GENE TRANSFER AND EXPRESSION

A Laboratory Manual

Sudhir Paul

GENE TRANSFER AND EXPRESSION

A Laboratory Manual

Michael Kriegler

W. H. Freeman and Company • New York

Library of Congress Cataloging-in-Publication Data

Kriegler, Michael P., 1954–
Gene transfer and expression : a laboratory manual / Michael Kriegler.
p. cm.
Reprint. Originally published: New York, N.Y. ; Stockton Press, 1990.
Includes bibliographical references and index.
ISBN 0-7167-7004-0
1. Genetic transformation—Laboratory manuals. 2. Genetic regulation—Laboratory manuals. I. Title.
[DNLM: 1. Gene Expression Regulation—laboratory manuals. 2. Transfection—laboratory manuals. QH 450 K92g 1990a]
QH442.K73 1991
574.37′322′078—dc20
DNLM/DLC
for Library of Congress
91-20303
CIP

2 3 4 5 6 7 8 9 KP 9 9 8 7 6 5 4 3 2 1

*This book is dedicated to my wife,
Jan, and my daughter, Sophie.*

Contents

	Acknowledgments	xi
	How to Use This Manual	xiii
PART I	GENE TRANSFER	1
1	**Eukaryotic Control Elements**	3
	Introduction	3
	Promoters and Enhancers	4
	Viral Enhancers	5
	Cellular Enhancers	10
	Inducible Promoters and Enhancers	16
	Mechanism of Enhancer Action	18
	Promoter and Intron Interactions	19
	Messenger RNA Degradation Signals and Polyadenylation	19
	Translational Control	20
	Related Phenomena	21
2	**Vectors**	23
	Introduction	23
	Transient Transfection	24
	Stable Transfection	24
	SV40-Based Vectors	25
	Polyoma-Virus-Based Vectors	29
	Adenovirus-Based Vectors	31
	Epstein-Barr-Virus-Based Vectors	35
	Herpes-Simplex-Virus-Based Vectors	36
	Vaccinia-Virus-Based Vectors	39
	Papilloma-Virus-Based Vectors	43
	Retroviral Vectors	47
	Retroviral Packaging Cell Lines	51
	Retroviral Vector Genomes	52

	Homologous Recombination and Gene-Replacement Vectors	56
	PCR-Based Expression Vectors	60
	PCR-Based Gene Assembly	60
	PCR-Based Expression	61
3	**Processing of Proteins Encoded by Transferred Genes**	**62**
	Introduction	62
	Protein Synthesis	62
	Processing	63
	Post-Translational Modification	65
	Part I References	**66**
PART II	**GENE TRANSFER METHODS**	**83**
4	**Cells and Cell Lines**	**85**
	Basic Tissue-Culture Techniques	85
	Propagation of Cell Lines	87
	Propagation of Embryonal Stem Cells	91
	Assays for Colony Formation, Anchorage-Independent Growth, and Focus Formation	94
5	**DNA Transfer**	**96**
	Calcium Phosphate Transfection Method	96
	DEAE Dextran Transfection Method	99
	Electroporation	101
6	**Selection and Amplification**	**103**
	Positive Selection	103
	Amplification	108
	Negative Selection	112
7	**Expression Cloning**	**114**
	Preparation and Technical Requirements	117
	Library Construction and Amplification	117
	Comparison of Construction Methodologies	117
	cDNA Library Construction	121
	Solid-State Amplification	131
	Transfection and SIB Selection	133
	Transfection	133
	SIB Selection	133

8	**Subtractive Hybridization**	**136**
	Generation of Subtracted Probes	139
	Production of First-Strand cDNA, High- or Low-Specific-Activity	139
	Subtracted cDNA Probe: Aqueous Hybridization and Hydroxylapatite (HAP) Chromatography	141
	Subtracted cDNA Probe: Aqueous Hybridization and Biotin/Phenol Extraction	144
	Subtracted cDNA(+) Probe: Phenol-Emulsion Reassociation Technique (PERT) and HAP Chromatography	146
	Generation of Subtracted Libraries	149
	Constructing a Subtracted cDNA Library: Biotin/Phenol Extraction	155
	Hybridization Analysis of Plasmid cDNA Libraries	158
9	**Retrovirus-Mediated Gene Transfer**	**161**
	Generation of High-Titer Helper-Virus-Free Recombinant-Retrovirus Stocks	161
	Titration and Analysis of Recombinant-Retrovirus Stocks	163
10	**PCR-Based Expression**	**165**
	PCR-Based Gene Assembly	165
	PCR-Based Gene Expression	171
	Part II References	**173**
PART III	**ASSAYS FOR GENE TRANSFER AND EXPRESSION**	**177**
11	**Assays for Gene Transfer**	**179**
	Preparation of Genomic DNA	180
	Southern-Type Techniques	181
	Method I	181
	Method II	182
	PCR-Based Techniques	184
	Cell-Lysate Work-Up	185
	PCR Conditions	186
	Nested-PCR Conditions	187
	Blotting Procedure	187
	Probe Synthesis	187
	Hybridization	188
	Detection	188
	Synthesis of Nucleic-Acid Probes for Hybridization Analysis	189
	Synthesis of Radiolabeled Nucleic-Acid Probes	189
	Non-Isotopically Labeled Probes	192
	Non-Isotopic Detection	199
	Hybridization Conditions for Nucleic-Acid Probes	201

12	**Assays for Gene Expression**	**205**
	Analysis of RNA (Northern- and PCR-Based Techniques)	205
	Chemical Precautions	206
	Total mRNA Isolation	206
	Northern Blotting	209
	PCR Analysis of RNA	211
	Analysis of Protein	213
	Immunoprecipitation	213
	Western-Type Techniques	216
	Immunofluorescence Analysis	219
	Analysis of Post-Translational Modification	222
	Analysis of N-Linked Glycosylation	222
	Analysis of O-Linked Glycosylation	224
	Analysis of Tyrosine Sulfation	225
	Analysis of Fatty Acid Acylation	226
	Part III References	**227**
	Appendix: Suppliers	**229**
	INDEX	**239**

Acknowledgments

The writing of this laboratory manual reflects the efforts of a number of dedicated individuals without whose assistance I could not have completed the task. Many of the protocols were developed and virtually all of the protocols were tested with the assistance of the personnel in my laboratory, Carl Perez, Kim DeFay, Iris Albert, and Sharon Hubby, as well as my colleagues from the PCR division, John Lyons and Ernie Kawasaki. M. J. Poklar and George McGregor assisted with the library research. Jan Tuttleman critically reviewed the manuscript and made many useful suggestions regarding content and organization. Denise Ramirez and Edna McCallan typed the manuscript and Ward Ruth drew the figures. I am deeply indebted to those individuals for their help. I am also indebted to my colleagues who encouraged me to take on this project, expressing the need for a manual of this type and to those individuals who developed many of the techniques contained herein and who shared their experiences with me. I also wish to thank Ingrid Krohn for her encouragement and support.

Michael Kriegler

How to Use This Manual

This manual is composed of three parts: Part I, Gene Transfer/Background Information; Part II, Gene Transfer/Methods; and Part III, Assays for Gene Transfer and Expression. Part I provides background information on three subjects: (1) eukaryotic regulatory elements, (2) gene transfer vectors, and (3) protein synthesis, processing and glycosylation. If you are unfamiliar with any of these topics, I suggest you familiarize yourself with the information in Part I. It reflects our current understanding of the variety and function of eukaryotic (mammalian) cis-acting regulatory elements, the variety of gene transfer vectors currently available, as well as what is currently understood about protein synthesis, processing and glycosylation. If you are familiar with these topics, proceed to Part II. The techniques described in this section will enable you to introduce the gene or genes of interest into cells in culture. Once gene transfer has been attempted, you will want to assay for transfer and expression; Part III describes how to accomplish this.

GENE TRANSFER

This section of the manual contains a description and analysis of the genetic components, molecular tools, and biochemical processes that you, the experimentalist, must manipulate and understand to design incisive and convincing gene transfer studies. I have included a description of *cis*-acting elements involved in the regulation of expression of eukaryotic genes, a description with diagrammatic representation of the various gene expression vector systems that currently exist as well as a discussion of their potential applications, and a description of the post-translational mechanisms that alter protein structure. Relevant references are included to serve as a starting point for further reading on the particular *cis*-acting element, vector system, or post-translational process you are interested in studying.

1

EUKARYOTIC CONTROL ELEMENTS

INTRODUCTION

As you attempt to express genes transferred into cells or into an animal, you must deal intelligently with a plethora of information about *cis*-acting DNA elements that, either directly or indirectly, affect gene expression. Each gene appears to have not only a promoter but in many cases a matched enhancer, as well as splice signals, polyadenylation signals, and signals that determine the messenger RNA half-life. These elements exert their influence in concert in a cell-type-specific manner and serve to determine where and when in an organism the gene is expressed. These elements can also be experimentally mixed and matched with each other, resulting in recombinant elements that manifest new biologies. Thus, before a single test tube is raised, you must familiarize yourself with these elements. Despite the apparent complexity of the interaction of these *cis*-acting elements during the regulation of gene expression, there are a number of rules you should keep in mind to guide genetic design.

The information that follows is provided to illustrate the range and behavior of *cis*-acting elements in viral and cellular cistrons available to you. In this background section, data arising from the experimental manipulation of regulatory sequences of eukaryotic genes are compiled. In most cases, specific sequence information is not included, but references are cited.

First, promoters are described, and a variety of enhancers, both viral and cellular, are discussed in detail. Then the effects of splicing on gene expression are described, mRNA stability elements are examined, and both *cis*- and *trans*-acting elements that affect translational control are discussed. At the end of each section, the mechanisms by which these elements exert their influences are explored. You should note that the section on enhancers is but a partial listing of the viral and cellular enhancers identified to date. New elements are described monthly, and the identifications of their *trans*-acting enhancer-binding protein congeners follow shortly thereafter. What should be apparent from this listing is that enhancers are highly varied and function in a variety of ways. The enhancer section contains an abundance of detail because changes in enhancer sequences, even single base-pair changes, can dramatically alter the function of the enhancer.

Adapted and reprinted by permission of the publisher from "Assembly of Enhancers, Promoters, and Splice Signals to Control Expression of Transferred Genes" by Michael Kriegler in GENE EXPRESSION TECHNOLOGY (METHODS IN ENZYMOLOGY, Volume 185) edited by David V. Goeddel. Copyright © 1990 by Academic Press, Inc.

Note also that promoters, enhancers, splice signals, and other *cis*-acting elements do not function as fully independent elements. It appears that there are preferred and nonfunctional combinations. In many cases, appropriate time- and tissue-specific expression is dependent on the assembly of an appropriate expression ensemble: ideally matching enhancer with promoter and promoter with splice signal and mRNA degradation signal. The selection and arrangement of *cis*-acting elements in a recombinant cistron is thus critical in achieving the desired expression phenotype. Therefore, prior to attempting gene transfer, take your time in engineering your recombinant molecule. Don't let convenient restriction sites dictate your design parameters. If necessary, destroy or create new sites via either site-directed mutagenesis or PCR-based gene assembly to ensure the inclusion or exclusion of any potentially important or detrimental DNA sequences in your construction. If you pay attention to these details, you can expect to succeed. Failure to do so may lead to an uninterpretable result and thus a waste of time and resources.

PROMOTERS AND ENHANCERS

The genetic dissection, through recombinant DNA techniques, of promoter elements, initially viral, led to a description of the molecular structure of eukaryotic promoters and to the discovery of a new class of *cis*-acting elements that appeared to modulate gene expression in a time- and tissue-specific fashion. These elements have come to be called enhancers. Enhancers appear to form part of the structural framework for DNA/protein interactions that may result in the formation of higher-order chromatin structures that may, in turn, serve to regulate transcription of subject genes from RNA polymerase II promoters. The identification of enhancer elements has led to the isolation and characterization of *trans*-acting factors that either bind to or in some other way affect the promoters or enhancers with which they interact. For a compilation of transcription-regulating proteins and an analysis of their mode of action, see Wingender (1988), Mitchell and Tjian (1989), and Ptashne and Gann (1990). What has become clear from site-directed mutagenesis of eukaryotic expression elements is that both enhancers and promoters play pivotal roles in the regulation of gene expression.

The elegant early studies of the herpes simplex thymidine kinase (HSVtk) promoter (McKnight and Kingsbury, 1982; McKnight et al., 1984) and the human β-globin promoter (Breathnach and Chambon, 1981) demonstrate that the structure of RNA polymerase II promoter elements is relatively set. One critical feature of all promoter elements is that they contain mRNA cap sites, the point at which the mRNA transcript actually begins. Yet another frequently occurring homology, the so-called TATA box, is centered around position -25, just upstream of the mRNA cap site. The TATA box accurately positions the start of transcription. Further upstream, the relative positions of promoter sequence homologies become more variable. They can, as is the case of the CCAAT homology in the HSVtk and human β-globin promoters, actually reside in different positions on different strands around 70 to 80 nucleotides upstream from the mRNA cap site. Another structural component of many housekeeping genes consists of multiple copies of GC-rich regions upstream of the mRNA cap site. There appears to be an absolute requirement for these sequences in eukaryotic promoters, so they must be included in all engineered recombinant cistrons.

With few exceptions, the most critical variable in the design of chimeric expression cistrons is the selection of an enhancer element or elements for inclusion in the recombinant molecule. First identified in the genomes of SV40 and murine retroviruses, enhancers are the most peculiar of all known expression elements. Little is known about their mechanism of action. The key properties that characterize an enhancer are as follows:

- They are relatively large elements and may contain repeated sequences that can function independently.
- They may act over considerable distances, up to several thousand base pairs (bp).
- They may function in either orientation.
- They may function in a position-independent manner and can be within or downstream of the transcribed region, but they can function only in *cis*. If several promoters lie nearby, the enhancer may preferentially act on the closest.
- They may function in a tissue-specific manner or at a particular stage of differentiation in a given cell.

Enhancers have been identified in association with both viral and cellular genes. To minimize confusion, I will treat these two classes separately, and in yet another section discuss the class of inducible enhancers and promoters. A listing of viral enhancers can be found in Table 1-1, cellular enhancers in Table 1-2, and inducible promoters and enhancers in Table 1-3.

Viral Enhancers

The SV40 Enhancer

The first enhancing DNA sequence identified was of viral origin and resides in the genome of SV40 near the origin of replication (Banerji et al., 1981; Moreau et al., 1981). The SV40 enhancer is composed of three functional units—A, B, and C—each of which cooperates with the others or with duplicates of themselves to enhance transcription. When element C, containing the so-called enhancer core consensus sequence, is inactivated by point mutations, revertants with reduced enhancer function can be isolated that contain duplications of either one or both of the elements A or B (Herr and Clarke, 1986; Firak and Subramanian, 1986). In addition, each element can act autonomously when present in multiple tandem copies. Amplified copies of the B and C elements exhibit different cell-specific activities (Ondek, et al., 1987).

An analysis of the effects that varying the position of the SV40 enhancer had on the expression of multiple transcription units in a single plasmid revealed two types of position effects. The first, promoter occlusion, results in reduced transcription at a downstream promoter if transcription is initiated at a nearby upstream promoter. This effect does not involve enhancer elements directly, even though the effect is most pronounced when the downstream promoter lacks an enhancer element. The second effect stems from the ability of promoter sequences to reduce the effect of a single enhancer element on other promoters in the same plasmid. This effect is mediated by either promoters adjacent to the enhancer element or promoters interposed between the enhancer element and the other promoters on the plasmid (Kadesch and Berg, 1986). Perhaps one of the more controversial observations regarding the SV40 enhancer is that enhancer-binding factors are required for the establishment but not the maintenance of enhancer-dependent transcriptional activation (Wang and Calame, 1986). This conclusion has been elegantly contested (Schaffner et al., 1988), and the disagreement has not yet been resolved. Such enhancer-binding factors may be induced or activated, either directly or indirectly, in a human hepatoma cell line by the tumor promoter TPA. In this cell line,

TABLE 1-1 Viral Enhancers

Enhancer	References
SV40	Banerji et al., 1981; Moreau et al., 1981; Sleigh and Lockett, 1985; Firak and Subramanian, 1986; Herr and Clarke, 1986; Imbra and Karin, 1986; Kadesch and Berg, 1986; Wang and Calame, 1986; Ondek et al., 1987; Kuhl et al., 1987; Schaffner et al., 1988
Polyoma	Swartzendruber and Lehman, 1975; Vasseur et al., 1980; Katinka et al., 1980, 1981; Tyndall et al., 1981; Dandolo et al., 1983; deVilliers et al., 1984; Hen et al., 1986; Satake et al., 1988; Campbell and Villarreal, 1988
Retroviruses	Kriegler and Botchan, 1982, 1983; Levinson et al., 1982; Kriegler et al., 1983, 1984a,b, 1988; Bosze et al., 1986; Miksicek et al., 1986; Celander and Haseltine, 1987; Thiesen et al., 1988; Celander et al., 1988; Choi et al., 1988; Reisman and Rotter, 1989
Papilloma Virus	Campo et al., 1983; Lusky et al., 1983; Spandidos and Wilkie, 1983; Spalholz et al., 1985; Lusky and Botchan, 1986; Cripe et al., 1987; Gloss et al., 1987; Hirochika et al., 1987; Stephens and Hentschel, 1987; Gius et al., 1988
Hepatitis B Virus	Bulla and Siddiqui, 1986; Jameel and Siddiqui, 1986; Shaul and Ben-Levy, 1987; Spandau and Lee, 1988; Vannice and Levinson, 1988
Human Immunodeficiency Virus	Muesing et al., 1987; Hauber and Cullan, 1988; Jakobovits et al., 1988; Feng and Holland, 1988; Takebe et al., 1988; Rosen et al., 1988; Berkhout et al., 1989; Laspia et al., 1989; Sharp and Marciniak, 1989; Braddock et al., 1989
Cytomegalovirus	Weber et al., 1984; Boshart et al., 1985; Foecking and Hofstetter 1986
Gibbon Ape Leukemia Virus	Holbrook et al., 1987; Quinn et al., 1989

TPA induces the transcriptional stimulatory activity of the SV40 enhancer (Imbra and Karin, 1986).

Similarly, F9 embryonal carcinoma (EC) cells, when forced to differentiate, exhibit a marked increase in transcription from a transiently transfected genome whose transcription is driven from an SV40 promoter. Deletion of the enhancer from this plasmid ablates the effect. Further, in both undifferentiated and differentiated F9 EC cell types, the level of transcription was found to be limited by the availability and/or activity of cellular factors necessary for enhancer function (Sleigh and Lockett, 1985). Chimeras constructed with the SV40 enhancer and the virus responsive elements (VREs) of the interferon α promoter indicate that the two subsequences, Rep A and Rep B of the VRE, when individually inserted between the TATA box of the α-interferon promoter and the SV40 enhancer, serve to silence that promoter. Such promoter silencing is fully reversible after induction. Silencing is not observed when the intact VRE is placed between the TATA box and the SV40 enhancer (Kuhl et al., 1987).

Thus, the SV40 enhancer element is a complex structure whose function is subject to some position effects, and whose cell-type-specific activation is dependent, in part, on the absence or presence of active cellular factors or proximal sequences.

The Polyoma Enhancer

The polyoma virus enhancer, also found near the viral origin of replication (Tyndall et al., 1981; Veldman et al., 1985), is similar in many ways to the SV40 enhancer. However, there are observations unique to this regulatory element. The polyoma virus enhancer is normally not active in undifferentiated F9 EC cells, but it is active in these cells after differentiation (Swartzendruber and Lehman, 1975). Nevertheless, polyoma virus mutants that can express and replicate their genomes in both cell types have been described (Katinka et al., 1980, 1981; Vasseur et al., 1980). The sequence alterations responsible for this phenotype consist of duplications and mutations around the polyoma virus origin of replication and enhancer. These mutations appear to dramatically increase the transcription of the viral genes (Dandolo et al., 1983). Recently, an enhancer point mutant has been isolated that is fully functional in both cell types. Mutant F441 carries a single base mutation at nucleotide 5233 of the viral genome. Enhancers of mutants displaying a similar phenotype contain a duplicated segment of viral DNA encompassing nucleotide 5233. Furthermore, duplication of the point-mutated segment results in an even higher level of expression in the undifferentiated cell type. In addition, co-transfection of the F441 oligonucleotide, but not the wild-type sequence, inhibits the activity of the enhancer fragment of F441 attached to a reporter gene. Thus, it appears that the point mutation is a target for a cellular factor or factors that act in a positive manner to increase the transcription of a gene in undifferentiated F9 EC cells (Satake et al., 1988).

Studies have shown that both the SV40 enhancer and the wild-type polyoma enhancer can be repressed by adenovirus E1A gene products. Other work has shown that although the wild-type polyoma enhancer cannot function in undifferentiated F9 EC cells, the SV40 enhancer can, at least to a limited extent. On the basis of these observations, it has been postulated that undifferentiated embryonal carcinoma F9 cells contain an E1A-like activity, and that this activity is responsible for the lack of polyoma virus enhancer activity in F9 EC cells. Hen et al. (1986) report that E1A gene products do not repress a point mutant of the polyoma virus enhancer that is active in undifferentiated F9 EC cells. Their result is consistent with the notion that undifferentiated F9 EC cells contain a cellular repressor that blocks the polyoma virus enhancer and that this repressor has the same target sequence as the E1A proteins.

However, the polyoma story is even more complex. The polyoma virus is normally not permissive for replication in most lymphoid lines. Nevertheless, deletion of the PvuII.D fragment from the wild-type polyoma genome (spanning the F441 point mutation) facilitated replication in some T-cells and mastocytoma cell lines; therefore, the PvuII D fragment can act as both a positive and a negative regulatory element. Substitution of the polyoma enhancer with the MuLV enhancer facilitates replication in 3T6 and B lymphoid cells. Substitution with the immunoglobulin heavy-chain enhancer facilitates replication in B lymphoid cells but not 3T6 cells (deVilliers et al., 1984) or mastocytomas (Campbell and Villarreal 1986, 1988).

Retroviral Enhancers

Retroviruses also carry potent transcriptional enhancers (Kriegler and Botchan, 1982, 1983; Levinson et al., 1982). I will consider the enhancers of the murine leukemia and sarcoma viruses simultaneously because it makes for a more interesting story when one compares their activities. The type of tissue in which these retroviruses induce disease often, though not always, reflects the cell-type specificity of the enhancers resident in the long terminal repeats (LTRs) of the integrated provirus. The enhancers in these viruses are quite similar to one another, so mix-and-match experiments between elements of their enhancers can reveal the nucleotide differences that determine tissue-type expression specificity. The transcriptional enhancers of Moloney murine leukemia virus (MoMuLV) and Moloney murine sarcoma virus (MoMuSV) exhibit different cell-type expression specificities from the enhancer of Friend murine leukemia virus. Although the three enhancers are approximately equally active in erythroid cells, the MoMuSV and MoMuLV enhancers are 20–40-fold more active than the Friend MuLV enhancer in T-lymphoid cells. There appears to be an element, repeated several times within these enhancers, that modulates the activity of the enhancer in T-cells without affecting it in erythroid cells. This element thus seems to be one of the determinants of the tissue specificity of the enhancer (Bosze et al., 1986; Li et al., 1987; Thiesen et al., 1988).

Enhancer elements within the sarcoma virus MoMuSV and the non-leukemogenic (AKV) and T-cell leukemogenic (SL3-3) murine retroviruses also exhibit strong cell-type preferences in transcriptional activity. These elements are additionally regulated by the glucocorticoid dexamethasone. Mapping studies in combination with DNAse I footprinting experiments define the presence of glucocorticoid regulatory elements at the promoter-proximal ends of each enhancer repeat. These elements behave like inducible enhancers. Their regulatory activity is independent of position and orientation when they are linked in *cis* to a heterologous promoter. The sequences required for dexamethasone regulation for both AKV and SL3-3 include a 17-nucleotide consensus sequence termed the glucocorticoid-responsive element (GRE), located at the promoter-proximal ends of each enhancer repeat. Although the GREs are identical for these enhancers the sequences surrounding these elements differ (Miksicek et al., 1986; Celander and Haseltine, 1987; Celander et al., 1988).

Our experiments with both infected and transfected retroviral genomes have led us to some interesting observations regarding promoter and enhancer strengths in specific cell types. First, we and others observed that infected retroviral genomes express their gene products at substantially higher levels than do the same transfected genomes present in the same copy number (Hwang and Gilboa, 1984). Second, comparison of the synthetic capabilities of a wild-type SV40 virus and an engineered murine retrovirus (MuLV) expressing SV40 T antigen revealed that, in mouse fibroblasts, an engineered SV40 retrovirus could produce approximately 10 times the amount of T antigen as its wild-type SV40 equivalent. These observations hold true for a variety of structural genes we have analyzed (Kriegler and Botchan, 1982, 1983; Kriegler et al., 1983, 1984a,b, 1988; Choi et al., 1988).

Papilloma Virus Enhancers

The papilloma viruses, both bovine and human, induce tissue-specific diseases in their hosts and have complex enhancer architectures. Cells transformed by these viruses maintain the viral DNA as nuclear plasmids. Three enhancer elements modulate the transcriptional activity of the bovine papilloma virus (BPV). They map to a 59 bp region 3' to the early polyadenylation sequence at the end of the early viral gene transcripts (Campo et al., 1983; Lusky et al., 1983), the 31 percent non-transforming, late region (Lusky and Botchan, 1986), and the 5', 900 bp non-coding region of the viral genome (Spalholz et al., 1985). The 59 bp 3' enhancer is host specific (Spandidos and Wilkie, 1983). The 5' enhancer is located in the vicinity of the viral promoter and can be *trans*-activated by the viral protein E2. Human papilloma virus (HPV) type 18 also contains three different enhancer domains, two of which are inducible and one of which is constitutive. The inducible enhancers are responsive to papilloma virus-encoded *trans*-acting factors, whereas the constitutive enhancer requires only cellular factors for activity. Inducible enhancer IE2 is located proximal to the E6 cap site and responds to the E2 *trans*-activator. Inducible enhancer IE6 is located 500 bp upstream of the E6 cap and responds to the E6 gene product. The third, constitutive enhancer lies between the two inducible enhancers. Each enhancer functions independently of the others and may function at different stages of the viral life cycle (Gius et al., 1988). Human papilloma virus type 16 (HPV16) contains a keratinocyte-dependent enhancer in a sequence 5' to the P97 coding region. The E2 *trans*-activator of HPV16, as well as that of the related bovine papilloma virus, further enhance HPV16 transcription. Similarly, the "short E2" (SE2) gene products of both viruses repress E2 activation and suppress activation in keratinocytes (Cripe et al., 1987). Furthermore, Gloss et al. (1987) report that the region upstream of P97 in HPV16 functions as a GRE and shares sequence homology to the consensus GRE. Elements of the enhancer behavior described above are shared by HPV types 1, 6b, 7, and 11, as well as by cottontail rabbit papilloma virus (Hirochika et al., 1987).

Hepatitis B Virus Enhancer

Another potentially tissue-specific viral enhancer has been described in hepatitis B virus. This enhancer is located 3' to the hepatitis B surface antigen coding sequences but is contained within the mature viral transcripts (Bulla and Siddiqui, 1986). Vannice and Levinson (1988) report that the HBV enhancer can dramatically increase expression levels of genes controlled by the SV40 enhancer/promoter, but only when the enhancer is located within the transcribed region of the gene. Further, this effect appears to be orientation-dependent, a violation of usual enhancer rules. Last, these authors report that when this enhancer is located either 5' to the promoter or in the 3' untranslated region of a recombinant cistron, its enhancing spectrum is not cell-type-specific. This finding is in striking contradiction to the study by Jameel and Siddiqui (1986), who report that the enhancer displays strict host and tissue specificity in that it is functional only in human liver cells. This specificity requires *trans*-acting factors, potentially those described by Shaul and Ben-Levy (1987). Additional studies support the contention that the X gene product of hepatitis B virus can *trans*-activate the HBV enhancer (Spandau and Lee, 1988).

Human Immunodeficiency Virus Enhancers

The human immunodeficiency viruses (HIVs) exhibit strikingly complex regulatory strategies, involving virally encoded *trans*-activating proteins (*tat*) and their *cis*-acting element congeners (*tar*). These congeners are not categorized formally as enhancers, but they are relevant to our discussion because the effects they manifest on gene expression are quite dramatic and are most useful in engineering gene expression. *Tat* acts on *tar* to increase viral mRNA and protein synthesis as much as 100-fold (Hauber and Cullen, 1988; Jakobovits et al., 1988). This increase can be divided roughly into a 20-fold increase in viral RNA synthesis and a five-fold increase in protein synthesized from each messenger RNA. The studies of Muesing et al. (1987) and Hauber and Cullen (1988) serve to distinguish *tar* from typical transcription-factor-binding sites. First, *tar* must be located immediately downstream of the site of transcription initiation by RNA polymerase II. Second, it is orientation-dependent: inversion of *tar* so that the complementary strand is transcribed into RNA eliminates the activity of *tar*. Further, *tar* activity does not require HIV regulatory sequences upstream of the site of transcription initiation. *Tar* deletion does not result in an increase in the basal level of transcription from the HIV promoter in the absence of *tat* and thus is not a negative regulatory element and does not serve to terminate transcription. The *tar* element has been extensively analyzed via site-directed mutagenesis, and it has been found that sequences capable of forming an RNA stem-loop structure are required for the activity of *tar* (Feng and Holland, 1988; Berkhout et al., 1989).

The experiments of Laspia et al. (1989) indicate that the major effect of *tat* on *tar* is to increase the rate of initiation of transcription through the recognition of *tar* sequences on viral mRNAs. It has been suggested that a newly synthesized

RNA could bind *tat* and one or more proteins, resulting in the formation of a complex with the ability to activate the formation of subsequent transcription complexes (Sharp and Marciniak, 1989). A secondary but notable effect of *tat* on *tar* is to increase the efficiency of translation of *tar*-containing mRNAs (Braddock et al., 1989). Surprisingly, pure *tat* protein *trans*-activates HIV expression when added to the medium of cultured cells; apparently, cells take up the *tat* protein intact.

A curious parallel can be drawn between the HIV *tar* sequence and sequences found in the R-U5 segment of the HTLV-I long terminal repeat, with one important exception. This HTLV-I sequence, which appears to function in a fashion analogous to *tar*, can, in combination with the SV40 promoter, increase the production of a variety of gene products 100-fold without the presence of any known virally encoded *trans*-acting factors. This sequence functions to up regulate gene expression when placed just downstream of the SV40 promoter and works in only one orientation (Takebe et al., 1988). Unlike the adenovirus VA gene products, this HTLV-I sequence does not *trans*-activate other promoters upon co-transfection, and the VA gene products do not *trans*-activate promoters to which this HTLV-I sequence is coupled (Kriegler et al., unpublished observations). Further, sequences within the HTLV-I R-U5 region of the viral LTRs can form stem-loop structures similar to those found in HIV-I. The incorporation of these sequences into an SV40 replicon containing a multiple cloning site results in the generation of a remarkably efficient expression-cloning vector called SRα (Takebe et al., 1988).

In addition, HIV carries an authentic enhancer, upstream of the SP1-binding sites within the U3 region of the LTR between nucleotides -137 and -17. The enhancer is non-specific in that it can be replaced by either the RSV or SV40 enhancer (Rosen et al., 1988).

Human Cytomegalovirus Enhancer

The human cytomegalovirus (HCMV) genome, 235 kb in size, has been shown to contain a strong transcriptional enhancer. The enhancer in this member of the herpesvirus group was identified through the application of an enhancer trap (Weber et al., 1984). With this approach, linearized SV40 DNA lacking its own enhancer is mixed with short, randomly fragmented pieces of the DNA from which the enhancer is to be isolated. This mixture is transfected into cells permissive for the replication of SV40. Recombination between the enhancerless SV40 DNA and fragments encoding enhancer activity derived from, in this case, HCMV results in the appearance of lytically growing SV40-type recombinants, and from these the new enhancer activity can be isolated. The HCMV enhancer is located upstream of the transcription initiation site of the major immediate-early gene, between nucleotides -524 and -118. This enhancer has little cell-type or species preference and is several fold more active than the SV40 enhancer (Boshart et al., 1985). When the HCMV enhancer-promoter is compared with the SV40 enhancer-promoter and the Rous sarcoma virus long terminal repeat, which contains its enhancer-promoter, the HCMV expression element is considerably stronger when transfected into a variety of cell types (Foecking and Hofstetter, 1986).

Cellular Enhancers

The initial demonstration of enhancers of cellular gene expression has led to the identification of a great number of cellular enhancers. For the purposes of these discussions, these enhancers will be divided into two groups, enhancers that affect immune-system genes and enhancers that affect other genes. The immune-system genes include those encoding the immunoglobulin heavy and light chains, the histocompatibility genes, the T-cell receptor genes, the β-interferon genes, and the interleukin-2 receptor gene. The other group of genes includes human β-actin, muscle creatine kinase, prealbumin, elastase I, metallothionein, collagenase, α-fetoprotein, γ-globin and β-globin, c-fos, c-ras, insulin, and the neural cell adhesion molecule (NCAM).

Immune System Enhancers

Sequence analysis of the regulatory domains of a variety of immune-system genes has revealed a common but not ubiquitous sequence motif that functions as a binding site for the nuclear regulatory factor NF-κ-B. NF-κ-B-binding sites have been identified in the regulatory regions of the IL-2 (Hoyos et al., 1989), IL-2 receptor (Leung and Nabel, 1988; Lowenthal et al., 1989),

TABLE 1-2 Cellular Enhancers

Enhancer	References
Immunoglobulin Heavy Chain	Banerji et al., 1983; Gilles et al., 1983; Grosschedl and Baltimore, 1985; Atchinson and Perry, 1986, 1987; Imler et al., 1987; Weinberger et al., 1988; Kiledjian et al., 1988; Porton et al., 1990
Immunoglobulin Light Chain	Queen and Baltimore, 1983; Picard and Schaffner, 1984
T–Cell Receptor	Luria et al., 1987; Winoto and Baltimore, 1989; Redondo et al., 1990
HLA DQ α and DQ β	Sullivan and Peterlin, 1987
β-Interferon	Goodbourn et al., 1986; Fujita et al., 1987; Goodbourn and Maniatis, 1988
Interleukin-2	Greene et al., 1989
Interleukin-2 Receptor	Greene et al., 1989; Lin et al., 1990
MHC Class II E_α^k	Koch et al., 1989
MHC Class II HLA-DRα	Sherman et al., 1989
β-Actin	Kawamoto et al., 1988; Ng et al., 1989
Muscle Creatine Kinase	Jaynes et al., 1988, Horlick and Benfield, 1989; Johnson et al., 1989a
Prealbumin (Transthyretin)	Costa et al., 1988
Elastase I	Ornitz et al., 1987
Metallothionein	Karin et al., 1987; Culotta and Hamer, 1989
Collagenase	Pinkert et al., 1987; Angel et al. 1987
Albumin Gene	Pinkert et. al., 1987; Tronche et al., 1989, 1990
α-Fetoprotein	Godbout et al., 1988; Campere and Tilghman, 1989
γ-Globin	Bodine and Ley, 1987; Perez-Stable and Constantini, 1990
β-Globin	Trudel and Constantini, 1987
c-fos	Treisman, 1985; Deschamps et al., 1985
c-HA-ras	Cohen et al., 1987
Insulin	Edlund et al., 1985
Neural Cell Adhesion Molecule (NCAM)	Hirsch et al., 1990
α_1-Antitrypsin	Latimer et al., 1990
H2B (TH2B) Histone	Hwang et al., 1990
Mouse α Type I Collagen	Ripe et al., 1989
Glucose-Regulated Proteins (GRP94 and GRP78)	Chang et al., 1989
Rat Growth Hormone	Larsen et al., 1986
Human Serum Amyloid A (SAA)	Edbrooke et al., 1989
Troponin I (TN I)	Yutzey et al., 1989
Platelet-Derived Growth Factor	Pech et al., 1989
Duchenne Muscular Dystrophy	Klamut et al., 1990

immunoglobulin κ (light chain) (Sen and Baltimore, 1986; Lenardo et al., 1987), β-interferon (Lenardo et al., 1989), serum amyloid A (Edbrooke et al., 1989), and IL-6 genes (Liberman and Baltimore, 1990; Shimizu et al., 1990). This observation suggests that the NF-κ-B-binding site may serve to induce the expression of numerous genes activated in response to trauma or infection. In some instances the NF-κ-B-binding site has been implicated in the regulation of some immune system genes through genetic analysis as well. A number of immune system enhancers, with and without NF-κ-B-binding sites, that have been identified through genetic analysis are described here.

Immunoglobulin Gene Enhancers. Publications by two groups of the existence of a tissue-specific enhancer in the immunoglobulin heavy-chain genes indicated that the enhancer resided in an intron, downstream of the joining region of the heavy-chain genes (Banerji et al., 1983; Gilles et al., 1983). Concurrently, an enhancer was identified in the major intron of the immunoglobulin κ (light-chain) genes (Queen and Baltimore, 1983; Picard and Schaffner, 1984). These enhancers are interesting for at least three reasons. The first is that they are located within introns, the second is that they appear to be lymphoid-specific, and the third is that enhancer activity may be only transiently required in the case of the κ gene (Atkinson and Perry, 1987).

The heavy-chain enhancer is considerably stronger than the κ (light-chain) enhancer (Picard and Schaffner, 1984). The tissue specificity of the heavy-chain enhancer appears to be determined in part by two distinct negative regulatory elements in and around the heavy-chain enhancer, such that, in non-lymphoid cells, activation by ubiquitous transcription factors is repressed (Imler et al., 1987). It has been shown that removal of this repressive or "silencing" sequence allows the heavy-chain enhancer to function in fibroblasts, cells in which the wild-type enhancer does not normally function (Weinberger et al., 1988). In fact, in the heavy-chain enhancer, three different domains serve to alter the expression of the immunoglobulin heavy-chain gene in a tissue-specific manner: the promoter, an enhancer, and a third element (Grosschedl and Baltimore, 1985).

Polymerization—the creation of multimers—of the entire heavy-chain enhancer, results in an increase in overall transcription. Polymerization of enhancer-DNA segments and subsequent analysis revealed two functional domains within the enhancer. In fact, greater-than-wild-type activity can be obtained through polymerization of these "sub-enhancers." One domain contains three regions thought to be involved in protein binding (E1, E2, E3), and the other contains the fourth E motif (E4) and the conserved oligonucleotide ATTTGCAT (Kiledjian et al., 1988). Each element is necessary but insufficient to direct high-level gene expression.

Thus, the immunoglobulin enhancers achieve their dramatic cell-type specificity through a combination of both positive and negative enhancer elements. Although there exists an abundance of evidence suggesting that enhancer-binding proteins exist and are important for the immunoglobulin enhancer to manifest its activity, there is some evidence that the immunoglobulin enhancer is not merely an entry site for RNA polymerase or a transcription complex. Atkinson and Perry (1986) dramatically demonstrated that tandem immunoglobulin promoters are equally active in the presence of the κ enhancer even when the promoters are located 7 kb from the enhancer. The observation that cell lines that lack the intron enhancer are still able to transcribe the immunoglobulin H (IgH) gene prompted the search for other Ig enhancers. In a recent report, Pettersson et al. (1990) identified a second B–cell-specific enhancer 3' of the IgH locus.

T–cell-Receptor Gene Enhancers. An enhancer of the α chain of the human T-cell receptor reportedly had been localized to a 1.1-kb BamHI-HindIII fragment located 5′ to the first exon of the C α gene. This enhancer's activity was said to be specific for lymphoid cells of both B- and T-cell origin (Luria et al., 1987). This report has been elegantly rebutted by Winoto and Baltimore (1989), and it now appears that an inducible, T-cell-specific enhancer is located at the 3′ end of the T-cell-receptor α locus. The exquisite expression specificity of the α locus is due to the action of a combination of this enhancer and a group of nearby transcriptional silencers (Winoto and Baltimore, 1989). A T-cell-specific enhancer has been identified within the human T-cell-receptor δ locus and has been localized to a 250-bp region (Redondo et al., 1990).

HLA DQ α and β Gene Enhancers. The HLA DQ α and DQ β genes have also been shown to possess enhancer elements. Two regions have been demonstrated in DQ α and one in DQ β (Sullivan and Peterlin, 1987).

β-Interferon Gene Enhancer. The β-interferon gene contains a virus-inducible enhancer between −65 and −109 relative to the cap site. This region is composed of a series of tandemly repeated 6-bp segments. Synthetic oligomers of the repeats also function as virus-inducible enhancers (Fujita et al., 1987). Deletion analysis within the enhancer region indicates that this enhancer is under negative control. Deleting sequences from the 3′ end of the enhancer leads to a dramatic increase in the basal level of β-interferon mRNA and a corresponding decrease in the induction ratio. The remaining 5′ region of the enhancer serves as a strong constitutive-expression element. Thus, in the case of β-interferon, derepression of a constitutive transcription element appears to play a key role in the regulation of the expression of the gene (Goodbourn et al., 1986; Goodbourn and Maniatis, 1988).

Interleukin-2 and Interleukin-2-Receptor Gene Enhancers. The interleukin-2 (IL-2) and interleukin-2-receptor (IL-2R) gene enhancers are subject to regulation by both cellular and viral *trans*-acting factors, including those encoded by HIV-I. The mobilization of the cellular *trans*-acting factors that stimulate expression of IL-2R is induced by a variety of immune stimuli, including antigens, cytokines, and various non-specific mitogens. Under normal circumstances, this induction is usually transient in nature. However, infection of human CD4$^+$ T-cells with HTLV-I results in persistent high-level expression of IL-2R, a consequence of the viral *trans*-acting factor *tax*. *Tax* induces transcription from the HTLV-I LTRs as well as from the IL-2 and IL-2R promoters. In contrast, an HIV-I *trans*-acting factor appears to subvert the expression of the genes induced by HTLV-I.

The *cis*-acting sequences involved in the regulation of IL-2R gene expression have been analyzed and serve to explain the complex behavior of this promoter in response to a wide variety of stimuli. Transient transfection of Jurkat cells with IL-2R-promoter constructs containing 471 nucleotides of 5′ flanking sequence, followed by stimulation with PMA, PHA, TNF α or *tax* results in a 5–10-fold increase in activity relative to unstimulated cells. Deletions beyond base 281 diminished and then destroyed activation mediated by various cytokines and mitogens. Promoter constructs terminating at nucleotide 266 remained highly inducible by HTLV-I *tax*. Thus, activation of IL-2R gene expression by PMA and by *tax* appear to occur by two fundamentally different mechanisms. It appears that the IL-2R promoter contains a NF-κ-B-like regulatory element between bases −267 and −256 that is strikingly similar to those found in κ (light-chain) immunoglobulin, major histocompatibility complex (MHC) Class I antigens, β-microglobulin, and IL-2 genes. It appears that HTLV-I deregulation of the IL-2R promoter occurs through the NF-κ-B-binding site either directly or indirectly (Greene et al., 1989).

Additional Cellular Enhancers

The second set of cellular genes is more diverse and includes genes encoding structural proteins, cellular enzymes, and secretory proteins.

β-Actin Gene Enhancer. β-actin is one of the most abundant proteins in eukaryotic cells and is expressed in a variety of cells. Kawamoto et al. (1988) constructed a series of deletion mutants in an attempt to localize the β-actin enhancer. They found that the first of five introns contains enhancer activity. Their results indicate that enhancer activity resides between positions +770 and +793 in the β-actin gene and contains a so-called enhancer "core." The β-actin promoter/enhancer assembly functions well in recombinant cistrons and when driving the expression of the neo^r gene. In transfection experiments, β-actin neo^r constructions consistently yield substantially greater numbers of stable transformants when transfected into murine cells than do equivalent constructions driven by either the MLV LTR or the SV40 promoter (Kriegler and Perez, unpublished observations).

Muscle Creatine Kinase Gene Enhancer. Muscle creatine kinase (MCK) is induced to high levels during skeletal-muscle differentiation. The MCK gene contains multiple upstream regulatory elements including an element, located between 1.031- and 1.190-kb upstream of the transcription start site, that has the properties of a transcriptional enhancer. It appears that even in the absence of this enhancer, low-level expression from the MCK promoter retains differentiation specificity (Jaynes et al., 1988). More recent evidence (Horlick and Benfield, 1989; Johnson et al., 1989a) suggests that, like the rat skeletal muscle α-actin gene and the rat myosin light-chain gene, the MCK enhancer is muscle-specific.

Prealbumin Gene Enhancer. Cell-specific expression of the transthyretin (prealbumin) gene has been localized to two upstream elements. One of these elements, located between 1.96- and 1.86-kb 5' to the mRNA cap site, behaves as an enhancer element. This element appears to be cell-specific because it can stimulate the β-globin promoter in HEP G2 cells but not in HeLa cells (Costa et al., 1988).

Elastase I Gene Enhancer. The elastase I gene promoter and enhancer have been analyzed and it is clear that the elastase enhancer, located between positions −205 and −73, exhibits tissue-type expression specificity for pancreatic acinar cells. The elastase enhancer can activate both the metallothionein and human growth hormone (hGH) promoters. Combinations of immunoglobulin and elastase enhancers with a heterologous promoter result in expression in all of the tissues predicted by the sum of each enhancer acting alone. Thus, these enhancers act independently of each other and, unlike the immunoglobulin genes, appear to lack silencing activity in cell types in which they do not normally function (Ornitz et al., 1987).

Metallothionein Gene Enhancer. The human metallothionein IIA gene contains two enhancer elements whose activities are induced by a variety of heavy-metal ions. Deletion of the metal-responsive elements (MREs) of the enhancers has no effect on the basal activity of the enhancer but prevents induction by heavy-metal ions. Replacement of the basal-level enhancer element with a DNA linker inactivates the enhancer both before and after induction, leading to the conclusion that the MREs act as positive regulators of metallothionein enhancer activity (Karin et al., 1987). The mouse metallothionein I (MT-1) MRE is 15–17 bp in length. A single copy works bidirectionally to yield a three- to four-fold induction of gene expression. Dual copies yield a 10–20-fold response (Culotta and Hamer, 1989). More information on the metallothionein promoter and enhancer can be found in the next part of this section, on inducible promoters and enhancers.

Collagenase Gene Enhancer. TPA-induced expression of the human collagenase gene is mediated by an enhancer element located 5' to the mRNA start site. The 32-bp enhancer is located between positions −73 to −42 relative to the transcription start site. When inserted next to the herpes simplex virus thymidine kinase promoter, the new enhancer/promoter combination can be induced by TPA (Angel et al., 1987a,b).

Albumin Gene Enhancer. Transgenic mice were utilized to identify an enhancer element in the albumin gene responsible for efficient, liver-specific gene expression. This enhancer does function in an orientation-independent manner but does not function well with a heterologous promoter, a phenomenon similar to that shown by some other enhancer/promoter pairs (Pinkert et al., 1987).

α-Fetoprotein Gene Enhancer. The regulatory region of the α-fetoprotein gene contains at least three distinct enhancer elements. Each enhancer directs gene expression in the appropriate tissues, including the visceral endoderm of the yolk sac, the fetal liver, and the gastrointestinal tract. The DNA elements responsible for directing the activation, repression, and re-induction of transcription are contained within, or 5' to, the gene itself (Hammer et al., 1987). These enhancers have been fine-structure mapped to regions 200–300 bp in length that are 2.5-, 5.0- and 6.5-kb upstream of the transcription start site (Godbout et al., 1988).

γ-Globin/β-Globin Gene Enhancers. Another embryonic gene product, human A γ-globin, carries an enhancer at the 3' end of the gene, approximately 400 bp downstream from the polyadenylation site. This enhancer is less than 750 bp in length and behaves in a non–tissue-specific manner in tissue culture cells (Bodine and Ley, 1987). However, in transgenic mice, a related gene, the G γ-globin gene, is active only in embryonic erythroid cells. Sequences between −201 and −136 are essential for the expression of G γ-globin. Another fragment from −383 to −206 of the G γ-globin expression element was similarly active in an identical assay. The combination of both sequences, −136 to −383, was most active of all (Perez-Stable and Constantini, 1990).

The β-globin gene is expressed only in fetal and adult erythroid cells. There exists an enhancer at the 3' end of the β-globin gene that, when attached to the G γ-globin gene in either orientation, activates transcription of the otherwise silent G γ-globin gene in fetal liver. This enhancer is within the region 600–900 bp 3' to the β-globin polyadenylation signal and obviously contributes to the stage-specific expression of the β-globin gene (Trudel and Constantini, 1987).

c-fos Gene Enhancer. A serum-responsive enhancer element has been found associated with the c-fos oncogene. The c-fos transcriptional enhancer is located between −332 and −276 relative to the mRNA cap site and does not appear to be tissue-specific in nature (Treisman, 1985; Deschamps et al., 1985). The function of this serum-responsive element (SRE) cannot be replaced by either the SV40 or the Moloney murine leukemia virus enhancers.

c-HA-ras Gene Enhancer. An enhancer that lies 3' to the human c-HA-*ras* I gene has been identified. This enhancer is a repetitive-sequence element, variants of which have been associated with malignant disease. This element can function in a somewhat position- and orientation-independent manner (Cohen et al., 1987).

Insulin Gene Enhancer. Two regulatory elements, one of which functions as an enhancer, have been identified 5' to the rat insulin gene. The enhancer activity is spread over the region −103 to −333. Internal portions of the enhancer element (−159 to −249), which showed only low activity, stimulated full activity when duplicated. The activity of the insulin enhancer appears to be confined to insulin-producing cells (Edlund et al., 1985).

Neural-Cell-Adhesion Molecule (NCAM) Gene Enhancer. One of the most prevalent cell-adhesion molecules in vertebrates, the expression of NCAM is subject to complex cell-type- and developmental-stage-dependent regulation. NCAM is believed to be critically involved in specifying cell patterning and movement in the embryo and thus the three-dimensional organization of the bodies of higher organisms. The NCAM promoter lacks a typical TATA box, and it appears that several transcription start sites are used indiscriminately by different cell types. Sequences responsible for both the cell-type-specific promotion and inhibition of transcription reside within 840 base pairs upstream of the main transcription start site. The sequences between positions −645 and −37 direct high-level expression in NCAM-expressing N2A cells. This same fragment is six times less active in L cells, and this activity can be repressed by the inclusion of an additional upstream segment. Deletion analysis reveals that the region of maximal promoter activity can be subdivided into two domains, one of which (−645 to −462) increases promoter efficiency two-fold whereas the proximal domain (−462 to −245) contributes most of the promoter activity (Hirsch et al., 1990).

Inducible Promoters and Enhancers

All genes that are not constitutively expressed must, at some time during the life of the individual within which they reside, be either induced or repressed. In this section, I will discuss but a subset of these regulated genes, those whose expression can be practically regulated in tissue culture. This means expression elements that can be regulated by either the addition of a stock reagent to the growth medium or, in a few cases, the transfection of a gene encoding an inducer molecule into the cell line carrying a recombinant cistron whose expression is driven by a regulatable element. The two most commonly employed regulatable elements in eukaryotic gene transfer experiments are (1) the metallothionein expression elements and (2) the glucocorticoid-responsive elements, such as those found in the LTRs of the mouse mammary tumor virus (MMTV). In this section, I will discuss these two elements. In addition, I have compiled Table 1-3, listing other regulatable elements along with their inducers or repressors, as well as references for further information.

The Metallothionein Expression Elements

The heavy–metal-inducible metallothionein promoter has been used to control the expression of genes both for protein production in cultured mammalian cells and in transgenic animals. Some of these promoters, derived from human, murine, and other sources, display relatively high levels of basal expression in the absence of heavy metals, although moderate increases can be achieved when inducer is added (Palmiter et al., 1982).

Detailed analysis of the human metallothionein IIA (hMTII$_A$) promoter has established at least nine sequence elements involved in the regulation of expression of that gene (Haslinger and Karin, 1985; Imagawa et al., 1987; Karin et al., 1987). In addition to the TATA box, the sequences include four metal-responsive

TABLE 1-3 Inducible Promoters and Enhancers

Element	Inducer	References
MT II	Phorbol Ester (TPA) Heavy metals	Palmiter et al., 1982; Haslinger and Karin, 1985; Searle et al., 1985; Stuart et al., 1985; Imagawa et al., 1987; Karin et al., 1987; Angel et al., 1987b; McNeall et al., 1989
MMTV(mouse mammary tumor virus)	Glucocorticoids	Huang et al., 1981; Lee et al., 1981; Majors and Varmus, 1983; Chandler et al., 1983; Lee et al., 1984; Ponta et al., 1985; Sakai et al., 1988
β-Interferon	poly(rI)X poly (rc)	Tavernier et al., 1983
Adenovirus 5 E2	E1a	Imperiale and Nevins, 1984
Collagenase	Phorbol Ester (TPA)	Angel et al., 1987a
Stromelysin	Phorbol Ester (TPA)	Angel et al., 1987b
SV40	Phorbol Ester (TPA)	Angel et al., 1987b
Murine MX Gene	Interferon, Newcastle Disease Virus	Hug et al., 1988
GRP78 Gene	A23187	Resendez et al., 1988
α-2-Macroglobulin	IL-6	Kunz et al., 1989
Vimentin	Serum	Rittling et al., 1989
MHC Class I Gene H-2kb	Interferon	Blanar et al., 1989
HSP70	E1a, SV40 Large T Antigen	Williams et al., 1989 Taylor et al., 1989; Taylor and Kingston, 1990a,b
Proliferin	Phorbol Ester-TPA	Mordacq and Linzer, 1989
Tumor Necrosis Factor	PMA	Hensel et al., 1989
Thyroid Stimulating Hormone α Gene	Thyroid Hormone	Chatterjee et al., 1989

elements (MREs), all within 150 bp of the transcriptional start point, as well as elements involved in basal-level expression (BLEs).

The MREs are capable of conferring metal responsiveness on heterologous promoters, with the degree of induction increasing with the number of such sequence elements present (Searle et al., 1985; Stuart et al., 1985). Nevertheless, maximal induction from these heterologous constructs is less than the native metallothionein promoter. Karin et al. (1987) reported that a promoter containing tandem repeats of both BLE and MRE sequences from the hMTII$_A$ gene demonstrated increased inducibility. Unfortunately, such increased inducibility was accompanied by a parallel increase in the basal level of expression.

This problem was in large part overcome by McNeall et al. (1989). In their expression vectors, a region from -70 to -129 containing a BLE was deleted from the hMTII$_A$ promoter. Multiple MREs were inserted to replace this deletion. The resulting vectors exhibit low-level basal expression that is hyperinducible upon treatment with heavy metals. Thus, altering the ratio of MREs to BLEs has a profound effect on a homologous hMTII$_A$ promoter. In the best case reported, induction resulted in a greater-than-200-fold increase in expression.

Mouse Mammary Tumor Virus (MMTV) Expression Elements

The observation that the expression of mouse mammary tumor virus is transcriptionally regulated by glucocorticoid hormones led to the utilization of the MMTV LTR to drive the expression of foreign genes in a regulatable manner (Huang et al., 1981; Lee et al., 1981). Both glucocorticoid-responsive elements (GREs) and basal-level elements (BLEs) have been identified in the MMTV expression elements (Majors and Varmus, 1983; Chandler et al., 1983). A fairly precise mapping of these elements has been determined by deletion analysis

(Lee et al., 1984; Ponta et al., 1985). Sequences sufficient for wild-type promoter function are contained downstream of residue −64. A region between −220 and −140 contains sequences essential for hormonal control. Deletion analysis of an MMTV enhancer/HSVTK promoter chimera served to demonstrate that the region of the MMTV LTR between −236 and −52 is sufficient to confer glucocorticoid responsiveness on a heterologous promoter. Recently, a negative glucocorticoid-responsive element (nGRE) has been described upstream of the bovine prolactin gene. This element serves to silence the bovine prolactin gene promoter as well as heterologous promoters to which it is attached. A region of the bovine prolactin promoter between −51 and −562 contains multiple footprinting sites for purified glucocorticoid receptor binding sites. A 34-bp subfragment containing a single receptor-binding site is sufficient for nGRE activity (Sakai et al., 1988).

Mechanism of Enhancer Action

What is the molecular basis of enhancement? We know that enhancers dramatically affect transcription even when they are placed at great distances from their cognate promoters. We also know that, in many cases, enhancer multimers function more efficiently than the same enhancer present only as a monomer (Schaffner et al., 1988). Enhancer-binding proteins have been isolated and we know, through the conduct of *cis/trans* tests, that these sequence-specific proteins are required for efficient enhancer function.

There are two types of explanations of enhancer activities. The simplest explanation for the action-at-a-distance effect is that the enhancer-binding protein/DNA complex directly interacts with the transcription complex, including the RNA polymerase and other promoter-binding proteins, at the promoter. The simplest mechanism by which this could be accomplished is to bend the DNA into a loop structure in a manner such that the enhancer-binding complex is brought into direct physical contact with the appropriate promoter-binding complex (Müller et al, 1989). Experimental observation tells us that the distance between transcription modules is too short to allow for spontaneous DNA looping, so such DNA bending must be induced by bound proteins. Such DNA bends must be aligned to allow the proteins to move together. Also, the proteins must lie on the correct side of the DNA helix. The experimental observation that transcription modules can tolerate small changes in spacing is difficult to reconcile with the requirement for correct alignment of bound proteins for proper protein/protein interaction: Small changes in spacing might rotate one or the other DNA-binding protein complex around the helix, thus preventing correct alignment with the other DNA-binding protein complex as well as changing the helical phasing of the DNA bends.

An alternative explanation involves the inclusion of an integrator molecule that serves to organize the proteins bound to the DNA into a macromolecular complex in a linear array without any requisite bending of the DNA molecule. This model predicts the existence of an as-yet-unidentified cellular component to function as the integrator or adaptor, a factor whose existence can probably be determined through an analysis of *in vitro* transcription experiments (Dynan, 1989; Lewin, 1990).

These models explain the method of integration of information within an enhancer or promoter. However, they do not address the mechanism by which distal promoters and enhancers interact. It has been proposed that such long-range signaling occurs through the generation of long DNA loops, the simplest explanation of the action-at-a-distance phenomenon of enhancement. Such an interaction might allow for the formation of a more stable, more easily regulatable multimolecular enhancer/promoter-binding complex.

PROMOTER AND INTRON INTERACTIONS

One of the least explicable observations about factors affecting the efficacy of recombinant cistrons is experimental evidence that indicates that the requirement for a functional intron in a recombinant cistron is promoter-dependent. This means that, upon transfection, certain promoters function more efficiently when accompanied by a gene containing an intron.

In transfection assays of recombinant cistrons, an intron is required when transcription is driven by an immunoglobulin μ promoter/enhancer combination, although the requirement is not specific for a particular intron (Neuberger and Williams, 1988). β-globin gene expression driven by the SV40 promoter appears to require sequence information from intervening sequences; however, this intervening sequence information cannot support the stable production of β-globin mRNA when it is placed downstream of the polyadenylation site. Further, sequences derived from the transcribed herpes simplex virus thymidine kinase gene can support substantial intron-independent expression (a greater-than-100-fold increase) of a β-globin cDNA whose expression is driven by the SV40 early promoter (Buchman and Berg, 1988).

The ramifications of intron inclusion appear vitally important in determining the level of accumulation of mRNAs in exogenous cistrons introduced into transgenic mice. Brinster et al. (1988) examined the behavior of enhancer/promoter/structural gene chimeras by fusing either the mouse metallothionein I or the rat elastase I enhancer/promoter to the rat growth hormone gene. The mouse metallothionein I/growth hormone chimera was assayed in fetal liver, whereas the rat elastase I/growth hormone chimera was assayed in the pancreas. The behavior of a mouse metallothionein gene mutagenized with a synthetic oligonucleotide to facilitate the discrimination of the mRNAs produced by the exogenous and endogenous alleles, and the human β-globin gene also introduced into transgenic mice was examined in parallel. In each case, there were 10- to 100-fold more mRNAs produced from the intron-containing constructs. Most surprising perhaps is the observation that both the intron-minus and intron-plus mouse metallothionein constructs function equally well when transfected into tissue culture cells. This inexplicable observation strongly suggests that introns play a role in facilitating transcription of microinjected genes and that this effect may be manifest only when those genes have been placed in a developmental setting. The molecular mechanism of promoter/intron cross-talk remains a mystery.

MESSENGER RNA DEGRADATION SIGNALS AND POLYADENYLATION

The stability of mRNAs in cells appears to vary widely (15 minutes to many hours), and it is now clear that rates of mRNA decay are important control points in the regulation of gene expression. The factors governing mRNA stability appear to be structural in nature. The selectivity of the decay process appears to be the consequence of interactions of endonucleases or other factors with specific internal structural features of the RNA molecule.

The exquisite regulation of mRNA degradation is exemplified by the coupling of histone H3 mRNA decay to DNA replication during the cell cycle, and by the rapid turnover of transferrin-receptor mRNA in the presence of its ligand, iron. Both events can be attributed to a stem-loop structure found at the 3' end of the mRNAs encoding both proteins (Brawerman, 1987, 1989).

Several short-lived RNA species have been shown to contain a group of AU-rich sequences in their 3' non-coding regions (Shaw and Kamen, 1986; Caput et al., 1986). These sequences have been shown to be responsible for the metabolic instability of these RNAs. However, a similar site is found in the 3' non-coding sequence of β-globin mRNA, which is highly stable, so there must be some selectivity in AU-rich destabilizing sequences. These sequences have been described in a variety of lymphokine genes, as well as in a number of oncogenes such as c-fos and c-myc. In the case of c-fos, mRNA decay may be due to multiple destabilizing elements. The deletion of the AU-rich region alone is insufficient to maintain mRNA stability: The c-fos coding region itself can incur instability on an otherwise stable molecule. The work of Shyu et al. (1989) suggests that these two destabilizing regions function by different decay processes. In fact, inhibitors of transcription interfere with the decay of an RNA bearing the c-fos AU region but do not interfere with the decay of a construct containing the c-fos coding region.

The role of polyadenylation in mRNA stability remains unclear. Most available data suggest that the poly-A tract can protect mRNAs from rapid destruction. Circumstantial evidence for this conclusion comes from experiments that demonstrated time-dependent shortening of poly-A tracts: Newly synthesized globin mRNAs contain a chain of approximately 150 adenylate residues, which is progressively shortened to 40–60.

The half-lives of rabbit globin mRNAs containing poly-A tracts of varying lengths were compared directly by co-injection with radiolabeled histidine. The functional stability in this heterologous system was measured over time. The results indicate that mRNA molecules containing 32 or more A residues have the same stability as those containing 150 A's. However, mRNAs with only 16 A's were 10-fold lower in number than those with 32–150 A's. This behavior appears to be a function of structural stability rather than translational efficiency, a conclusion supported by *in vitro* decay experiments. In such experiments, the half-lives of polyadenylated histone and globin mRNAs are at least 10-fold longer than their de-adenylated counterparts (Bernstein and Ross, 1989). In no situation does it appear that polyadenylation has a negative effect on gene expression. Therefore, recombinant cistrons should carry polyadenylation signals to facilitate gene expression.

There is some indication that the molecular mechanism of mRNA stability/instability involves the interaction of the 3' untranslated AU-rich instability sequence with the poly-A tail, complexed with so-called poly-A–binding proteins (PABPs). Stable mRNAs, lacking the AU-rich instability sequences, maintain stable poly-A/PABP complexes, whereas the poly-A/PABP complexes of unstable mRNAs are somehow disrupted by the 3' untranslated instability sequences, and such destabilization leads to nucleolytic attack (Bernstein and Ross, 1989).

TRANSLATIONAL CONTROL

Several types of *cis*- and *trans*-acting elements affect the efficiency of translation of eukaryotic mRNAs and thus can be exploited in the generation of recombinant cistrons. Jang et al. (1988) described a sequence from an encephalomyocarditis virus that facilitates the translation of both coding sequences of a bi-cistronic mRNA. This sequence functions as a ribosome entry site. When this sequence is found downstream of the coding sequence of one gene but upstream of the coding sequence of another gene on the same mRNA, both coding regions will be translated. The insertion of this sequence into a recombinant DNA molecule at the appropriate location can allow for the efficient translation of both genes of a bi-cistronic mRNA, and could conceivably function with a tri-cistronic mRNA as well.

Another element, a *trans*-acting one, can function to dramatically affect mRNA translation: the so-called adenovirus VA genes (Kaufman, 1985). Upon co-transfection into COS cells, these gene products appear to enhance the translation of most but not all mRNAs. One such non-reactive cistron is that of the SRα vector described in the following chapter.

The putative mechanism of action of the adenovirus VA genes on translation involves a protein kinase known as the dsRNA-activated inhibitor (DAI). After viral infection, the two VA RNA polymerase III transcripts are expressed at very high levels, and they are required for efficient RNA translation of adenovirus mRNAs late in infection. VA RNA can bind DAI kinase and inhibit its activation by dsRNA *in vitro*. It has been proposed that the appearance of dsRNA, produced perhaps from asymmetric transcription of the adenoviral genome, activates DAI kinase to phosphorylate the α-subunit of active eukaryotic initiation factor 2 (eIF2), resulting in the inhibition of translation initiation. VA RNA inhibits DAI kinase and so prevents the inhibition of translation initiation, an effect that can be mimicked by the DAI kinase inhibitor 2-aminopurine (Kaufman, 1987). Kaufman and colleagues (1989) have created a similar situation in translation by expressing a serine to alanine mutant of eIF-2, which renders it phosphorylation resistant in transfected cells. Thus, the VA RNAs regulate the level of active eIF2. Surprisingly, in transient transfection experiments, the effects of VA RNA and 2-aminopurine reveal that the dramatic translational enhancement one observes is restricted to the transfected plasmid DNA; the endogenous chromosomal genes are unaffected by the inducers. Thus, in particular applications, especially in COS-type transient replication experiments, the inclusion of the VA genes in the transfection mix, either in *cis* or *trans*, may serve as a useful tool in increasing the expression of transfected genes from most but not all promoters.

RELATED PHENOMENA

The initial discovery of enhancers was motivated, in part, by the search for eukaryotic origins of replication as *cis*-acting elements that could enhance the transforming efficiency of a marker gene. In fact, numerous studies link enhancer function with DNA replication. Several recent studies of the effects of DNA replication on transcription and presumably enhancer function indicate that such effects are cell-specific (Grass et al., 1987; Enver et al., 1987; Campbell and Villarreal, 1988). Such effects should not be ignored.

Other factors may also affect the function of a recombinant cistron introduced into cells and animals. For instance, the pattern of DNA methylation may affect the developmental regulation of enhancer function within retroviral genomes integrated into the germ line of mice (Jahner and Jaenisch, 1985), as well as the developmental regulation of the immunoglobulin κ gene in tissue culture (Kelley et al., 1988). In addition, the site of integration of the recombinant cistron into the genome of the host cell can dramatically alter the "expressibility" of that cistron. This phenomenon is referred to as position effect. Recent experimental evidence indicates that random position effects can be overcome by exploiting a methodology that facilitates targeted integration of the recombinant molecule into the host genome through homologous recombination. In such experiments, the neo^r gene, driven by the SV40 promoter and flanked by β-globin sequences, was homologously recombined into the chromosomal β-globin locus. After integration, the foreign cistron was subject to the same transcriptional inducers as the natural β-globin allele (Nandi et al., 1988). Implementation of this technique should enable one to retain authentic biological regulation of a recombinant cistron after integration. This methodology is discussed in greater detail in the section dealing with homologous recombination and gene-replacement vectors in Part II.

2
VECTORS

INTRODUCTION

Much information on *cis*-acting elements has been exploited in the creation of a battery of readily available gene-transfer vectors, each with its own strengths and weaknesses, many of which are presented in abbreviated form in Table 2-1. In my review of these creations, I will illustrate how they can be used, as well as provide sufficient information to allow you to incorporate your own gene or

TABLE 2-1 Expression Levels and Utilities of Various Mammalian Gene Transfer and Expression Systems

Cell Line	Mode of DNA Transfer	Optimal Expression Level ($\mu g/ml$)	Primary Utility
Monkey cells/Human cells			
CV-1	SV40 virus infection	1–10	Expression of wild-type and mutant proteins
CV-1/293	Adenovirus infection	1–10	
COS	Transient DEAE-dextran	1	Cloning by expression in mammalian cells; rapid characterization of cDNA clones; expression of mutant proteins
CV-1	Transient-DEAE-dextran DNA transfection	0.05	
Murine-fibroblasts			
MOP	Transient DEAE Dextran	NA	Rapid characterization of cDNA clones; expression of mutant proteins
C127	BPV stable transformant	1–5	High-level constitutive expression
3T3	Retrovirus infection	0.1–0.5	Gene transfer into animals; expression in different cell types
CHO-DHFR	Stable DHFR + transformant	0.01–0.05	
	Amplified MTX	10	High-level constitutive expression
Primate/rodent	Vaccinia virus	1	Vaccines
	EBV vector	NA	Cloning by expression
Neurons	HSV vector	NA	Gene transfer

Source: Adapted from Kaufman, *Methods Enzymol.* Vol. 185, Table 1, p. 509. Copyright © 1990 by Academic Press, Inc.

cDNA inserts into these vectors. Before I describe this collection of gene-transfer vectors, it is important to discuss two fundamentally different approaches to gene transfer: transient transfection and stable transfection.

Transient Transfection

Transient transfection is an excellent method for obtaining very high levels of expression of a transferred gene in tissue culture cells, second only to the generation of cell lines containing stably integrated copies of transferred DNA that have undergone substantial amplification via co-selection with an amplifiable marker. In transient transfection, one introduces either the recombinant DNA molecule or the recombinant viral particle to the cell by either DNA-mediated gene transfer techniques or viral infection. The three most commonly used transient DNA-mediated gene transfer techniques incorporate either DEAE dextran or calcium phosphate in the transfection mixture to facilitate DNA uptake by the recipient cells, or use the technique of electroporation, wherein the recipient cells are mixed in suspension with the DNA to be transferred, after which the mixture is exposed to a transient electric pulse. A fourth type of transient expression methodology involves the generation of infectious recombinant retroviruses. This approach will be discussed in detail in the section on retrovirus-mediated gene transfer in Part II.

The object of transient transfection is to obtain a burst of gene expression, after which the cultures are disposed of. It is most commonly used when the vectoring system employed is lytic upon transfection and therefore stable expression is impossible, or when transient expression of a particular gene product is all that is required for the experimenter's needs. Generally speaking, transient transfection is the method of choice when speed of execution is a primary consideration in the experiment. This is often the case when one is confronted with numerous DNA samples to evaluate, as in expression cloning experiments. Transient transfection is, in fact, the only practical alternative when the experimenter is confronted with a large number of samples that must be assayed in a short period of time. This is because the generation and maintenance of stable transfectants or infectants require not only the generation of the transfected plates but subsequent maintenance (e.g., splitting for propagation and drug selection) of all of the cultures. This task is tremendously labor-intensive, time-consuming, and costly when handling tens or hundreds of plates.

The major drawback of transient transfection is that gene expression occurs in an explosive burst and then disappears, as the cells either lyse or fail to stably integrate and express the transfected or infected genetic information. Therefore, transfected cultures must be generated over and over again to maintain a continuous source of fresh material for analysis.

Stable Transfection

Stable transfection is the best method for obtaining moderate expression levels from a transferred gene in a situation where multiple transfections need not be carried out. It is also the best method, when combined with selective gene amplification, for obtaining very high levels of expression in continuous culture.

In this method, you again introduce either the recombinant DNA molecule or the recombinant viral particle to the cell by DNA-mediated gene transfer techniques or viral infection. In this case, the most commonly used gene transfer techniques are calcium-phosphate-mediated gene transfer, electroporation, and viral infection. In general, the establishment of stably transformed cells occurs one or even two orders of magnitude less frequently (depending on the cell type under investigation) than does the establishment of a transiently expressing recombinant plasmid in a cell during transient transfection.

Identification of a rare stable transfectant among a population of untransfected cells requires either a phenotypic change in the cell, such as morphologic transformation, or the inclusion of a drug-selectable marker in the recombinant cistron or co-transfection with a drug-selectable marker on a separate DNA molecule. Execution of the drug-selection protocol, followed by the isolation and characterization of individual transfectant or infectant clones, is quite labor-intensive and may require several weeks of tissue culturing—several months or more if the transfectant genes are to be co-amplified by drug selection. This is the major drawback to stable transfection. However, drug selection may not always be necessary when transferring genes via infectious recombinant retroviruses. Infection of cells in culture with high-titer recombinant retroviruses can result in infection of virtually all members of the population of recipient cells. If you intend to carry out an in-depth, long-term analysis of the expressed product, then stable expression will save many weeks of work in the long run.

SV40-BASED VECTORS

This chapter begins with gene transfer vectors based on the simian virus 40 (SV40) genome (Figure 2-1) because, as far as transfer vectors are concerned, this is where it all began. In fact, the basic gene transfer methodologies—DEAE-dextran-mediated gene transfer, calcium-phosphate-mediated gene transfer, and DNA microinjection—were developed, at least in part, with the analysis of wild-type and mutant SV40 genomes in mind.

Studies of SV40 biology led to the removal of SV40 late-gene sequences from the genome and subsequent replacement by genes encoding tRNAs and other sequences. Once these constructions were shown to replicate in simian cells (CV-1), SV40 genomes were inserted into bacterial plasmids such as pBR322 to facilitate further manipulation. These chimeric molecules could be propagated in *E. coli* and subsequently transferred to tissue culture cells where, in the appropriate cell type, they could express the early viral genes and replicate both the viral and bacterial DNA as part of a single molecule. These recombinant DNAs became the basis of so-called *shuttle vectors*, DNA molecules that could be experimentally manipulated to carry foreign DNA and could replicate in both bacterial and animal cells. While conducting experiments with these shuttle vectors, it became apparent that the rates of replication of these viral/bacterial plasmid chimeras relative to wild-type SV40 DNA were considerably lower. Excision of the viral DNA from the plasmid component followed by recircularization and transfection indicated that the bacterial component substantially depressed the replication of the chimera in *cis* but not in *trans*. It was ultimately determined that plasmid sequences around the NIC/BOM site of pBR322 were "poisonous" to viral DNA replication in animal cells, and viral DNA/plasmid chimeric molecules lacking these sequences replicated as efficiently as wild-type SV40 DNA. Such "poison-minus" constructions serve as the basis of all modern SV40 gene transfer vectors (Lusky and Botchan, 1981).

For purposes of discussion, our definition of an SV40-based vector is a vector that contains the SV40 origin of replication (ori); it may or may not contain functional SV40 early or late promoters. Such a vector is capable of replicating in a cell permissive for SV40 viral replication when the SV40 large T (tumor) antigen is present. The gene encoding the SV40 large T antigen can either be provided in *cis* as part of the shuttle vector or in *trans* as it is in COS-type cells. In fact, the most useful SV40-based gene transfer vectors do

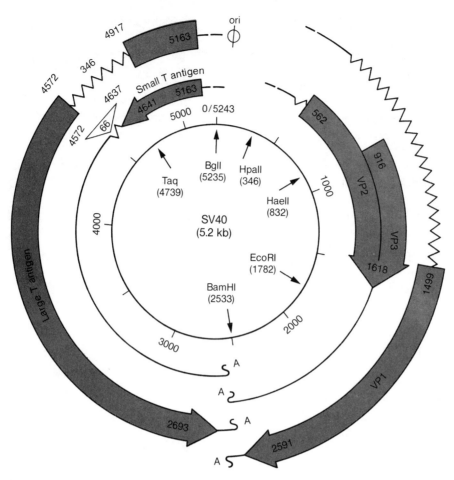

FIGURE 2-1 *SV40 genome.* The genome of simian virus 40 (SV40) is depicted here. The early region, encoding the T(umor) antigens, lies to the left. The late region, encoding the viral coat proteins, lies to the right. The origin of replication (ori) is found between the early and the late regions. The ori and early region functions play key roles in the expression of genes from SV40-based vectors and in COS cells. See the text for details. [Adapted from Tooze (1980), *DNA Tumor Viruses*, Part 2, Cold Spring Harbor Press, p. 813.]

not encode their own SV40 large T antigens but are COS-dependent for DNA replication. As such, the COS expression system warrants description at this point.

The COS-type cell is a simian cell that is permissive for SV40 replication and has been genetically engineered to contain all viral and cellular *trans*-acting factors necessary to support the replication of a recombinant DNA molecule containing an SV40 ori. The development of COS was based on the notion that one could destroy the origin of replication of SV40 without destroying the function of the SV40 early promoter responsible for driving the expression of the viral T antigens. Such a mutant viral genome, when transfected into a permissive simian cell, would produce the SV40 T antigens, would not replicate, would stably integrate into the host chromosome, and presumably would not lyse the transfected cell. Such a cell could support runaway replication of an exogenously introduced recombinant molecule carrying an SV40 ori.

This reasoning proved correct and led to the creation of COS cells, including some producing a temperature-sensitive mutant of the SV40 large T antigen. Recombinant molecules bearing an SV40 replicon transfected into COS-type cells undergo runaway replication and can lead to very high levels of expression of exogenously introduced genes carried by the replicon (see section on expression cloning).

All SV40 vectors are similar in nature, with the fundamental differences being (1) the manner in which the DNA is inserted into the vector and (2) the promoter that serves to drive the expression of the gene of interest. SV40 vectors have been most effectively used to drive the expression of cloned cDNAs. The prototype for modern SV40-based cDNA cloning vectors was developed by Okayama and Berg (1983). In this vector, expression of the cDNA is driven from the SV40 early promoter. To insert the genes of interest, the synthesis of the cDNA is vector-primed. In this method, the first cDNA strand is primed by poly-dT covalently coupled to one end of the linearized vector. After synthesis of the cDNA first strand, the plasmid vector is circularized with a linker DNA segment that binds one end of the plasmid vector to the 5' end of the cDNA sequence. This method ensures that all of the cDNAs have the same orientation, an orientation appropriate for obtaining expression of the cDNA sense strand from the resident SV40 early promoter. The result is termed *directional cloning*. DNA fragments can be inserted into this vector by more conventional methods such as tailing, annealing, and linker ligation. The advantages and disadvantages of these various joining methodologies are discussed further in the section on expression cloning in Part II.

The Okayama and Berg vector has since been modified in a variety of ways to either alter its replication-host range or to increase its expression efficiency in Cos cells. Margolskee et al. (1988) modified this vector by inserting an Epstein-Barr virus ori in place of the SV40 ori, thus enabling the vector to replicate in special types of human cells (see Epstein Barr virus vectors). Takebe et al. (1988) invented an Okayama and Berg variant that represents a significant advance in gene expression technology. Their vector is called SRα (Figure 2-2). The key improvement is in the insertion of a *cis*-acting element derived from the R-U5 segment of the LTR of HTLV-I (see previous chapter) just downstream of the SV40 early promoter in the Okayama and Berg vector. This element serves to up-regulate expression from the SV40 promoter. Its function is orientation dependent, and it must be positioned downstream of the transcription start site. Therefore this element exists as a component of the mRNA transcript and may function in a manner analogous to HIV-I *tar*. In all other respects, SRα is identical to its Okayama and Berg counterpart. A side-by-side comparison of these two vectors reveals that SRα is 50–100 times more efficient in driving the expression of a variety of exogenously introduced genes.

A rather clever, if somewhat complicated, combination COS-based Chinese Hamster Ovary (CHO)-based expression vector has been developed by Kaufman (1985). He has assembled a vector, pg1023(B), and a derivative, pMT2 (Figure 2-2), both of which contain an SV40 origin of replication and a transcriptional enhancer segment. Expression is driven from the adenovirus major late promoter (MLP) combined with the tripartite leader and a hybrid intron consisting of a 5' splice site from the first exon of the tripartite leader and a 3' splice site from a murine immunoglobulin gene. Expression from these vectors is augmented by the *trans*-acting adenovirus VA I RNA. Downstream of the EcoRI site used for cDNA insertion is the murine dihydrofolate reductase (DHFR) coding sequence, which is itself followed by the SV40 early polyadenylation signal. Expression of DHFR from the resulting bicistronic transcripts facilitates direct selection of transfectant DHFR-CHO cells. Exposure of such transfectants to increasing concentrations of methotrexate facilitates direct selection for amplified copy number of the transfected gene. The adenovirus tripartite leader sequence and VAI gene product both enhance the translatability of the mRNA of interest. This vector has been used to expression-clone a number of cytokine genes.

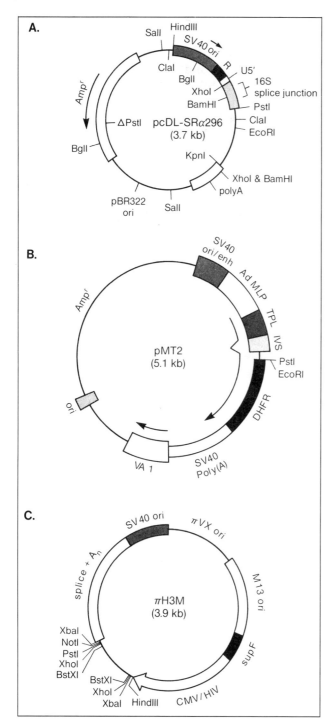

FIGURE 2-2 *SV40-based expression vectors.* These three SV40-based expression vectors have been used extensively to express a number of exogenous genes in COS cells. **Panel A.** SRα296. The direction of transcription from the eukaryotic promoter is indicated by the arrow on the right. The vector is composed of three segments. Segment one (Hind III–Pst I fragment) contains the SRα promoter and the SV40 late-gene splice junction. Segment two (Pst I–Kpn I fragment) contains an intervening DNA piece between the plasmid linker and the vector-primer segments. Segment three (Kpn I–Hind III fragment) contains the vector primer that includes the Kpn I-oligo(dT) priming site, the SV40 late region polyadenylation signal, the pBR322 replication origin, and β-lactamase gene lacking a Pst I site. **Panel B.** pMT2. The vector is composed of three segments. Segment one contains the SV40 origin of replication and enhancer, the adenovirus major late promoter (MLP), a majority of the adenovirus tripartite leader sequences, and an intervening sequence. Segment two contains the dihydrofolate reductase (DHFR) coding region, the SV40 polyadenylation sequence (poly A), and sequences encoding the adenovirus VA1 RNA. Segment three contains the bacterial origin of replication and the bacterial β-lactamase gene conferring ampicillin resistance (Amp^r) derived from the pUC18 plasmid. **Panel C.** πH3M. The direction of transcription from the eukaryotic promoter is indicated by the arrow. The vector is composed of seven segments. Segment one is the pBR322 origin of replication (πVX ori). Segment two is the M13 origin of replication. Segment three is the supF gene (the selectable marker for this plasmid in the appropriate bacterial host). Segment four is a chimeric cytomegalovirus/human immunodeficiency virus promoter. Segment five is a stuffer fragment with which you replace the cDNAs of interest. Segment six contains splice and polyadenylation signals derived from the plasmid pSV2. Segment seven contains the SV40 origin of replication. [Adapted from Takebe et al. (1988), *Mol. Cell. Biol.*, Vol. 8, p. 469, Fig 3, and Kaufman (1990), *Methods Enzymol.*, Vol. 185, D. Goeddel Ed., Academic Press, p. 500, Fig. 1.]

A non-COS-based two-part glucocorticoid inducible expression system has been developed from the pMT2 vector. Part I of the expression system is a DHFR-CHO cell line, GRA, that has been genetically engineered to overexpress the rat glucocorticoid receptor. Part II of the system is a pMT2 derivative, pMG18PC, into which multiple copies of the MMTV glucocorticoid-responsive element have been inserted in such a manner that they can affect the expression of the inserted gene (Israel and Kaufman, 1989). Transfection of this plasmid into this cell line, followed by treatment of the transfected cells with glucocorticoids, results in a sevenfold increase in expression relative to the already high basal level of expression of the uninduced vector. The utility of this vector has been further increased by removing the constitutive SV40 enhancer from the construct. In this vector, pMG18δS, basal expression has been eliminated, but the high level of glucocorticoid induction has been retained.

A slightly different approach to developing a more useful vector has been adopted by Aruffo and Seed (1987). Their vector, πH3M (Figure 2-2), has been used in the expression cloning of a variety of cell-surface molecules. Three features of this vector make it particularly useful for COS-type transfection. First, the eukaryotic transcription unit, a chimeric cytomegalovirus/human immunodeficiency virus promoter, is well-suited to COS-based expression. Second, the small size and particular placement of the vector sequences allow for high-level replication in COS cells. Third, the presence of two identical BstXI sites in inverted orientations and separated by a short replaceable fragment allows the use of an efficient oligonucleotide-based strategy to promote cDNA insertion into the vector—in essence, linker insertion. The plasmid backbone contains a πVX origin of replication and a suppressor tRNA gene to facilitate replication and selection in *E. coli* strain MC1061 (p3) for tetracycline and ampicillin resistance because the suppressor tRNA complements the amber mutations in the tetracycline resistance and beta-lactamase genes carried on the p3 plasmid. As is the case with SR-α and pMT2, πH3M is a shuttle vector and can, after transfection of tissue-culture cells, be rescued by Hirt extraction (Hirt, 1967) and subsequently transferred to the appropriate *E. coli* host for further propagation and subsequent analysis. πH3M has been used to expression-clone a remarkable number of cell-surface molecules. In comparison to the two other expression-cloning vectors described above, πH3M and its relative, CDM8, have one minor drawback. When cloning into πH3M, one must excise and remove the short stuffer fragment from the vector. In addition, the vector is a defective M13 molecule, and therefore a second DNA, p3, must be grown concurrently. The vector must be purified from this DNA as well.

We have developed an improved cDNA-cloning technique that can be applied to all of the expression vectors described and, in combination with our solid-state plasmid-library amplification technique, results in cDNA-library-generation-efficiencies of 1×10^7 CFU/μg RNA, considerably better than any published reports. With this procedure, vector background is less than 1 percent, which represents a significant improvement in cDNA cloning for all purposes. The method is described in detail in the section on expression cloning in Part II.

POLYOMA-VIRUS-BASED VECTORS

The development of polyoma-virus-based vectors parallels the development of SV40 vectors. Wild-type polyoma virus replicates efficiently in murine cells. The organization of the polyoma virus genome is quite similar to that of SV40 with one striking exception: Whereas SV40 encodes two early-region gene products (T antigens), polyoma virus encodes three T antigens—small, middle, and large (Figure 2-3).

The polyoma virus large T antigen is both necessary and sufficient, in a permissive environment, to support the replication of polyoma virus genomes carrying a polyoma ori. This observation led to the development of polyoma-virus-equivalents of SV40 COS cells. These new lines were generated by transfecting NIH 3T3 cells with an originless polyoma virus, resulting in the

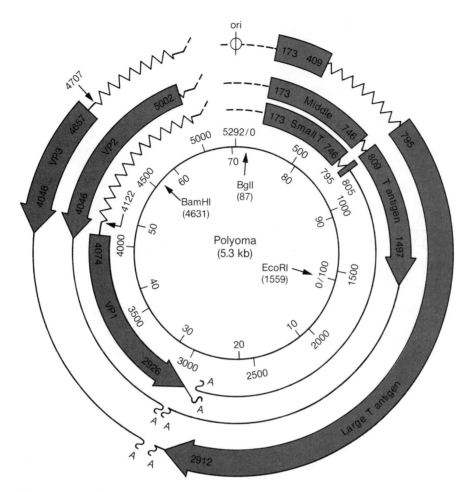

FIGURE 2-3 *Polyoma virus genome.* The genome of polyoma virus is depicted here. The early region, encoding the T(umor) antigens, lies to the right. The late region, encoding the viral coat proteins, lies to the left. The origin of replication (ori) is found between the early and the late regions. The ori and early region functions play key roles in the expression of genes from polyoma-virus-based vectors and in MOP cells. See the text for details. [Adapted from Tooze (1980), *DNA Tumor Viruses*, Part 2, Cold Spring Harbor Press, p. 907.]

creation of mouse originless polyoma virus (MOP) cell lines (Müller et al., 1984). Shuttle vectors carrying the polyoma virus ori inserted into a poison-minus derivative of pBR322 can replicate with wild-type efficiency in such cell lines. In fact, replication efficiencies are virtually identical to COS/SV40 plasmid experiments (Müller et al., 1984). Thus, MOP cell lines provide a workable alternative to COS when it is important to express a gene of interest in a murine cell.

An equivalent cell line, WOP 32-4, which encodes a temperature-sensitive polyoma virus large T antigen from an originless plasmid, has been described as well (Kern and Basilico, 1986). This cell line can be stably transfected with a plasmid containing a functional polyoma virus ori, a dominant selectable marker gene, and the gene of interest, as long as the line is maintained at 39°C, the non-permissive temperature for the temperature-sensitive large T antigen. After shifting such cultures to 33°C, the polyoma-virus-ori-containing plasmids excise due to characteristic instability, replicate the excised genome, and express the foreign gene, which is now carried by the replicating nuclear plasmid.

Unfortunately, the development of such cell lines has not led to the development of expression vectors as sophisticated as those described for SV40-based COS transfection. The result of such neglect is that COS-based transfection is the overwhelming favorite for transiently expressing genes at high levels for purposes such as expression cloning. Nevertheless, the MOP and WOP cell lines may be appropriate for your use, particularly if you intend to construct your own polyoma virus expression vector. In addition, as mentioned previously, point-mutated polyoma virus expression elements have been identified and described that are competent for replication and gene expression in embryonic cells, a claim that cannot be made for any SV40-based expression system (see previous chapter).

ADENOVIRUS-BASED VECTORS

There are a number of features of adenovirus biology that have been exploited to make adenovirus-based vectors important to consider when attempting to transiently express a gene of interest at very high levels. The notion that one might construct, with recombinant techniques, adenoviral vectors arose from the observation that co-infection of simian cells with adenovirus and SV40 often results in the formation of novel adeno-SV40 hybrid viruses. This observation indicated a certain level of plasticity in the adenoviral genome. Adenovirus normally grows poorly in simian cells. Only in the presence of the SV40 T antigen (a product of the SV40 A gene), which provides the so-called helper function, can adenovirus replicate well in simian cells. In an attempt to develop adenovirus into a useful expression vector, investigators inserted the SV40 A gene into the adenovirus genome downstream from a number of adenoviral promoters to monitor adenoviral expression from these promoters. Adenoviral promoters are categorized into two classes, early (ORE) and late (ORL). Early promoters are expressed prior to the onset of viral DNA synthesis, and late promoters are expressed after. Thummel et al. (1983) were able to, with a fair degree of precision, position the SV40 A gene close to and downstream from what is known as the adenovirus major late promoter (MLP) (Figure 2-4). Their experiments demonstrated the importance of the tripartite leader adjacent to the MLP in obtaining maximal adenovirus-directed gene expression, a key step in the development of adenovirus as an expression vector.

The strategy just described worked because adenovirus/SV40 recombinants replicate preferentially in simian cells. However, the substitution of other, non-selectable, foreign genes for the SV40 T antigens placed new requirements on the vector system. To address this problem, a generalized adenovirus vector system was developed. The vector carried a complete structural gene encoding the SV40 T antigen whose expression was driven by its own promoter. The foreign genes of interest, either the polyoma T-antigen genes or the HSVtk gene, were inserted just upstream of the SV40 T-antigen cistron, and just downstream from the adenovirus MLP and tripartite leader. Neither the polyoma T antigens nor the HSVtk gene provides adenovirus helper function. Therefore, in this strategy, one does not select directly for the gene product of interest (Mansour et al., 1985; Yamada et al., 1985). In these experiments, which incorporated the HSVtk gene, as much as 10 percent of the protein synthesized late in infection and as much as 1 percent of the total cell protein was found to be the HSVtk gene product (Yamada et al., 1985).

To use these vectors, one must insert the non-selectable gene into a type of transfer vector (Figure 2-5). In this vector, the non-selectable gene is flanked

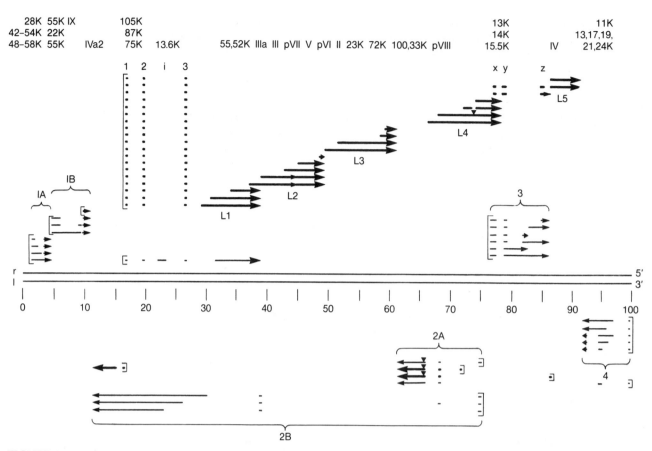

FIGURE 2-4 *Adenovirus genome*. The transcriptional map of the adenovirus-2 genome, a major component of the recombinant, hybrid adenoviruses that are used in the generation of the adenovirus-based expression vectors described in the text, is shown here. The 36,500-bp genome is divided into 100 map units. Arrows indicate the direction of transcription of the r and l strands of the DNA. Several *cis*-acting elements important in the function of both adenovirus-based expression vectors and the SV40-based expression vector, pMT2, are noted. They are the early regions eIA and eIB and the tripartite leader found in transcripts of the r strand denoted 1, 2, and 3. The major late promoter is found just to the left of the first segment of the tripartite leader.[Reprinted from Tooze (1980), *DNA Tumor Viruses*, Part 2, Cold Spring Harbor Press, p. 995.]

at the 5' end by adenoviral DNA sequences homologous to the desired site of insertion of the transfer vector into the adenoviral genome. The non-selectable gene is flanked at the 3' end by an intact SV40 A gene whose expression is driven by its own promoter. After digestion of both the transfer vector and an adenovirus 2/5 recombinant viral genome (Ad 1 × 51) with a restriction endonuclease (BamHI), the DNAs are mixed and ligated to one another. This ligation mixture is mixed with intact adenoviral 1 × 51 DNA, which serves as helper, and this mixture is co-transfected into human 293 cells. Three days post transfection, viral lysates are prepared. During infection of the 293 cells, which are fully permissive for wild-type adenoviral replication, either intermolecular or intramolecular recombination occurs between the duplicated regions of the recombinant adenoviral genome present in the ligation mixture. This results in the accurate alignment of the inserted, non-selectable gene with the MLP and tripartite leader. This recombination event results in optimal positioning of the inserted gene for maximal gene expression. The viruses arising from co-transfection of the 293 cells are harvested and amplified by two cycles of infection on CV-1 (simian) cells, thus selecting for any adeno/SV40 recombinant viruses. In these experiments, adenovirus 1×51 is again used as a helper. At this

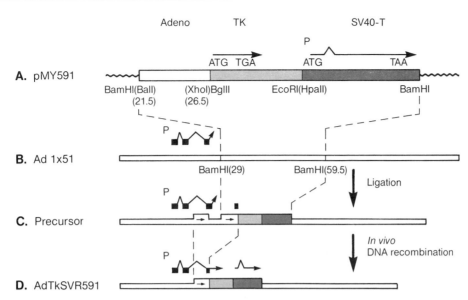

FIGURE 2-5 *Adenovirus-based expression vector.* The structure and mechanism of generation through recombination of non-defective adenovirus expressing the HSVtk gene described in the text are depicted here. **A**. pMY591. The tk gene was first inserted between adenovirus sequences (21.5–26.5 map units) containing most of the third segment of the tripartite leader at its distal end and an SV40 segment which contains the entire T-antigen coding sequence, the early promoter (P), the enhancer, replication origin, and polyadenylation signal. This entire assembly was inserted into pBR322, denoted by the wavy line. The direction of transcription of the tk gene, the SV40 T-antigen gene, and the adenoviral sequences is co-linear. **B**. This is a map of the adenovirus vector 1 × 51 showing the position of the major late promoter (P) and the tripartite leader (black boxes). **C**. Both pMY591 and vector 1 × 51 are digested with Bam HI, mixed, and ligated to form the precursor depicted in C. This precursor carries a duplication of adenoviral DNA sequences from map positions 21.5 to 26.5 that is depicted by the open boxes with arrows. Post-transfection, these duplicated regions are resolved by either intermolecular or intramolecular recombination between the duplicated regions forming the genome of AdTkSVR591 depicted in **D**. [Reprinted from Yamada et al. (1985), *Proc. Natl. Acad. Sci. U.S.A.*, Vol. 82, p. 3568, Fig. 1.]

point, individual plaques are isolated and again amplified by two passages on CV-1 cells.

The apparent high levels of expression delivered by this adenoviral vector system can be attributed to three characteristics of the system: (1) the copy number of the tk gene is high because it is efficiently introduced into cells by replicating hybrid adenoviruses; (2) transcription of the non-selected gene is under the control of the highly efficient adenovirus MLP; and (3) translation of the recombinant mRNA is highly efficient because it contains the adenovirus tripartite leader. A major advantage of this vector system is that the recombinant adenovirus stock, once generated, can serve as a handy reagent for large-scale infection of cells in culture, leading to high-level production of the protein of interest.

Helper-virus-independent adenoviral vectors have also been developed in a number of laboratories (Van Doren et al., 1984; Haj-Ahmad and Graham, 1986; Massie et al., 1986; Davidson and Hassell, 1987; Berkner et al., 1987). In these vectors, the adenovirus early region has been deleted to create space for exogenous DNA insertion. The E1a functions are provided in *trans* by human 293 cells, which constitutively express the adenoviral early-gene products. Thus, the cell line complements the defect in the viral genome.

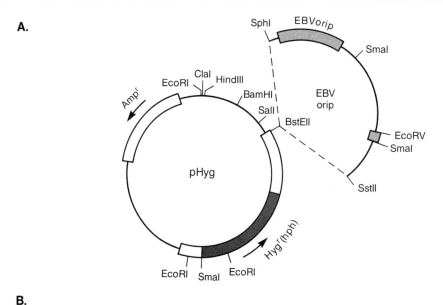
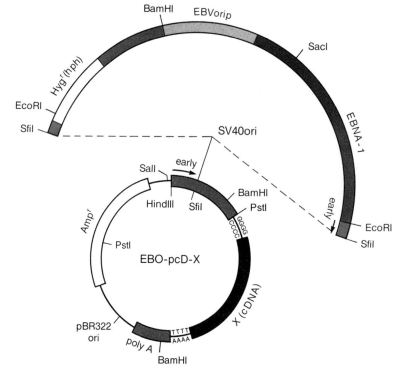

FIGURE 2-6 *Epstein-Barr-virus (EBV)-based expression vectors*. **Panel A**. Vector pHEBO. This plasmid, described in the text, can replicate in EBV transformed B-lymphoblasts. pHEBO was constructed by insertion of sequences derived from the Epstein-Barr virus genome containing the two regions of the genome that constitute the viral origin of replication (indicated by the two boxes in the orip fragment) into the plasmid pHyg (hygromycin). pHyg contains both the bacterial origin of replication and the bacterial β-lactamase gene as well as the hygromycin phosphotransferase (hph) gene of *E. coli*, which encodes resistance to hygromycin B, flanked by the promoter and polyadenylation signals derived from the herpes simplex virus thymidine kinase gene, which serve to drive the expression of the Hyg^r gene.
Panel B. Vector EBO-pcD-X. This plasmid, described in the text, was constructed by linearizing the plasmid pcD-X, the original Okayama and Berg expression cloning vector, with Sfi I. Similarly, a derivative of pHEBO into which the gene encoding the Epstein-Barr virus nuclear antigen (EBNA) had been inserted (EBO) was linearized, and the EBO linear molecule was inserted into the linearized pcD-X to form

EPSTEIN-BARR-VIRUS-BASED VECTORS

Epstein-Barr Virus (EBV) is a herpes virus that contains a 135-kb double-stranded DNA genome and transforms human B-lymphocytes into proliferating blasts, which can be efficiently established into continuous cell lines. Such cell lines contain the viral DNA as a nuclear plasmid. A *cis*-acting element, termed orip, facilitates plasmid maintenance in substrate-attached human, monkey, and dog but not rodent cells transformed with wild-type EBV DNA. Sugden and colleagues (1985) initiated the development of Epstein-Barr Virus nuclear plasmid vectors carrying orip by constructing the vector pHEBO (Figure 2-6). Transfection of cells with the orip-containing plasmid is 1×10^5 times more efficient than transformation with the orip-minus congener.

In addition to the orip element, this vector carries the *E. coli* hph gene, which confers resistance to hygromycin B when the vector is transfected into mammalian cells including human lymphoblasts. Introduction of exogenous genes into pHEBO and subsequent transfection of such derivatives into cells followed by hygromycin selection permits the isolation and expansion of human lymphoblastoid cell lines that carry multiple copies of an autonomously replicating nuclear plasmid bearing the gene of interest. Although the copy number of the plasmid DNAs in such transformants can vary from 1–60 copies, the average number of plasmid copies per cell is 10.

The virally encoded *trans*-acting factor necessary to drive the replication of the orip-containing plasmids is the Epstein-Barr Virus nuclear antigen (EBNA) (Reisman and Sugden, 1986). Yates et al. (1985) incorporated the EBNA gene into pHEBO and thus generated a plasmid that carried the virally encoded *trans*-acting factors normally provided by the wild-type *trans*-acting helper EBV genome. This plasmid was further modified by Kioussis et al. (1987) into an EBV-based shuttle cosmid vector. This modification facilitated the inclusion of relatively large stretches of DNA into the nuclear plasmids. Therefore, one can construct genomic libraries that can be transfected into human cells and screened for a particular phenotype. The insertion of large stretches of genomic DNA into this cosmid vector reduces the efficiency of transformation with this vector by two orders of magnitude. It appears that insertion of large DNA fragments into this EBV cosmid shuttle vector adversely affects the expression of the EBNA gene resident in the plasmid. However, the transformation efficiency of the

---

EBO-pcD-X (the addition of the EBNA gene to pHEBO and EBO-pcD-X dramatically expands the host range of replication of these vectors because the EBNA gene product is required for EBV replication and is normally produced by EBV-transformed cells). This vector is composed of seven segments. Segment one contains the SV40 origin of replication and the early-region promoter (early) oriented in the clockwise direction joined to a fragment containing the junctions of the 19S and 16S SV40 late-region pre-mRNA intervening sequences. Segment two contains the insertion site for the various cDNA sequences to be expressed and the flanking dG/dC and dA/dT tails with which the cDNAs have been inserted. Segment three contains the SV40 late-region polyadenylation signal (poly A) and the bacterial origin of replication and ampicillin resistance gene derived from pBR322. The semicircular arc above contains segments 4, 5, 6, and 7. Segment four contains the SV40 regulatory region including the origin of replication, the early-region promoter oriented in the clockwise direction, and the hph gene. Segment five contains the SV40 small t-antigen intervening sequence and polyadenylation signal. Segment six contains the EBV origin of plasmid replication (orip), and segment seven contains the EBNA-1 gene, whose expression is driven by the SV40 late-region promoter oriented in the clockwise direction and depicted at the extreme right end of the semicircular arc. The applications of this vector are described in the text. [Adapted from Sugden et al. (1985) *Mol. Cell. Biol.*, Vol. 5, p. 411, Fig. 1; Margolskee et al. (1988), *Mol. Cell. Biol.*, Vol. 8, p. 2839, Fig. 1.]

genomic-DNA-containing molecules can be restored by again providing the EBNA gene product in *trans*.

Hambor et al. (1988) compared the behavior of five eukaryotic promoters carried by an Epstein-Barr-virus-based vector and transfected into human T-cell clones. They found that, in their constructions, the RSV LTR functioned most efficiently. This EBV/RSV LTR construction was used to achieve selective, reversible anti-sense-RNA-mediated gene inhibition of the T-cell associated molecule CD8. The experimenters achieved a similar level of expression with the promoter from the lymphopapilloma virus 5′ LTR. The rat GRP78 gene calcium ionophore-inducible promoter (GRP78) and the human heavy-metal-inducible metallothionein II_A promoter (HMT II_A), when inserted into the EBV vector, displayed high levels of basal activity upon transfection; however, although these promoters were still inducible, the levels of expression at maximal induction were far below that of the RSV construction. A construction whose expression was driven by the SV40 promoter was least efficient of all.

An Epstein-Barr-virus-based shuttle vector, EBO pcDX, for cloning cDNAs has been developed by Margolskee et al. (1988) (Figure 2-6). It is quite similar to the Okayama and Berg SV40 cDNA cloning vectors; however, in this instance, the SV40 replication machinery has been replaced by the EBV oriP and EBNA gene product. Reconstruction experiments indicate that one can isolate one plasmid encoding HPRT per 10^6 plasmid clones and one plasmid encoding LEU-2 per 10^4 plasmid clones upon transfection of human lymphoblastoid cells with EBO pcD libraries followed by appropriate drug selection.

Heinzel et al. (1988) introduced a different wrinkle into the EBV-based vector. They inserted an SV40 ori into their EBV vector in an attempt to boost episomal copy number in transfected cells. They achieved an amplification of EBV copy number by transiently introducing a plasmid encoding the SV40 T antigen but lacking an origin of replication of its own. The amplification they achieved with this procedure was, on the average, 10-fold. Such amplification should result in a concomitant increase in the expression of genes carried by these vectors.

HERPES-SIMPLEX-VIRUS-BASED VECTORS

The motivation for developing herpes-simplex-virus-based vectors stemmed from the perceived need to develop viral vectors that could carry more than 5 kb of exogenous DNA. Defective herpes simplex virus (HSV) vectors were ideal for this purpose for a variety of reasons. First, the HSV virion can accommodate 150 kb of DNA. Second, the 150 kb-DNA molecules present in serially passaged virus stocks are composed of multiple iterations of sequences, so-called repeat units, arranged in a head-to-tail tandem array (Figure 2-7). The unit repeat sizes range from 3–30 kb. Thus, inclusion of a foreign DNA molecule into a unit repeat could lead to substantial amplification of the foreign molecule in the recombinant virion. Third, defective viral genomes appear to be relatively stable during serial undiluted virus propagation. Fourth, regenerated concatameric viral genomes can be regenerated from individual monomeric repeats following co-transfection of cells with helper virus DNA. This allows for implementation of the amplification strategy mentioned above. Fifth, HSV-1 can be used to infect a wide variety of host-cell species.

There appear to be two separate sets of sequences required for the propagation of defective HSV genomes. The first is the so-called S-component of

HERPES-SIMPLEX-VIRUS-BASED VECTORS

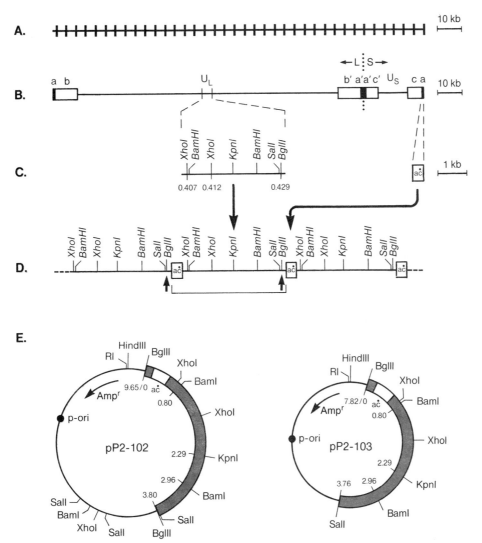

FIGURE 2-7 *Defective herpes-simplex-virus (HSV-1)-based expression vector.* **Panel A.** A schematic representation of the HSV-1 (Patton) defective genomes containing approximately 38 reiterations of 3.9-kb repeat units. The defective genomes terminate at one end with sequences corresponding to the ac terminus of standard virus DNA. **Panel B.** A schematic representation of the standard HSV DNA, displaying the arrangement of the unique and inverted repeat sequences of the S and L components. **Panel C.** The arrangement of the Xho I, Bam HI, Kpn I, Sal I, and Bg1 II restriction enzyme sites in the U_L segment of the standard HSV DNA bounded within the map coordinates shown. ac represents the portion of the S inverted repeat sequences present in the defective genome repeat units. This region has a maximum size of 500 bp. There is some uncertainty in the amount of c sequence present in the 3.9-kb Patton defective genome repeat units. **Panel D.** The arrangement of the U_L and ac sequences within the Patton defective genomes depicted in panel A. **Panel E.** The recombinant HSV-derived plasmid amplicons described in the text, pP2-102 and pP2-103, are depicted here. These vectors contain two segments. Segment one contains the HSV-derived sequences (denoted in the boxes). Segment two contains the bacterial origin of replication and the bacterial β-lactamase gene (Amp^r). pP2-103 was derived from pP2-102 after digestion of pP2-102 with Sal I. [Reprinted from Spaete and Frenkel (1982), *Cell*, Vol. 30, p. 296, Fig. 1.]

standard viral DNA. This region contains the recognition sequence for the cleavage of the HSV concatamers and the subsequent packaging of the viral DNA into nucleocapsids. The second set of sequences is contained within the U_L region of the viral genome. This sequence may represent a replication origin.

Spaete and Frenkel (1982) exploited these properties in the development of an HSV amplicon that carries these sequences. They began their work with a defective genome derived from the Patton strain of HSV-1. The selected genome consisted of a repeat size of 3.9 kb. They coupled the bacterial plasmid pKC7 to this repeat and demonstrated that this amplicon carrying foreign DNA could be propagated in serially passaged virus populations. Similar results have been obtained by Barnett et al. (1983). Kwong and Frenkel (1985) report efficient expression of an ovalbumin gene into such defective amplicons.

Two major limitations of this vectoring system are that it is lytic and therefore only useful for transient expression of a given gene, and that there appears to be an upper size limit, approximately 15 kb, on the efficient replication of HSV-1 amplicons (Kwong and Frenkel, 1984).

HSV-1 infection results in the shutdown of host-cell protein synthesis, apparently at the level of translation. Of course, viral gene expression proceeds in a normal, albeit regulated, fashion.

In an attempt to circumvent the effects of shutdown of host-cell protein synthesis, Shih et al. (1984) developed a non-defective HSV-1 vector (Figure 2-8) and placed the exogenous gene under the control of a viral promoter and other viral *cis*-acting elements in a transfer vector and after co-transfection with wild-type HSV-1 DNA obtained expression of the hepatitis B virus surface antigen (SAg) gene. In these experiments, the SAg gene was fused to the

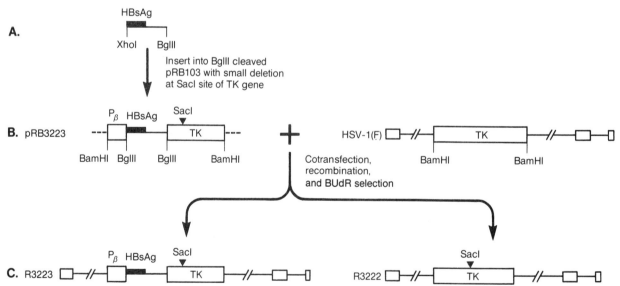

FIGURE 2-8 *Non-defective Herpes-simplex-virus-based expression vector.* The construction of an HSV-1 replication competent recombinant virus containing a chimeric β-tk promoter regulated hepatitis B virus surface antigen (HBV Sag) gene is depicted here. **Panel A**. An Xho I-Bgl II fragment containing the coding sequence for the HBV Sag gene was inserted into a bacterial plasmid, pRB103, containing the Bam HI Q fragment of HSV-1, which includes the HSV-1 tk gene and a 200-bp deletion around the unique Sac I site introduced to inactivate the tk gene as shown in **Panel B**. The resultant plasmid, PRB3223, was linearized and co-transfected with wild type HSV-1 DNA into 143tk$^-$ cells. After recombination and selection for the tk$^-$ phenotype in BUdR, two types of recombinant viruses were identified. The structures of their genomes are shown in **Panel C**. R3223 contains the HVB Sag gene expressed from the β-tk promoter (P_β) and a mutant tk gene. R3223 lacks the recombinant HBV Sag cistron but carries the mutated tk gene. [Adapted from Shih et al. (1984), *Proc. Natl. Acad. Sci.*, Vol. 81, p. 5868, Fig. 1.]

promoter/regulatory region of the β-thymidine kinase and α-4 gene of HSV-1. HSV-1 can undergo homologous recombination; therefore, co-transfection of the HSV-1 promoter/hepatitis B SAg gene chimera with intact HSV-1 DNA results in the appearance of recombinant HSV-1 genomes in which the viral tk gene has undergone recombination with the homologous flanking region of the chimeric molecule. This recombination event results in the appearance of tk⁻ viruses carrying the exogenous gene. Cells infected with and producing these viruses can be selected by incubation with bromodeoxyuridine. Shih et al. (1984) found that the foreign gene incorporated into the viral genome was not shut off during viral replication; rather, the expression of this gene was regulated in a manner identical to that of the authentic viral gene driven by that viral promoter, either β-thymidine kinase or α-4.

Similarly, Whealy et al. (1988) reported the development of a swine herpes virus, pseudorabies virus (PRV), as an expression vector. They generated chimeric genes containing portions of the PRV virion-envelope glycoprotein gIII and the human immunodeficiency virus type I (HIV-I) envelope glycoproteins gp120 and gp41 and obtained expression of these chimeric envelope glycoproteins. Rather than select for viral recombinants with bromodeoxyuridine, they implemented the black-plaque test to identify gIII-positive and negative viruses, an immunostaining method that facilitates detection of disruption of the gIII gene. Similarly, pseudorabies virus has been used to express tissue plasminogen activator (tPA) (Thomsen et al., 1987). A further development of the pseudorabies virus vector is the incorporation of the Cre-*lox* site-specific recombination system of coliphage P1 (Sauer et al., 1987). With this system, DNA can be readily inserted into the pseudorabies virus vector PRV42, which contains the *lox* site within the non-essential gIII gene. Incubation in vitro of PRV42 DNA with Cre protein and a circular plasmid containing a *lox* site generated approximately 5 percent recombinant molecules in which the plasmid integrated into the PRV42 genome at the *lox* site. After transfection, the virus can be propagated and viral DNA prepared and, if necessary, incubated with Cre, which facilitates the excision of the inserted plasmid from the viral genome. This plasmid can then be amplified by transformation of *E. coli*.

Recently, two groups have exploited the neurotropic nature of HSV-1, and both defective (Geller and Breakefield, 1988) and non-defective (Palella et al., 1988) HSV-1 vectors have been used to transfer and express genes into neuronal cells in culture. Geller and Breakefield transferred the *E. coli* beta-galactosidase gene into cultured peripheral neurons by co-infecting the neurons with the defective virus carrying beta-galactosidase and a temperature-sensitive mutant of HSV-1 (tsK) providing helper virus functions. Palella et al. transferred the human hypoxanthine-guanine phosphoribosyltransferase (HGPRT) gene to an HGPRT-deficient rat neuroma cell line with their non-defective HSV-1 vector. In this case, the HGPRT gene was inserted, via homologous recombination, into the HSV-1 tk gene. The potential applications of the HSV-1 vector systems have just begun to be explored.

VACCINIA-VIRUS-BASED VECTORS

The primary use of vaccinia virus vectors has been in the generation of recombinant vaccinia viruses for use as experimental vaccines. However, the biology of this poxvirus makes it a promising candidate for use as a more general expression vector as well. This is due to the fact that viral infection shuts down host-cell protein synthesis while retaining high levels of viral gene

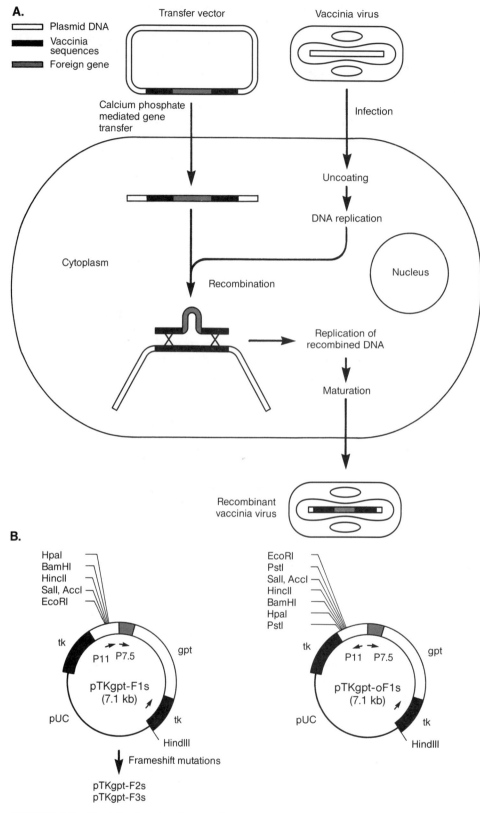

FIGURE 2-9 *Vaccinia-virus-based expression system.* **Panel A.** The protocol and mechanism for the insertion of foreign genes into infectious vaccinia virus is depicted here. First the gene to be expressed is inserted into a subgenomic fragment of vaccinia virus carried by a plasmid vector, creating the transfer vector. This

expression, an important consideration when employing vaccinia virus vectors for transferring and expressing foreign genes.

The development of vaccinia virus as an expression/gene-transfer vector was initiated in earnest by Panicali and Paoleiti (1982) and Mackett and colleagues (Mackett et al., 1982). What has arisen out of their original efforts is a highly flexible vector system that has been used to express a number of exogenous genes employing a number of very creative genetic-engineering tricks. The general method employed for the production and selection of infectious vaccinia virus recombinants is quite straightforward. With this method one constructs a plasmid transfer vector that, due to its unique design, undergoes homologous recombination, post transfection into vaccinia virus infected cells, with the 185-kb vaccinia virus genome, replacing the homologous endogenous viral sequences in the process. The transfer vectors contain all or part of the vaccinia virus thymidine kinase (tk) gene interrupted by the insertion of a multiple cloning site placed adjacent to, just downstream of, the tk promoter or another promoter inserted into the tk gene. The foreign gene to be expressed is inserted adjacent to the promoter, just downstream of the transcriptional start site. The features of this system are displayed in Figure 2-9. When the interrupted tk gene of the transfer vector recombines with the intact, viral genomic tk gene, the viral tk gene is replaced by the transfer vector tk gene interrupted by the foreign gene to be expressed, rendering the recombinant virus tk^-. Such recombinant viruses are selected on the basis of their tk^- phenotype.

Recombinant vaccinia viruses have been used to express hepatitis B surface antigen (Paoletti et al., 1984), influenza A virus haemagglutin (Coupar et al., 1986a), herpes simplex virus (HSV) type I D glycoprotein (Paoletti et al., 1984), the rabies virus G glycoprotein (Kieny et al., 1984), vesicular stomatitis virus G glycoprotein (Stephens, et al., 1986), polyoma virus virion proteins (Stamatos et al., 1987), streptococcal M protein (Hruby et al., 1988), nerve growth factor, (Edwards and Rutter, 1988), MHC antigen H-2k^d (Coupar et al., 1986b), and the cytokine interleukin-2 (Flexner et al., 1987; Ramshaw et al., 1987).

Falkner and Moss (1988) introduced an important improvement in their vaccinia virus vectors: they incorporated the *E. coli* gpt gene and a multiple cloning site into their transfer vector. The multiple cloning site allows for the insertion of a foreign gene into all three reading frames between the translation-initiation codon of a strong viral late promoter and the synthetic translation-termination sequence. After integration of this transfer vector into the genome

---

recombinant plasmid is transfected into tissue culture cells. The cell is coinfected with rescuing vaccinia virus. *In vivo* recombination occurs between the vaccinia virus DNA sequences flanking the foreign DNA insert and homologous DNA sequences in the replicating vaccinia genome forming a novel recombinant DNA molecule, which in turn can be replicated and packaged into infectious recombinant vaccinia virus. **Panel B.** The gpt-encoding, dominant selectable, vaccinia virus open reading frame vectors described in the text are depicted here. These vectors contain vaccinia virus tk gene sequences flanking the *E. coli* gpt gene, the promoters of the genes encoding the 11K polypeptide (P11) and the 7.5K polypeptide (P7.5), and the pUC-derived ampicillin resistance gene and origin of replication. The unique cloning sites into which foreign DNA can be inserted in both plasmids downstream of the initiation codon of the gene encoding the 11K peptide are indicated. Two derivatives of pTKgpt-F1s that shift the reading frame of the inserted sequence one nucleotide (pTKgpt-F2s) and two nucleotides (pTKgpt-F3s) relative to the 11K initiation codon have been generated that permit the insertion of foreign sequences into all three possible reading frames. [Adapted from Piccini et al. (1987), *Methods Enzymol.*, Vol. 153, Academic Press, p. 547; Falkner and Moss (1988), *J. Virol.*, Vol. 62, p. 1853, Fig. 3.]

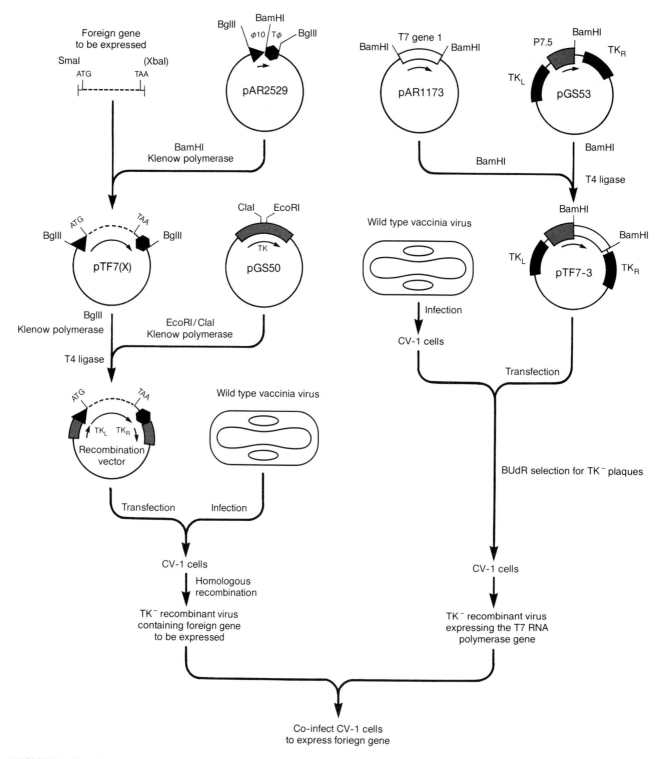

FIGURE 2-10 *Vaccinia virus T7-based expression system.* The construction of the two recombinant vaccinia viruses required for the expression of an exogenous gene from a bacteriophage T7 promoter carried by a recombinant vaccinia virus is depicted here. The left portion of the diagram explains the generation of a vaccinia virus recombinant expressing a foreign gene from the T7 promoter. First, the foreign structural gene is inserted into the plasmid pAR2529 between the T7 promoter ($\phi10$) and the T7 terminator sequences ($T\phi$) to form pTF7(x), where (x) is the foreign gene. The expression cassette, $\phi10$-(x)-$T\phi$, is excised from pTF7(x) with Bgl II and inserted into and interrupting the tk gene in pGS50. The resulting recombination vector is transfected into cells that are infected with wild-type vaccinia virus, and after homologous recombination is allowed to occur, tk⁻ recombinant viruses are selected. The left portion of the diagram explains the generation of a second vaccinia virus recombinant expressing

of the vaccinia virus, one can employ dominant selection for the gpt gene with mycophenolic acid (MPA) and thus select for cells infected with the recombinant virus presumably expressing the exogenous gene of interest. When Falkner and Moss inserted *E. coli* β-galactosidase into this vector, 24 hours post infection with the recombinant virus, approximately 3.5 percent of the total infected cell protein was found to be β-galactosidase.

Fuerst and colleagues also developed two interesting variations of their vectors (Fuerst et al., 1987, 1989). The first involves the generation of two distinct recombinant vaccinia virus populations. One population carries the gene encoding the bacteriophage T7 RNA polymerase under the control of a vaccinia virus promoter. The second population carries a gene of choice inserted between a bacteriophage T7 promoter and termination sequences. Thus, upon co-infection, the RNA polymerase encoded and overproduced by one vaccinia virus drives the transcription of the gene of interest carried by the other virus (Figure 2-10). Maximum expression of the exogenous gene product appears to occur when cells are infected with 10 PFU of each recombinant virus. This approach has proven successful for the expression of β-galactosidase, hepatitis B virus surface antigen, and human immunodeficiency virus envelope proteins. These studies indicated that the level of synthesis achieved was greater than that employing the more-conventional recombinant vaccinia virus expression system.

Then, Fuerst et al. (1989) introduced the *cis*- and *trans*-acting elements of the *E. coli* lac operon into vaccinia virus in an attempt to regulate gene expression. In this construct, a strong vaccinia virus late promoter was modified by insertion of the lac operator (lac O) at various positions. Once the optimal position for insertion had been determined, a single vaccinia virus stock containing lac I and the β-galactosidase gene under the control of the optimal lac O promoter was constructed. In the absence of the inducer, isopropyl-β-D-thiogalactoside, cells infected with this virus synthesize little or no detectable β-galactosidase. Addition of the inducer restores expression to approximately 20 percent of the unrepressed level. Under optimal conditions, employing the appropriate vector, vaccinia virus based expression can result in the foreign gene's producing up to 3 percent of the total cell protein.

PAPILLOMA-VIRUS-BASED VECTORS

Transfection of selected cells with papilloma virus DNA results in the appearance of transformed foci in tissue culture dishes. Examination of the DNA in the transformed foci reveals that the papilloma virus genome is maintained as a multicopy nuclear plasmid whose replication is non-lytic. This characteristic has driven the development of papilloma virus DNA as a eukaryotic gene-transfer vector. Two types of papilloma viruses have been extensively studied, the bovine papilloma viruses (BPV) and the human papilloma viruses (HPV).

---

the T7 RNA polymerase gene. First, the T7 gene 1 sequence, encoding T7 RNA polymerase, is excised from pAR1173 and is subsequently inserted into the plasmid pGS53 just downstream from the vaccinia virus P 7.5 promoter, which itself interrupts the vaccinia virus tk gene to form pTF7-3. This recombination vector is transfected into cells that are infected with wild-type vaccinia virus, and after homologous recombination is allowed to occur, tk⁻ recombinant viruses are selected. Once isolated and expanded, the virus encoding the foreign gene and the virus encoding the T7 RNA polymerase can be used in a mixed infection to drive the expression of the foreign gene. [Adapted from Moss et al. (1986), *Proc. Natl. Acad Sci. U.S.A.*, Vol. 83, p. 8123, Fig. 1; Fuerst and Moss (1987), *Mol. Cell. Biol.*, Vol. 7, p. 2540, Fig. 1.]

The BPV genome is a double-stranded DNA molecule of 7945 nucleotides (Figure 2-11). Lowy et al. (1980) showed that a 69 percent BamHI-HindIII fragment of the viral genome is capable of inducing cellular transformation. The open reading frames (ORFs) in this fragment are referred to as early ORFs (Danos et al., 1983). This fragment appears to produce 5 mRNA species in

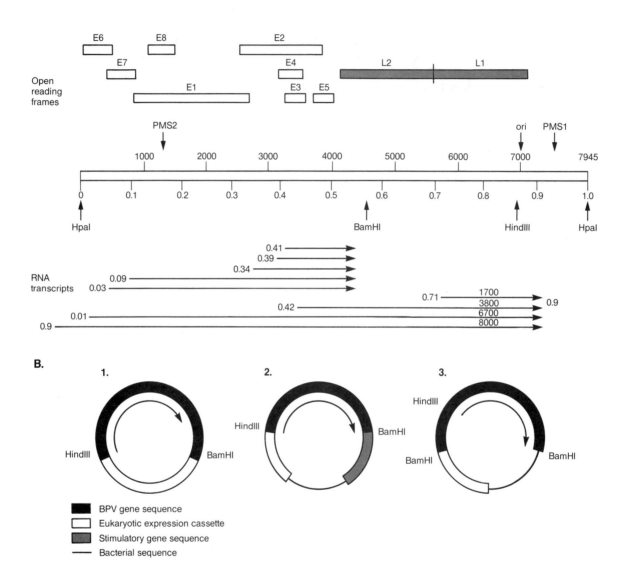

FIGURE 2-11 *Bovine papilloma virus genome and expression vectors.* **Panel A.** The circular bovine papilloma virus (BPV) genome is shown as a linear map digested at the unique Hpa I site. The position of the eight early open reading frames (ORFs) and two late ORFs are shown above the linear map. The RNA transcripts corresponding to the ORFs, both early and late, are shown below the linear map. The viral origin of replication and the two plasmid maintenance sequences are also indicated. **Panel B.** The genetic components of the three types of bovine papilloma virus vectors capable of replicating as nuclear plasmids are shown. Type 1 contains the basic 69% transforming fragment plus the gene of interest. Type 2 contains the basic 69% transforming fragment, a stimulating DNA sequence, bacterial sequences necessary for propagation as a plasmid, and the gene of interest. Type 3 contains the entire BPV genome, bacterial sequences necessary for propagation as a plasmid, and the gene of interest. The arrows indicate the direction of BPV transcription. [Adapted from Stephens and Hentschel (1987), *Biochem. J.,* Vol. 248, p. 2, Fig. 1, and p. 5, Fig. 2.]

transformed mouse cells in culture (Stenlund et al., 1983). RNA transcripts that hybridize with the 31 percent fragment are only found in tumors and fibropapillomas and coincide with the late-region ORFs L1 and L2 (Amtmann and Sauer, 1982; Engel et al., 1983). A region lying between the early and late ORFs of about 900 base pairs (7093–48) is known to contain several regulatory sequences required for transcription and regulation. For the purposes of our discussion, I will only describe the early ORFs, those ORFs that are functionally important to BPV-based vectors (Stephens and Hentschel, 1987).

The E1 ORF has the largest coding capacity of the early ORFs and overlaps with the E2, E7, and E8 ORFs. E1 appears to be involved in the control of viral replication. Mutations in E1 lead to a phenotype in which the BPV fails to replicate and becomes integrated into the genome.

The E2 ORF overlaps with the E3 and E4 ORFs. The E2 product was initially thought to be a transforming protein but now has been shown to be a transactivator of the BPV regulatory unit.

The E5 ORF has a transforming function. Expression of this ORF alone is sufficient to cause transformation of C127 and NIH 3T3 cells.

The E6 ORF overlaps with the E7 ORF and also appears to have a transforming function; however, its transforming function appears to be more cell-specific than its E5 counterpart. E6 can transform C127 cells but not 3T3 cells. The E6 product has many properties characteristic of nucleic-acid-binding proteins.

The E7 ORF overlaps both the E6 and E1 ORFs. Mutations in the E7 ORF maintain the BPV genome as an extrachromosomal element in transformed cells but at a greatly reduced copy number (1–5 copies per cell). The studies of Berg and co-workers (1986) suggest that there exists a temporal need for the expression of E7 for the initial-establishment phase of BPV replication.

The E3 and E4 ORFs overlap with each other and E2. The E8 ORF overlaps with E1. No clear functions have been assigned to these ORFs to date. The E3 ORF lacks an in-phase methionine codon and thus may not encode a polypeptide.

As was the case with SV40, the insertion of the 69 percent fragment of BPV into pBR322 resulted in decreased transformation efficiency. Similar efficiency was observed with the poison-minus pML plasmid. However, when Campo et al. (1983) linked the plasmid pAT153, a pBR322 derivative containing a deletion not as extensive as that in pML, to the 69 percent transforming fragment, transformation competence was restored. The block in transformation can also be restored by the insertion of foreign genes such as β-globin or preproinsulin, which function as stimulatory elements. When the entire BPV genome is linked to either pML or pAT153, replication competence is restored as well. In contrast, linkage of the entire BPV genome to pBR322 results in a drop in transformation competence of at least two orders of magnitude.

These studies have led to the development of three types of BPV expression vectors (Figure 2-11). The first type consists of the 69 percent transforming sequence, no bacterial sequences, and a eukaryotic expression cassette. This type of vector has three distinct disadvantages: the number of restriction sites available for engineering is limited, transfection of tissue culture cells is often performed with linear DNA and complex rearrangements can occur upon transfection, and the lack of bacterial replication sequences prevents the shuttling of DNA between bacteria and mammalian cells (Zinn et al., 1982).

The second and third type of vector incorporate a number of improvements over the first. The second type consists of the 69 percent transforming region of BPV, bacterial sequences including an ori and a selectable marker, one of a number of stimulatory segments, and the gene of interest (Zinn et al., 1983). The third type of vector contains the entire BPV genome, bacterial sequences as before, and the gene of interest. The incorporation of these other sequence elements in the second and third types of BPV vectors eliminates the need to transfect with linear DNA and facilitates the shuttling of DNA between bacteria and mammalian cells. Thus, for most purposes, these vectors are preferred over the first type.

Identification procedures for successful BPV-vector-mediated gene-transfer include focus formation as well as selection for thymidine kinase (tk), xanthine-guanine phosphoribosyltransferase (gpt), amino glycoside phosphotransferase (neo), and metallothionein (MT). There is no reason to believe that other markers, such as multiple drug resistance (MDR-1), will not work as well. Selection for tk+ or gpt+ has been problematic, and the preferred selection procedures employ neo^r and MT. In fact, these selection protocols result in an approximately 10-fold increase in copy number relative to that found by a focus-forming assay.

BPV vectors have been used with a high degree of success in two experimental areas. They have been used to study gene regulation, and they have been used to obtain high-level expression of foreign genes. The principal argument for employing BPV vectors to study gene regulation is that they, unlike vectors that stably integrate into the chromosome of a transfected cell, are not subject to the chromosomal position effects discussed in the previous chapter. Experimental systems that have been investigated include the induction of metallothionein gene expression by heavy metals and glucocorticoids, the regulation of human β-interferon, and the cell-cycle-dependent expression of histone genes.

Studies initiated to examine the behavior of foreign promoters inserted into BPV vectors generated some peculiar results. The chromosomal mouse metallothionein promoter can be induced by cadmium, zinc, and the glucocorticoid dexamethasone. Insertion of this promoter into a BPV vector in which the promoter is positioned to drive the expression of a foreign gene alters the response of the promoter to these inducers. In such constructs, the mMTI promoter could be induced by cadmium but not by zinc or dexamethasone. Furthermore, the level of induction by cadmium was low relative to the endogenous mMTI gene (Pavlakis and Hamer, 1983). A similar effect has been reported for the human metallothionein promoter hMTIa. In contrast, the hMTIIa promoter retained zinc and glucocorticoid induction (Richards et al., 1984; Karin et al., 1983). The molecular basis of the altered inducibility of MTI promoters inserted into BPV vectors remains a mystery.

Insertion of the human β-interferon gene under the control of its own promoter into a BPV vector results in apparently normal regulatory control. β-interferon expression can be induced by both double-stranded RNA and inactivated Newcastle Disease Virus (Zinn et al., 1982; Maroteaux et al., 1983; Mitrani-Rosenbaum et al., 1983). Zinn et al. (1982) observed that a number of β-interferon-BPV episomes underwent some rearrangement upon transfection. In these experiments, they transfected the recipient cells with a linear BPV molecule that had been excised from the bacterial plasmid. Insertion of the β-globin gene into this vector and simultaneous replacement of the original bacterial sequences with bacterial sequences that did not require excision resulted in the generation of a BPV vector that did not rearrange (Zinn et al., 1983).

Green et al. (1986) studied the cell-cycle regulation of the human histone H4 gene linked to the 69 percent transforming fragment of BPV. When transfected cells were synchronized by two cycles of thymidine block, expression of the H4 histone gene carried by the episome was regulated coordinately with DNA replication.

BPV vectors have been used to generate stable cell lines expressing over 30 different proteins, including chloramphenicol acetyltransferase (Matthias et al., 1986); human growth hormone (Pavlakis and Hamer, 1983); tissue plasminogen activator (Reddy et al., 1987); α-, β-, and γ-interferon (Zinn et al., 1982; Fukunaga et al., 1984; Maroteaux et al., 1983); hepatitis B surface antigen (Stenlund et al., 1983; Hsiung et al., 1984; Wang et al., 1983; Denniston et al., 1984); influenza virus haemagglutinin and CAP-recognizing proteins (Sambrook et al., 1985; Braam-Markson et al., 1985); type I human T-cell leukemia virus small-envelope protein (Eiden et al., 1985); HLA heavy chain (DiMaio et al., 1984); human interleukin-2 receptor (Cosman et al., 1986); rat preproinsulin (Sarver et al., 1985); bovine growth hormone (Reitz et al., 1987); factor VIII (Sarver et al., 1987); class II MHC (A_B^d) gene (Fukushima et al., 1988); immunoglobulin κ light chain (Baker et al., 1988); and interleukins 2, 3, 4, and 5 (Karasuyama and Melchers, 1988).

The use of BPV as an expression vector can be quite complex, and the episomal nature of the vector/gene constructs is not predictable. The viral DNA can exist as episomal monomers, episomal concatamers, concatenates, integrated concatamers, or multiple integrated monomers. One interpretation of this complex behavior is an as-yet-unexplained interaction between viral and inserted *cis*-acting expression and replication elements. Therefore, the construction of BPV vectors that express foreign proteins at high levels is still a trial-and-error exercise.

RETROVIRAL VECTORS

Retroviruses were envisaged as potentially useful gene-transfer devices for a variety of reasons. First, the retroviral genome stably integrates into the host chromosome of the infected cell and is therefore transmitted from generation to generation. Second, integration is site-specific with respect to the viral genome at the extreme ends of the proviral long terminal repeats (LTRs). Thus, post integration, the structures of the viral genes that lie within the LTRs are preserved intact. Third, as a consequence of the reverse transcription of RNA to DNA, it was perceived that retroviruses could be developed into cDNA-cloning machines (Kriegler et al., 1984a). Fourth, retroviruses display a wide infectivity and expression host range. Fifth, their genomes are highly plastic and manifest a high degree of natural size variation. Experimentally, they appear to tolerate a tremendous amount of manipulation.

The list of cellular genes rescued as recombinant retroviruses continues to grow and includes selectable markers, cellular oncogenes, non-retroviral viral gene products, cytokines, and cytotoxins. By and large, the choice of gene has little bearing on the success to rescue infectious recombinant viruses. This observation reflects the dramatic plasticity just mentioned.

Before I launch into a discussion of vectors and packaging cell lines designed for retrovirus-mediated gene transfer, it will be useful to acquire a basic understanding of the lifecycle of the simplest retroviruses, as well as the viral genes required for retrovirus rescue and infectivity. Retrovirally infected cells shed a membrane virus containing a diploid RNA genome. The virus, studded with an envelope glycoprotein (which serves to determine the host range of infectivity), attaches to a cellular receptor in the plasma membrane of the cell to be infected. After receptor binding, the virus is internalized and uncoated as it passes through the cytoplasm of the host cell. Either on its way to the nucleus or in the nucleus, the reverse transcriptase molecules resident in the viral core drive the synthesis of the double-stranded DNA provirus, a synthesis that is primed by the binding of a tRNA molecule to the genomic viral RNA. The double-stranded DNA provirus is subsequently integrated in the genome of the host cell, where it can serve as a transcriptional template for both mRNAs encoding viral proteins and virion genomic RNA, which will be packaged into viral core particles. On their way out of the infected cell, core particles move through the cytoplasm, attach to the inside of the plasma membrane of the newly infected cell, and bud, taking with them tracts of membrane containing the virally encoded envelope glycoprotein gene product. This cycle of infection—reverse transcription, transcription, translation, virion assembly, and budding—repeats itself over and over again as infection spreads.

The viral RNA and, as a result, the proviral DNA encode several *cis*-acting elements that are vital to the successful completion of the viral lifecycle. The virion RNA carries the viral promoter at its 3' end. Replicative acrobatics place the viral promoter at the 5' end of the proviral genome as the genome is reverse-transcribed (Figure 2-12). Just 3' to the 5' retroviral LTR lies the viral packaging site or so-called *psi site* (ψ). Initially mapped to the region downstream of the 5' LTR and upstream of the splice donor in the viral genome, there is now a good deal of evidence that suggests the ψ site extends further downstream and apparently right into the coding region of the group-specific antigen (*gag*) gene in non-defective retroviruses. (Bender et al., 1987; Adam and Miller, 1988).

If the virus encodes both spliced and unspliced transcripts, as all replication-competent (non-defective) retroviruses do, the viral genome will encode a splice donor in the midst of the ψ site. Further downstream, the viral genome will encode a splice acceptor as well. Last, both spliced and unspliced transcripts will encode polyadenylation signals, vital to the efficient utilization of the transcripts as mRNAs.

The retroviral lifecycle requires the presence of virally encoded *trans*-acting factors. The viral *gag* molecules, which form the structural framework of the viral core, also appear to interact with the ψ site and play a key role in the incorporation of the viral RNA into the virion core during the packaging process. The viral-RNA-dependent DNA polymerase (*pol*)—reverse transcriptase—is also contained within the viral core and is vital to the viral life cycle in that it is responsible for the conversion of the genomic RNA to the integrative intermediate proviral DNA. The viral envelope glycoprotein, *env*, is required for viral attachment to the uninfected cell and for viral spread. *env* gene products differ in different strains of viruses and, as mentioned previously, *env* can serve to determine infectivity host range. Recent research on human immunodeficiency and T-cell leukemia viruses has revealed the existence of several virally encoded transcriptional *trans*-activating factors, so called *trans-activators*, that can serve to modulate the level of transcription of the integrated parental provirus. Some of these factors are discussed in the previous chapter.

Typically, replication-competent (non-defective) viruses are self-contained in that they encode all of these *trans*-acting factors. Their defective, often acutely transforming, counterparts are not self-contained and are, in the normal chain of events, usually accompanied by a non-defective congener. They are able to spread via a parasitic mechanism. The non-defective virus provides the defective virus with *gag*, *pol*, and *env*. The end result is a defective genome packaged with all the biochemical machinery required to successfully infect a cell and to integrate its defective provirus into the host genome.

The host range of infectivity of a retrovirus is determined in large part by the type of *env* gene product encoded by the viral genome. The mechanism that confers specificity on a particular cell type is the presence of a particular receptor on the surface of a cell with which a specific *env* gene product can bind. *env* gene products fall into three basic classes: ecotropic, amphotropic,

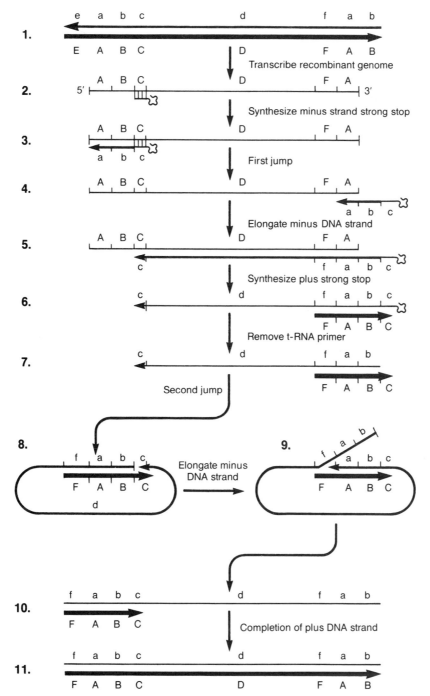

FIGURE 2-12 *Mechanism of reverse transcription and the immortalization of mutations introduced into the 3' LTR in the 5' LTR.* The mechanism of reverse transcription and the mechanisms by which a mutation (either a deletion, an insertion, or a substitution) introduced by recombinant DNA techniques into the 3' LTR of the recombinant provirus are immortalized in both LTRs after reverse transcription are depicted here. **Step 1**: Diagram of the recombinant retroviral genome into which the 3' long terminal repeat (sequences FAB) have been mutated in the U3 region by the substitution of F sequences for the E sequences originally present in both the 5' and 3' LTRs. **Step 2**: The recombinant genome is transcribed and the proline tRNA primer binds the genomic transcript. **Step 3**: The proline tRNA primes the synthesis of minus strand strong stop DNA. **Step 4**: Minus strand strong stop makes the first jump. **Step 5**: The minus DNA strand is elongated to the full length of the genome. **Step 6**: Plus strand strong stop is synthesized. **Step 7**: The tRNA primer is removed from the DNA minus strand. **Step 8**: Plus strand strong stop makes the second jump. **Steps 9 and 10**: The plus strand strong stop is elongated to the full length of the genome. It is at this step that the F mutation introduced into the 3' LTR is immortalized in the 5' LTR of the newly synthesized provirus. **Step 11**: The recombinant provirus is completed. [Adapted from Gilboa et al. (1979), *Cell*, Vol. 18, p. 94, Fig. 1.]

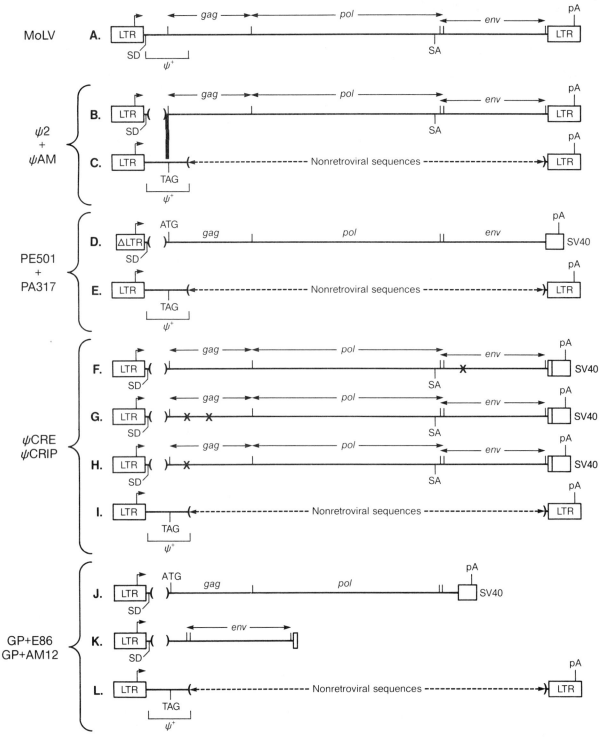

FIGURE 2-13 *Recombinant retroviral packaging line genomes.* The structures of the recombinant retroviral genomes carried by the packaging cell lines described in the text are depicted here. **A.** Wild-type murine leukemia virus genome. The extended packaging signal sequence described in the text is denoted +. **B.** $\psi2$ and ψAM genomic structure. **C.** The basic genomic structure of the LN type vectors described in the text. The vertical bar joining **B** and **C** denotes the regions of the two genomes between which recombination must occur to result in the generation of an infectious replication competent retroviral genome. **D.** PE501 and PA317 genomic structure. In addition to the deletion of the ψ region, these genomes carry a deletion in the 5' LTR. In addition, the 3' LTR has been replaced, in part, by an SV40 polyadenylation signal. **E.** The basic genomic structure of the LN type vectors described in the text is provided for comparison. **F, G,** and **H.** ψCRE and ψCRIP genomic structure. This packaging system involves two different

and xenotropic. Ecotropic viruses can infect their natural hosts, mice and rats. Amphotropic viruses can infect both their natural hosts and a number of other species, including *homo sapiens*. Xenotropic viruses cannot infect their natural hosts but can infect a number of other species, including *homo sapiens*.

Retroviral Packaging Cell Lines

One can engineer cell lines that do not produce a replication-competent retrovirus but can provide the *trans*-acting factors required by a replication-defective retrovirus. In such a situation, a replication-competent retroviral genome is enfeebled in a manner such that it is unrescuable. This means that, although the genome can produce the *trans*-acting factors required to rescue a defective viral genome present in a cell containing the enfeebled genome, the enfeebled genome cannot rescue itself.

The classic example of such a cell line is known as ψ 2, which was generated by Mulligan and colleagues (Mann et al., 1983). To create this line, a 3T3 cell line was transfected with an ecotropic Moloney murine leukemia virus proviral genome from which the packaging site—ψ site—had been deleted with restriction endonucleases (Figure 2-13). Transfection of this cell line with genetically engineered defective retroviruses containing reporter genes results in the production of infectious retroviruses carrying the reporter gene and transmissible for but a single round of infection, because after infection of the next cell, they lack the *trans*-acting factors provided by ψ 2.

Miller and colleagues have generated another cell line carrying a considerably more enfeebled non-packaging genome called PE501 (Miller and Buttimore, 1986). PE501 was generated by transfecting a tk$^-$ 3T3 cell line with a recombinant murine leukemia virus genome that had been enfeebled in a number of ways. First, the 5' sequences from the 5' LTR were deleted. Second, the ψ site upstream of the viral splice donor was deleted. Third, the 3' LTR was deleted and replaced with an SV40 polyadenylation signal (Figure 2-13).

Markowitz et al. (1988) have generated a similar cell line, GP&E-86, in which gag and pol are expressed from one plasmid and env is expressed from another plasmid (Figure 2-13). Danos and Mulligan (1988) developed another line, ψ CRE, employing a similar strategy (Figure 2-13). Lines such as ψ 2, PE501 GP&E-86, and ψ CRE are known as *packaging lines*.

Other similar packaging lines have been created that provide, in *trans*, the factors necessary to encapsidate the defective virus containing the re-

---

defective genomes to provide the necessary packaging functions. The first genome, **F,** provides the *gag* and *pol* gene products; however, the *env* gene has been mutated at the position marked with an X. The second genome, either G for ecotropic packaging or H for amphotropic packaging, carries a non-functional *gag* gene mutated at the position(s) marked with an X. Thus, both genomes are required to provide *gag, pol,* and *env* functions required for packaging. In all three cases the 3' LTR has been replaced with an SV40 polyadenylation signal. **I.** The basic genomic structure of the LN type vectors described in the text is provided for comparison. **J** and **K.** GP+E86 and GP+AM12 genomic structures. This packaging system also involves two different defective genomes to provide the necessary packaging functions. The first genome, **J,** provides *gag* and *pol* gene functions but lacks an *env* gene. The second genome, encoding either an ecotropic or amphotropic *env* gene product, lacks *gag* and *pol* sequences. Thus, both genomes are required in order to provide *gag, pol,* and *env* functions required for packaging. **L.** The basic genomic structure of the LN type vectors described in the text is provided for comparison. [Adapted from Miller and Rosman (1989), *Biotechniques*, Vol. 7, p. 981, Fig. 1; p. 982, Fig. 2; and p. 984, Fig. 3.]

porter gene in an amphotropic coat. The most commonly used lines of this type are ψ AM, the amphotropic equivalent of ψ 2, developed by Cone and Mulligan (1984); PA317, the amphotropic equivalent of PE501, developed by Miller and colleagues (1986); and ψ CRIP, the amphotropic equivalent of ψ CRE, developed by Danos and Mulligan (1988).

Retroviral Vector Genomes

Structure function studies of both non-defective and defective retroviruses led to the development of the first retroviral vectors. In 1982, Tabin et al. inserted an intact HSVtk gene into the backbone of a Moloney murine leukemia virus proviral clone and rescued an infectious retrovirus that conferred hypoxanthine, aminopterin, and thymidine (HAT) resistance on tk$^-$ cells. Mulligan and colleagues then reported the rescue of the GPT gene and the subsequent mapping of a retroviral packaging signal, the ψ site (Mann et al., 1983). Miller and colleagues and Temin and colleagues reported the rescue of a variety of gene sequences, all in vectors of similar design. We reported that the overlapping genes of the SV40 early region, encoded by differentially spliced mRNAs, could be separated by a retroviral vector (Kriegler et al., 1984a). We further observed that an unpaired functional retroviral splice donor located upstream of an unspliced DNA sequence could have a dramatic effect on the expression of that gene in a retroviral vector: Mutation of the donor splice signal greatly enhanced expression of our tester gene (Kriegler et al., 1984a, 1984b). The next major advance was achieved by Cepko et al. (1984), who reported the construction of a retroviral vector, pZIP-NEOSV(X)1, that could drive the expression of two inserted genes, one the dominant selectable marker neo^r and the other a gene of choice. The expression of both genes was driven from the promoter in the 5′ LTR. These vectors offered experimenters the opportunity to monitor the transmission of a non-selectable gene by following the neo^r phenotype, resistance to the antibiotic G418 (Figure 2-14). These vectors, when used in combination with the ψ 2 and ψ AM packaging cell lines (Mann et al., 1983; Cone and Mulligan, 1984) have proven to be powerful experimental tools.

As mentioned previously, initial studies of the retrovirus packaging site suggested that a non-coding sequence between the splice donor (of Moloney murine leukemia virus) and the beginning of the structural region of the *gag* gene was sufficient to drive the packaging of genomic retroviral RNA. In the last two years, it has become apparent that sequence elements within the structural region of the *gag* gene can dramatically increase the efficiency of packaging of retroviral RNA. As a result, a new type of vector, the *gag*+ vector, has been developed. This represents a real improvement in the effective titer of recombinant viruses that can be rescued. In fact, an increase in titer of one order of magnitude is not unusual. As far as maximizing titer of rescued viruses is concerned, *gag*+ vectors represent the current state of the art (Miller and Rosman, 1989).

These vectors and their progenitors can be rescued with or without internal exogenous promoters between the viral LTRs and downstream of the packaging site driving the expression of the transferred gene. Viral genomes encoding and expressing two different genes from two different promoters (one of them in the 5′ viral LTR) can be efficiently rescued as well.

A group of *gag*+ retroviral vectors that represents a substantial improvement over most earlier retroviral vectors are the *gag*+ vectors LN and derivatives designed by Miller and colleagues (1989). These vectors are ideally suited to be used in combination with the packaging lines PE501 and PA317. LN-type vectors transfected in these lines have been shown to result rarely, if ever, in the generation, through recombination of the vector with the endogenous en-

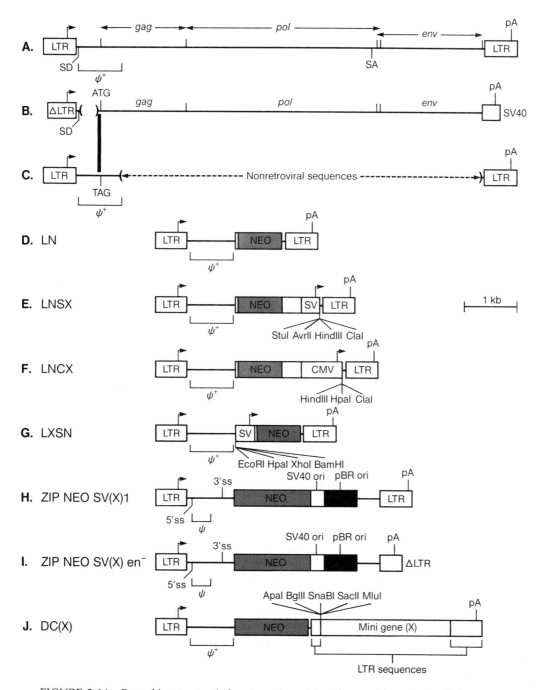

FIGURE 2-14 *Recombinant retroviral vectors.* A variety of recombinant retroviral vectors described in the text are depicted here. **A.** Wild-type murine leukemia virus genome. **B.** Genomic structure of PE501/PA317 packaging line genome. **C.** Genomic structure of the LN type vectors described in the text is provided for comparison. **D, E, F, G, H, I,** and **J.** The characteristics of these vectors are described in the text. [Adapted from Miller and Rosman (1989), *Biotechniques*, Vol. 9, p. 982, Fig. 2, and p. 984, Fig. 3; Cone et al. (1987), *Mol. Cell. Biol.*, Vol. 7, p. 889, Fig. 1; and Hantzopoulos et al. (1989), *Proc. Natl. Acad. Sci. U.S.A.*, Vol. 86, p. 3521, Fig. 2.]

feebled proviral genome, of a replication-competent retrovirus. Recombination would have to occur just upstream of the gag initiation codon within a 53-bp region that the vector and the enfeebled genome have in common. A second recombination event would also have to occur to replace the LTR missing from the PE501 or PA317 genome, and the LN vectors have been designed to preclude this (Figures 2-13 and 2-14).

The LN vectors are of two types (Figure 2-14). The first is composed of two LTRs, the *gag+* packaging region, and a multiple cloning site. The second type is composed of two LTRs, the *gag+* packaging site, a multiple cloning site, an internal promoter, and a neomycin resistance gene (Miller and Rosman, 1989). In these constructs the neomycin resistance gene is expressed from either the 5′, LNSX, LTR or an internal SV40 promoter LXSN. In one construct, where the internal SV40 promoter is replaced with a cytomegalovirus promoter, the expression of the neo^r gene is driven from the 5′ LTR LNCX. Just downstream of the promoter, not driving the expression of the neo^r gene, is a convenient multiple cloning site. We have inserted and rescued a number of genes in these vectors including those encoding TNF, IL-2, and the multiple-drug resistance gene. Analysis of the rescued viral supernatants revealed that they are (1) of high titer, 1×10^5 to 8×10^6 CFU/ml; (2) helper-virus-negative when analyzed by a battery of tests; and (3) relatively stable. Cells infected with constructions encoding two genes co-express those genes with high efficiency (> 95 percent).

Viral rescue/expression schemes are becoming ever more elaborate. In an attempt to achieve high-level expression of genes transferred into embryonal carcinoma cells by a retroviral vector, numerous schemes have been employed with some degree of success. Investigators have inserted internal promoters into their vectors to drive the expression of transferred genes in an effort to overcome the transcriptional silencing of the expression elements within the retroviral LTR commonly observed in EC cells (Wagner et al., 1985; Stewart et al., 1985; Boulter and Wagner, 1987). They have exploited naturally occurring retroviral variants, such as the myeloproliferative sarcoma virus (MPSV), whose LTRs bear mutations in the U3 region of the LTR (Stacey et al., 1984). MPSV displays an expanded host range that includes cells of the hematopoietic compartment as well as EC cells (Franz et al., 1986). Recently, they have generated hybrid LTRs that incorporate expression elements from viruses permissive for expression in EC cells, such as the polyoma virus enhancer mutant pyF101, which was selected for growth on EC cells (Valerio et al., 1989). This hybrid LTR, when incorporated into a retroviral vector, displays an expression host range reminiscent of the polyoma virus donor. Investigators have achieved the regulated expression of human globin genes transferred with retroviral vectors carrying both the globin coding sequences and *cis*-acting globin regulatory sequences between the LTRs. After infection of mouse erythroleukemia (MEL) cells with a recombinant virus carrying the human β-globin gene and flanking *cis*-acting elements, Cone et al. (1987) reported correct regulation of the exogenous allele after induction with dimethyl sulfoxide (DMSO). They reported the somewhat surprising observation that viral RNA expression was also inducible by DMSO. In similar experiments performed by Karlsson et al. (1987), viral RNA expression was not inducible by DMSO although expression of the transferred globin genes was inducible. This discrepancy is largely unresolved.

As mentioned previously, the retroviral virion genomic RNA carries the promoter sequence of the virus at the 3′ end of the RNA, yet there are promoter sequences in both the 5′ and 3′ LTRs of the integrated proviral DNA. As a consequence of the mechanism of the retroviral replication scheme, the sequences at the 3′ end of the virion RNA, encoded by the 3′ LTR, are immortalized in both the 3′ and 5′ LTRs of the next viral generation. A mutation in the promoter of the 3′ LTR of a recombinant provirus will serve to drive expression from the 5' LTR of the next viral generation. This peculiar circumstance permits the experimenter to construct a provirus whose expression is driven by the wild-type LTR in the first generation but, after viral rescue and infection of another cell, is driven by a mutated or defective LTR (Figure 2-12). As a result, viral vectors lacking an expression element in the 3' LTR have been developed. These vectors have been termed *suicide vectors* and have been developed in a number of laboratories.

Such vectors have been generated for two different reasons, both bearing on the desire to generate recombinant retroviruses that exhibit a tissue- or cell-type-specific host range. The native promoter within the retroviral LTR exhibits some specificity. Unfortunately, there are cell types in which it functions poorly, as well as cell types in which it functions in a constitutive, dramatically unregulated fashion. Furthermore, the expression host range of the retroviral LTR can influence the expression of flanking promoters. In fact, multiple promoters within the retroviral genome can alter each other's ability to express (Emerman and Temin, 1984, 1986a, 1986b). The experimental notion arose that deleting the retroviral expression elements and substituting their function with an exogenous internal promoter, complete with all the necessary *cis*-acting elements to drive tissue- or cell-type-specific expression, might facilitate appropriate regulation of the exogenous, internal promoter. After a good deal of effort, it appears that this strategy can work. Yu et al. (1986) reported the development of a suicide or self-inactivating (SIN) vector from which the enhancer and promoter had been deleted from the 3' LTR. They inserted a hybrid gene in which the human metallothionein promoter drove the expression of the c-fos gene. Following transfection, rescue of this recombinant genome, and reinfection of recipient cells, the transferred retroviral genomes synthesized RNA transcripts from the metallothionein promoter and not the now-mutated 5' LTR. In some cases, this promoter could be induced by cadmium. Similarly, Yee et al. (1987) rescued the human hypoxanthine phosphoribosyltransferase gene (HPRT), whose expression was driven by either the human metallothionein (MT-11$_A$) promoter or the human cytomegalovirus (hcMX) immediate early promoter. Lim et al. (1987) inserted into a SIN vector the adenosine deaminase (ADA) gene, whose expression was driven from the human phosphoglycerate kinase promoter. They obtained reasonable expression levels of ADA in murine hematopoietic cells in culture and spleen-colony-forming units (CFU-S) *in vivo*. Cone et al. (1987) exploited a self-inactivating derivative of pZIP-NEOSV(X)1 called pZIP-NEOSV(X)$_{en-}$ in their studies of β-globin gene expression (Figure 2-14). Guild et al. (1988) generated a *gag*+SIN vector that could be rescued at high titers and was shown to be transcriptionally active *in vivo* in both embryonal carcinoma (EC) cells and hematopoietic cells. A *gag*+ construct in which neo^r gene expression was driven by the human histone H4 promoter functioned most efficiently in murine hematopoietic cells both *in vitro* and *in vivo*, even when reconstituted bone marrow was analyzed 2.5 months post transplantation.

Recently, this self-inactivating technology has been given a new twist: A new double-copy (DC) vector has been developed. Rather than delete the enhancer and promoter from the 3' LTR, Hantzopoulos et al. (1989) inserted an adenosine deaminase minigene, complete with its own promoter, into the U3 region of the 3' LTR. Upon rescue of this recombinant genome and subsequent infection of recipient cells, the ADA minigene was duplicated in each LTR. The authors report a 10–20-fold increase in ADA expression when compared to vectors carrying the same minigene between the two LTRs.

Yet another horizon in retroviral vectorology is the experimental specification of cell- or tissue-type-*infectivity* host range. The object of such research is to generate, through the manipulation of proviral DNA, novel retroviral pseudotypes in which the native envelope gene product has been substituted with components of another viral envelope protein, a cell receptor, an immunoglobulin molecule, or some other membrane-bound ligand. One might substitute a hybrid gene encoding the transmembrane and related domains of the wild-type retroviral envelope gene product fused to the portion of the exogenous molecule necessary for recognition of the cell type to be infected. A basic understanding of the structure and functions of the multiple domains of the retroviral envelope proteins has facilitated such experiments, and it appears that efforts to generate functional hybrid molecules is beginning to bear fruit.

Recently, the first experiments in gene transfer into a human were conducted with a recombinant retrovirus encoding the neomycin drug resistance gene (Kasid et al., 1990). In these experiments, the neo^r gene was transferred *ex vivo* into human tumor-infiltrating lymphocytes (TILs), which were subsequently re-implanted into cancer patients. It was not expected that the therapeutic efficacy of the TILs would be enhanced by the transfer of the neo^r gene. The purpose of the experiment was to track the migration of the engineered TILs *in vivo* and to test the feasibility of this gene-transfer concept. Similar experiments with potentially therapeutic genes have been initiated, and the future of this technology appears quite bright.

HOMOLOGOUS RECOMBINATION AND GENE-REPLACEMENT VECTORS

One of the most exciting areas of gene-transfer research is the study and exploitation of homologous recombination in eukaryotic cells. The observation that homologous recombination occurs in other than meiotic cells has moved from a laboratory curiosity to that of a readily applicable technique for the deliberate inactivation of genes in somatic cells and embryonal stem cells, and ultimately in the inactivation of selected genes in whole animals (Kucherlapati et al., 1984, 1985; Smithies et al., 1985; Ayares et al., 1986; Song et al., 1987; Nandi et al., 1988; Doetschman et al., 1988; Koller et al., 1989; 1990). This technology has been used to create germ-line chimeras containing targeted disruptions in HPRT, ABL, EN-2, n-myc, β-2 microglobulin, 1gf-2, and INT-1 genes (Capecchi, 1990). Non-expressed genes in embryonic stem cells can be targeted as well (Johnson et al., 1989b). In addition, Ellis and Bernstein (1989) reported homologous recombination between an integration-defective retrovirus and chromosomal sequences in both rodent and human cells.

For each experiment, the particular genetic designs and appropriate protocols vary; however, the underlying concept is the same. It is based on the observation that DNAs transfected into cells in culture can recombine in a homologous fashion with their chromosomal equivalents. Although the frequency of such homologous recombination events is quite low, one can, by employing the appropriate selection or screening protocols, detect the homologous recombination event and subsequently isolate cellular clones in which the homologous recombination event has occurred. Such isolated clones can be further analyzed or manipulated. A diagrammatic representation of the method is shown in Figure 2-15.

Essentially, one constructs a recombinant DNA molecule wherein the mutagenic change one desires to make in the gene of interest is flanked by regions of complete homology, often of considerable length (Thomas and Capecchi, 1987), of that same gene. Such a recombinant molecule, upon transfection into a recipient cell, undergoes homologous recombination in the two flanking regions

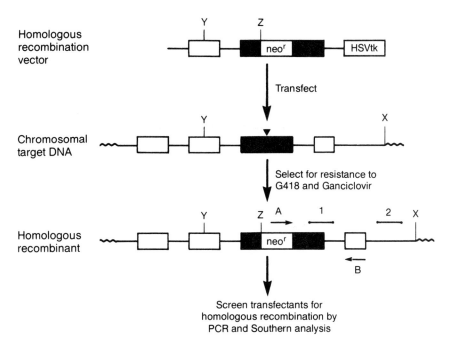

FIGURE 2-15 *Homologous recombination vector and mechanism of selection and verification.* The features of the vector structure appropriate for positive/negative selection for the introduction of exogenous DNA through homologous recombination are depicted here. The homologous recombination vector is composed of a fragment of the genomic DNA derived from the gene to be mutated. A neomycin resistance gene (neo^r), with or without a polyadenylation signal but carrying its own promoter, is inserted into an exon of the gene to be mutated to serve as a dominant selectable marker and to create the unique restriction site Z. An intact HSVtk gene is subsequently attached either 5' or 3' to the interrupted exon, leaving a sufficient amount of genomic DNA sequence between them to allow for efficient homologous recombination between this vector and the host cell chromosomal target DNA. Prior to transfection the homologous recombination vector is linearized. Post transfection, the cells are placed in media containing G418 and ganciclovir to select for neo^r and against HSVtk gene expression. The correct homologous recombinant appears as shown in the bottom diagram of the figure and after drug selection can be detected and verified by a variety of methods, two of which are described here. PCR analysis of transfected populations employing the amplimers labeled A (complementary to the neo^r gene) and B (complementary to the chromosomal DNA of the target gene but not included in the original homologous recombination vector sequences) will result in the appearance of a fragment of unique size that can be subjected to Southern hybridization analysis with the probe labeled 1. Alternatively, genomic DNA derived from transfected, selected pools can be digested with the restriction endonucleases Y and X, Y and Z, and X and Z. After electrophoresis of the digestion products and transfer to a nitrocellulose filter, the Southern blot can be probed with the hybridization probe labeled 2. When probed with 2, the X and Y digestion should reveal two bands, one identical to that found in DNA from the non-transfected control cells and one higher than that found in the non-transfected control population and characteristic of a homologous recombinant. The X and Z digestion should display a novel low molecular weight band again characteristic of a homologous recombinant.

such that the interior portion of the gene of interest is recombined out and replaced by its mutated equivalent. The recombinant molecule will integrate into other chromosomal positions in the cell population as well, and therefore the application of powerful selection or screening techniques is required.

Selection is first employed to identify those members of the transfected cell population that have stably incorporated the exogenous recombinant molecule into their chromosomal DNA at any location. This is most easily accomplished by incorporating a selectable marker in the recombinant molecule, preferably between the flanking regions of exact homology of the gene of interest.

To identify a subset of transfected cells that might contain a homologously recombined allele, Mansour and colleagues (1988) devised a protocol they call *positive/negative selection*. With this method, one includes a dominant, positive-selectable marker *between* the flanking region of homology. The positive marker they exploit is neo^r, which they select with the antibiotic G418. The negative marker they select against is the herpes simplex virus thymidine kinase (HSVtk) gene placed at the end of the construct outside the region of homology. The drug ganciclovir, which kills cells expressing HSVtk, is used to select against cells carrying the HSVtk gene. Thus, the researchers select for integration of the recombinant molecule and against the integration of the HSVtk gene. One manner in which such an integration can occur is if the regions flanking the dominant, positive-selectable marker undergo homologous recombination, stabilizing the positive marker but releasing the negative-selectable marker outside the regions of homology. Predictably, such a cell will be both G418- and ganciclovir-resistant.

It is conceivable that integration of the positive marker and non-integration of the negative marker can occur by means other than homologous recombination, and, in fact, they do. The frequency of the homologous recombination event varies from experiment to experiment. After positive/negative double-drug selection, Johnson et al. (1989b) reported homologous recombination frequencies between 1 in 5×10^7 and 1 in 6.6×10^6 ES cells electroporated. Approximately 1 in 1,000 electroporated cells were both G418- and ganciclovir-resistant; of those, one cell had undergone homologous recombination, the exact frequency depending on the allele disrupted. Considerably higher frequencies, almost one order of magnitude better, have been reported by Zijlstra et al. (1989). The reason for their higher efficiencies is unclear.

Sedivy and Sharp (1989) developed a technique that resulted in a frequency of targeted gene disruption of 1 in 10,000 cells incorporating exogenous DNA. Their procedure employs a targeting vector encoding neo^r that lacks a translation-initiation codon and thus selects for in-frame insertions within exons of active genes transcribed by RNA polymerase II. Frequencies of 1 in 1,000 to 1 in 10 have been reported by a number of groups, but frequencies of 1 in 1,000 or greater fairly represent the experiences of numerous investigators in this area.

In order to discriminate between bona fide homologous recombinants and random recombinants manifesting the identical drug-resistant phenotype, specific, sensitive detection methodologies must be employed. Two nucleic-acid-detection methods are most appropriate for this purpose: Southern-blot analysis, and application of the polymerase chain reaction with or without subsequent Southern-blot analysis. These methods are discussed in Part III of this manual. In fact, the detection methodology selected will have a significant impact on the design of the recombinant molecule used to introduce the mutation into the gene of interest via homologous recombination. To gain a greater appreciation of this point, I will run through the design of a plasmid for homologous recombination as well as a strategy for its subsequent detection (Figure 2-15). Although there are a number of locations one might select to inactivate a gene, I will, for the purposes of this discussion, inactivate the gene of interest within an exon. I will be discussing a gene for which a genomic clone has been characterized, partially sequenced, and comprehensively mapped with restriction endonucleases.

The first order of business is to identify a restriction site within an exon. This site will be used to insert the dominant selectable marker gene driven by an exogenous promoter. Insertion of this gene serves to disrupt the function of the gene of interest as well as enables one to identify tissue culture transfectants that have stably integrated the recombinant molecule. If one has chosen to exploit the positive/negative selection technique mentioned earlier, the next order of business is to identify restriction sites, either 5' or 3' to the interrupted exon, in which to insert the negative-selectable HSVtk gene, again driven by an exogenous promoter. There is strong evidence to support the notion that the length of the sequences flanking the dominant selectable marker—that is, the gene sequences available to undergo homologous recombination with their chromosomal complement—serve to determine, at least in part, the probability of creating a successful homologous recombinant. The thinking is that the greater the length of the sequence homology, the larger the recombination target. Thus, the distances flanking the mutational insertion in the gene-transfer vehicle should be maximized (Thomas and Capecchi, 1987), as should the distance from the mutational insertion to the negative-selectable HSVtk gene, which one is selecting against.

However, such decisions have a significant impact on the choice of the screening method employed to detect a successful homologous recombinant. Screening for a successful homologous recombinant requires the characterization of the distance, post recombination, between the insertional mutation in the recombinant molecule and a known location proximal to, but outside, the regions of homology, and therefore of the potential regions of recombination between the exogenous plasmid and the chromosomal DNA. Restriction endonuclease digestion and Southern hybridization analysis to detect a homologous recombination event is compatible with the measurement of large distances, greater than 1.5 kb. PCR-based detection and measurement of homologous recombination events is compatible with shorter flanking regions, less than 1.5 kb. Thus, although PCR detection may be simpler, the choice of PCR as a detection method will influence the design of the recombinant plasmid carrying the mutagenic insertion because it will limit the size of at least one of the flanking homologous regions of the genomic DNA in the plasmid. This decision may, as mentioned earlier, affect the frequency of homologous recombination events.

Yet another requirement for the efficient detection of homologous recombinants in mixed populations by either Southern blot or PCR analysis is the generation of a nucleic-acid probe complementary to a sequence *flanking* the gene sequence subcloned into the recombinant plasmid used for transfection. This probe must lie within the sequence containing the distal site encoding either: (1) the restriction endonuclease site used to free the fragment of chromosomal DNA in which the homologous recombination event has occurred, or, (2) the distal PCR amplimer used to drive the PCR-based generation of a similar fragment (the other amplimer lying in and characteristic of the insertional mutation of the gene of interest engineered into the recombinant plasmid).

In addition, the insertional mutation in the recombinant plasmid to be transfected should be engineered to encode a new restriction endonuclease site that, post digestion, will result in the appearance of a fragment spanning one of the flanking regions of the recombinant plasmid as well as containing the sequences homologous to the diagnostic probe described above. The inclusion of this feature in the genetic design of the recombinant plasmid will enable the investigator to visualize, after Southern-blot analysis, both the exogenous, wild-type allele of the gene of interest and the now-mutated counterpart of that gene migrating at a lower molecular weight.

In addition, the entire homologous recombination assembly engineered into the bacterial plasmid should be designed in such a manner that it can be excised intact, upon digestion with a restriction endonuclease, from the bacterial component of the vector. This linear DNA, lacking the bacterial sequences, appears to function as a more efficient homologous recombination substrate (Thomas and Capecchi 1986).

Although several lines of data support the notion that the extent of sequence homology flanking the nucleotide sequences to be inserted may have a profound effect on the frequency of homologous recombination into the chromosome (Thomas and Capecchi, 1987), at least one line of evidence suggests that, beyond a certain point, the size of the chromosomal target for homologous recombination is not important. In these experiments, cell lines containing the normal complement of the dihydrofolate reductase (DHFR) gene and an amplified array of the DHFR gene were evaluated as recipients of a transfected molecule designed to undergo homologous recombination within DHFR upon transfection. Zheng and Wilson (1990) reported that even a 200-fold increase in the size of the chromosomal target had no effect on the frequency of homologous recombination.

The elegant studies of Zijlstra et al. (1989, 1990) and Koller et al. (1990) serve to illustrate the power of this technology. In their studies, they generated embryonal stem cells in which one allele of the β-2 microglobulin gene had been disrupted. β-2 microglobulin has been shown to play a key role in the development of the immune response. It has been proposed that these gene products, expressed very early in embryogenesis, have non-immune functions as well. Zijlstra et al. and Koller et al. generated germ-line chimeric mice carrying the β-2 microglobulin defect. These mice were intercrossed, and offspring homozygous for the disrupted gene were identified. Although they lack MHC class I antigen and normal immune function, the mice develop normally. Thus, it appears that β-2 microglobulin does not play a key role in the early development of tissues other than those of the immune system.

PCR-BASED EXPRESSION VECTORS

The technical revolution created by the development of the polymerase chain reaction has led to a new gene-transfer methodology called PCR-based expression. The driving force behind the development of this method was the desire to enable the experimenter to generate, in a matter of hours, microgram quantities of DNA in which the expression of a given gene is driven by a selected promoter without any molecular cloning steps. Such a PCR-assembled construction could be used for transfection analysis immediately. This has, in fact, led to the development of two PCR-based methodologies: PCR based gene synthesis and assembly, and PCR-based expression. Both techniques are described here.

PCR-Based Gene Assembly

The inavailability of a given DNA sequence, be it the structural region of a cDNA or a *cis*-acting expression element, should not inhibit your ability to carry out experiments with that sequence. If the sequence of the DNA of interest has been published, you can use PCR to assemble the gene even if a source of mRNA is unavailable. We have used this technique to assemble a variety of genes including that encoding an artifical hybrid, the human leukocyte interferon A-D hybrid. To assemble this gene, we synthesized 16 overlapping 51-mers and carried out PCR on subsets of these amplimers, subsequently assembling the subsets into a completed gene. The procedure is described in detail in the section on PCR-based expression in Part II.

PCR-Based Expression

An alternative approach to insertion of a cDNA or gene of interest into an expression vector is the attachment, by a PCR reaction, of a DNA fragment encoding an expression element to the cDNA or gene of interest. In such a situation, sequences from the expression element and sequences from the cDNA or gene of interest are employed as amplimers to drive the synthesis of the coupled fragments. Upon completion of the PCR, the amplified DNA can be used directly in the transfection of cells in culture, eliminating the time-consuming process of plasmid subcloning and subsequent plasmid purification. The procedure is described in detail in the section on PCR-based expression in Part II.

3

PROCESSING OF PROTEINS ENCODED BY TRANSFERRED GENES

INTRODUCTION

Once you have assembled your expression construct or recombinant virus and have either transfected or infected the appropriate cells, you will want to study the gene product of interest. If that gene product is a protein molecule, you have a substantial armamentarium of methods at your disposal. In this section, I briefly describe (1) factors that can be experimentally manipulated to enhance mRNA translation, (2) the mechanism of protein processing and secretion, and (3) the biochemistry of post-translational modifications. The experimental methodology for the manipulation and analysis of these processes is described in Part III of this manual.

PROTEIN SYNTHESIS

The rate-limiting step in protein synthesis is the initiation of translation of the mRNA. Translation is initiated by the formation of a complex composed of eukaryotic initiation factor 2 (eIF-2), GTP, and an initiator methionine tRNA. This complex is bound to a 40S ribosomal subunit.

This complex binds to the 5' end of an mRNA and subsequently migrates in the 3' direction until it encounters an AUG (methionine) initiation codon. A 60S ribosomal subunit then binds this complex, and translation is initiated. The elegant studies of Kozak (1986) served to determine the optimal sequence surrounding the intitiator methionine for efficient mRNA translation. This consensus DNA sequence is 5'-CCA/GCC*ATG*G-3'. The two most important residues are the purine at the -3 position and the G in the $+4$ position. This sequence can, of course, be engineered into your recombinant cistron to enhance the translational efficiency of your transferred gene.

The presence of any AUG codons upstream of the authentic initiation codon can substantially reduce the translation of the mRNA. Whenever possible, such codons, either in frame or out of frame, should be removed.

Secondary structure in the mRNA can often, but not always, interfere with translation. If the expression of a given gene is poor in your expression system, you may be well served to examine the sequence of the transcript for significant secondary structures, keeping in mind that, in the case of some mRNAs, specific secondary structures may be beneficial. One clear example of this is the HTLV-1 transcribed expression element that has been incorporated into the SV40-based expression vector SRα. This contains a stem-loop structure that dramatically enhances gene expression (see the previous section on SV40 vectors and the section on expression cloning in Part II).

In attempts to enhance the expression of transferred genes, *trans*-acting factors can be recruited to enhance the translational efficiency of mRNAs derived from plasmid DNAs in COS transfections. As described in the previous sections on translational control and SV40-based vectors, cotransfection, in either *cis* or *trans* of adenovirus VA RNA with your expression vector can dramatically increase the translational efficiency of most COS-based expression vectors. VA RNA serves to stimulate translation by preventing the phosphorylation of the α subunit of eIF-2 by the double-stranded RNA-dependent protein kinase DAI, which normally induces translation arrest (Kaufman, 1985; Kitajewski et al., 1986; O'Malley et al., 1986; Svensson and Akusjarvi, 1985; Kaufman and Murtha, 1987; Mellits et al., 1990).

PROCESSING

The primary sequence of a protein serves to determine where that protein will ultimately reside in the cell. Although all proteins are synthesized in the cytoplasm of the cell, many function in specialized compartments within the cell, reside in the plasma membrane of the cell, or are secreted. This discussion will be limited to the processing of membrane and secretory molecules. For information on protein trafficking within the cytoplasm between organelles, see Meyer (1988); Verner and Schatz (1988); Bradshaw (1989); and Wickner (1989).

Studies of the mammalian endoplasmic reticulum (ER) suggest that protein translocation across translocation-competent membranes is receptor-mediated. In the case of the mammalian ER, four components of this receptor system have been identified. They are (1) a ribonucleoprotein particle (the signal recognition particle, or SRP), which binds (2) the protein precursor's signal sequence as it emerges from the ribosome, which then binds (3) a 72-kD integral ER protein (the SRP receptor, or docking protein), which facilitates the release of the SRP, which then allows the signal sequence of the precursor to interact with (4) a 35-kD integral ER glycoprotein (the signal-sequence acceptor) (Verner and Schatz, 1988).

As this process is completed, post-translational modification of the protein, as described below, begins. If the protein of interest is a cell-surface membrane protein, it will remain anchored in the membrane as it is transported to the Golgi complex (GC), then into vesicles, and ultimately to the cell surface. If the protein of interest is to be secreted, then it will follow a similar course. If, as is the case with tumor necrosis factor, the protein is synthesized as a cell-surface membrane protein that is also cleaved to function as a secreted molecule, that cleavage occurs en route to the cell membrane (Kriegler et al., 1988). We have been able to experimentally modify the secretory pathway of the tumor necrosis factor precursor. We have blocked the cleavage and release of the secretory form of tumor necrosis factor but retained the biological activity of the molecule by introducing small deletions around the cleavage sites of the precursor. Such a mutant tumor necrosis factor is transported to the cell membrane but is not cleaved and released into the medium. Similarly, we have substituted the signal peptide derived from γ-interferon for the transmembrane domain of the tumor necrosis factor precursor to create a tumor necrosis factor that cannot exist as a membrane protein. Upon transfection of the gene encoding this mutant into tissue culture cells, this altered tumor necrosis factor molecule is secreted as any normal secretory protein that lacks a precursor (Perez et al., 1990).

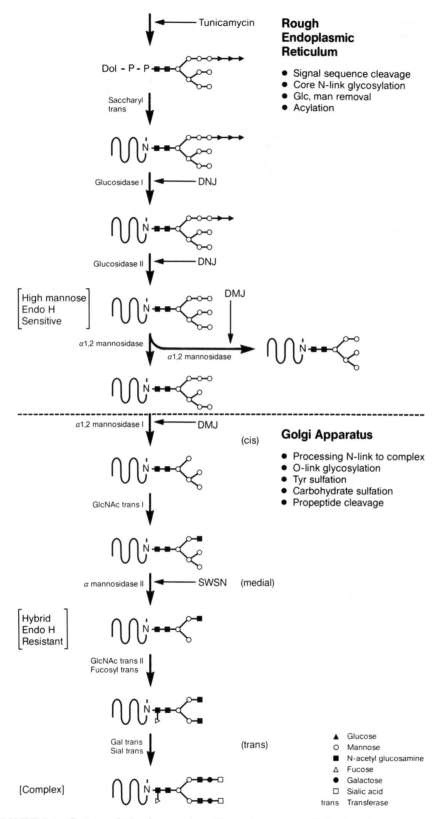

FIGURE 3-1 *Post-translational processing.* The order, type, cellular location, enzymes involved in, and inhibitors of various post-translational processing steps are depicted here. [Adapted from Dorner and Kaufman (1990), *Methods Enzymol.*, Vol. 185, Academic Press, p. 578, Fig. 1.]

POST-TRANSLATIONAL MODIFICATION

Post translation, proteins can be modified by glycosylation, sulfation, and acylation. These processes are represented diagrammatically in Figure 3-1. Upon translocation of newly synthesized membrane or secretory proteins into the ER, a high-mannose oligosaccharide core is added to the asparagine residue in the sequence Asn-X-Ser/Thr, where X can be any amino acid except proline. Next, terminal glucose residues are removed by glucosidases I and XI and at least one α-1,2-linked mannose residue is removed by an α-1,2 mannosidase. Acylation of proteins with a long-chain fatty acid may also occur in the ER, usually through the addition of a myristate or a palmitate to either serine or cysteine residues.

Further modification of proteins may occur as the molecules are transferred to the Golgi complex (GC). The systematic processing of proteins in the GC involves the removal of mannose residues by mannosidases I and II and the addition of N-acetylglucosamine, fucose, and galactose to silalic-acid residues by specific transferases to modify high-mannose carbohydrates to more complex forms.

In addition to N-linked glycosylation, carbohydrates can also be added to threonine and serine residues, a process termed O-linked glycosylation. Unlike N-linked glycosylation, no consensus sequence for O-linked glycosylation has been found. O-linked glycosylation is initiated with the addition of N-acetylgalactosamine to the protein. It is also in the GC that sulfation of tyrosine and carbohydrate residues occurs.

In general, expression of heterologous proteins in mammalian cells results in appropriate post-translational modifications. For more information on this subject, see Yan et al. (1989). The best method for determining if your protein falls into this category is empirical analysis. Specific experimental details on the analysis of post-translational modifications of proteins, including glycosylation, sulfation, and acylation, is presented in Part III of this manual.

PART I REFERENCES

ADAM, M. A. and MILLER, A. D. 1988. Identification of a signal in a murine retrovirus that is sufficient for packaging of nonretroviral RNA into virions. *J. Virology* **62**: 3802.

AMTMANN, E. and SAUER, G. 1982. Bovine papilloma virus transcription: polyadenylated RNA species and assessment of the direction of transcription. *J. Virol* **43**: 59.

ANGEL, P., BAUMAN, I., STEIN, B., DELLUS, H., RAHMSDORF, H. J., and HERRLICH, P. 1987a. 12-0-tetradecanoyl-phorbol-13-acetate induction of the human collagenase gene is mediated by an inducible enhancer element located in the 5′ flanking region. *Mol. Cell. Biol.* **7**: 2256.

ANGEL, P., IMAGAWA, M., CHIU, R., STEIN, B., IMBRA, R. J., RAHMSDORF, H. J., JONAT, C., HERRLICH, P., and KARIN, M. 1987b. Phorbol ester-inducible genes contain a common cis element recognized by a TPA-modulated trans-acting factor. *Cell* **49**: 729.

ARUFFO, A. and SEED, B. 1987. Molecular cloning of a CD28 cDNA by a high-efficiency COS cell expression system. *Proc. Natl. Acad. Sci. U.S.A.* **84**: 8573.

ATCHISON, M. L. and PERRY, R. P. 1986. Tandem kappa immunoglobulin promoters are equally active in the presence of the kappa enhancer: implications for models of enhancer function. *Cell.* **46**: 253.

ATCHISON, M. L. and PERRY, R. P. 1987. The role of the kappa enhancer and its binding factor NF-kappa B in the developmental regulation of kappa gene transcription. *Cell.* **48**: 121.

AYARES, D., CHEKURI, L., SONG, K. Y., and KUCHERLAPATI, R. 1986. Sequence-homology requirements for intermolecular recombination in mammalian cells. *Proc. Natl. Acad. Sci. U.S.A.* **83**: 5199.

BAKER, M. D., MURIALDO, and SHULMAN, M. J. 1988. Expression of an immunoglobulin kappa light-chain gene in lymphoid cells using a bovine papilloma virus-1 (BPV-1) vector. *Gene* **69**: 349.

BANERJI, J., OLSON, L., and SCHAFFNER, W. 1983. A lymphocyte-specific cellular enhancer is located downstream of the joining region in immunoglobulin heavy-chain genes. *Cell* **35**: 729.

BANERJI, J., RUSCONI, S., and SCHAFFNER, W. 1981. Expression of a beta-globin gene is enhanced by remote SV40 DNA sequences. *Cell* **27**: 299.

BARNETT, J. W., EPPSTEIN, D. A., and CHAN, H. W. 1983. Class I defective herpes simplex virus DNA as a molecular cloning vehicle in eukaryotic cells. *J. Virology* **48**: 384.

BENDER, M. A., PALMER, T. D., GELINAS, R. E., and MILLER, D. 1987. Evidence that the packaging signal of moloney murine leukemia virus extends into the gag region. *J. Virology* **61**: 1639.

BERG, L. J., SINGH, K., and BOTCHAN, M. 1986. Complementation of a bovine papilloma virus low-copy-number mutant: evidence for a temporal requirement of the complementing gene. *Mol. Cell. Biol.* **6**: 859.

BERKHOUT, B., SILVERMAN, R. H., and JEANG, K. 1989. Tat trans-activates the human immunodeficiency virus through a nascent RNA target. *Cell* **59**: 273.

BERKNER, K. K., SCHAFFHAUSEN, B. S., ROBERTS, T. M., and SHARP, P. A. 1987. Abundant expression of polyoma virus middle T antigen and dihydrofolate reductase in an adenovirus recombinant. *J. Virology* **61**: 1213.

BERNSTEIN, P. and ROSS, J. 1989. Poly (A), poly (A) binding protein, and the regulation of mRNA stability. *TIBS* **14**: 373.

BLANAR, M. A., BALDWIN, JR., A. S., FLAVELL, R. A., and SHARP, P. A. 1989. A gamma-interferon-induced factor that binds the interferon response sequence of the MHC class I gene, H-2Kb. *EMBO J.* **8**: 1139.

BODINE, D. M. and LEY, T. J. 1987. An enhancer element lies 3' to the human A gamma globin gene. *EMBO J.* **6**: 2997.

BOSHART, M., WEBER, F., JAHN, G., DORSCH-HASLER, K., FLECKENSTEIN, B., and SCHAFFNER, W. 1985. A very strong enhancer is located upstream of an immediate early gene of human cytomegalovirus. *Cell* **41**: 521.

BOSZE, Z., THIESEN, H. J., and CHARNAY, P. 1986. A transcriptional enhancer with specificity for erythroid cells is located in the long terminal repeat of the Friend murine leukemia virus. *EMBO J.* **5**: 1615.

BOULTER, C. A. and WAGNER, E. F. 1987. A universal retroviral vector for efficient constitutive expression of exogenous genes. *Nuc. Acids Res.* **15**: 7194.

BRAAM-MARKSON, J., JAUDON, C., and KRUG, R. M. 1985. Expression of a functional influenza viral cap-recognizing protein by using a bovine papilloma virus vector. *Proc. Natl. Acad. Sci. U.S.A.* **82**: 4326.

BRADDOCK, M., CHAMBERS, A., WILSON, W., ESNOUF, M. P., ADAMS, S. E., KINGSMAN, A. J., and KINGSMAN, S. M. 1989. HIV-I tat activates presynthesized RNA in the nucleus. *Cell* **58**: 269.

BRADSHAW, R. A. 1989. Protein translocation and turnover in eukaryotic cells. *TIBS* **14**: 1989.

BRAWERMAN, G. 1987. Determinants of messenger RNA stability. *Cell* **48**: 5.

BRAWERMAN, G. 1989. mRNA decay: finding the right targets. *Cell* **57**: 9.

BREATHNACH, R. and CHAMBON, P. 1981. Organization and expression of eukaryotic split genes coding for proteins. *Ann. Rev. of Biochem.* **50**: 349.

BRINSTER, R. L., ALLEN, J. M., BEHRINGER, R. R., GELINAS, R. E., and PALMITER, R. D. 1988. Introns increase transcriptional efficiency in transgenic mice. *Proc. Natl. Acad. Sci.* **85**: 836.

BUCHMAN, A. R. and BERG, P. 1988. Comparison of intron-dependent and intron-independent gene expression. *Mol. Cell. Bio.* **8**: 4395.

BULLA, G. A. and SIDDIQUI, A. 1986. The hepatitis B virus enhancer modulates transcription of the hepatitis B virus surface-antigen gene from an internal location. *J. Virol* **62**: 1437.

CAMPBELL, B. A. and VILLARREAL, L. P. 1986. Lymphoid and other tissue-specific phenotypes of polyoma virus enhancer recombinants: positive and negative combinational effects on enhancer specificity and activity. *Mol. Cell. Biol.* **6**: 2068.

CAMPBELL, B. A. and VILLARREAL, L. P. 1988. Functional analysis of the individual enhancer core sequences of polyoma virus: cell-specific uncoupling of DNA replication from transcription. *Mol. Cell. Biol.* **8**: 1993.

CAMPERE, S. A. and TILGHMAN, S. M. 1989. Postnatal repression of the α-fetoprotein gene is enhancer independent. *Genes and Dev.* **3**: 537.

CAMPO, M. S., SPANDIDOS, D. A., LANG, J., and WILKIE, N. M. 1983. Transcriptional control signals in the genome of bovine papilloma virus type 1. *Nature* **303**: 77.

CAPECCHI, M. 1990. Tapping the cellular telephone. *Nature* **344**: 105.

CAPUT, D., BEUTLER, B., HARTOG, K., THAYER, R., BROWN-SHIMER, S., and CERAMI, A. 1986. Identification of a common nucleotide sequence in the 3' untranslated region of mRNA molecules specifying inflammatory mediators. *Proc. Natl. Acad. Sci. U.S.A.* **83**: 1670.

CELANDER, D. and HASELTINE, W. A. 1987. Glucocorticoid regulation of murine leukemia virus transcription elements is specified by determinants within the viral enhancer region. *J. Virology* **61**: 269.

CELANDER, D., HSU, B. L., and HASELTINE, W. A. 1988. Regulatory elements within the murine leukemia virus enhancer regions mediate glucocorticoid responsiveness. *J. Virology*, **62**: 1314.

CEPKO, C., ROBERTS, B., and MULLIGAN, R. 1984. Construction and applications of a highly transmissible murine retrovirus shuttle vector. *Cell* **37**: 1053.

CHANDLER, V. L., MALER, B. A., and YAMAMOTO, K. R. 1983. DNA sequences bound specifically by glucocorticoid receptor in vitro render a heterologous promoter hormone responsive in vivo. *Cell* **33**: 489.

CHANG, S. C., ERWIN, A. E., and LEE, A. S. 1989. Glucose-regulated protein (GRP94 and GRP78) genes share common regulatory domains and are coordinately regulated by common trans-acting factors. *Mol. Cell. Biol.* **9**: 2153.

CHATTERJEE, V. K., LEE, J. K., RENTOUMIS, A., and JAMESON, J. L. 1989. Negative regulation of the thyroid-stimulating hormone alpha gene by thyroid hormone: receptor interaction adjacent to the TATA box. *Proc. Natl. Acad. Sci. U.S.A.* **86**: 9114.

CHOI, K., CHEN, C., KRIEGLER, M., and RONINSON, I. B. 1988. An altered pattern of cross-resistance in multi-drug-resistant human cells results from spontaneous mutations in the Mdr-1 (P-glycoprotein) gene. *Cell* **53**: 519.

COHEN, J. B., WALTER, M. V., and LEVINSON, A. D. 1987. A repetitive sequence element 3′ of the human c-Ha-ras1 gene has enhancer activity. *J. Cell Physiol.* **5**: 75.

CONE, R. and MULLIGAN, R. 1984. High-efficiency gene transfer into mammalian cells: generation of helper-free recombinant retrovirus with broad mammalian host range. *Proc. Natl. Acad. Sci. U.S.A.* **81**

CONE, R. D., WEBER-BENAROUS, A. W., BAORTO, D., and MULLIGAN, R. C. 1987. Regulated expression of a complete human β-globin gene encoded by a transmissible retrovirus vector. *Mol. Cell. Biol.* **7**: 887.

COSMAN, D., WIGNALL, J., LEWIS, A., ALPERT, A., CERRETTI, D. P., PARK, L., DOWER, S. K., GILLIS, S., and URDAL, D. L. 1986. High-level stable expression of human interleukin-2 receptors in mouse cells generates only low-affinity interleukin-2 binding sites. *Mol. Immunol.* **23**: 935.

COSTA, R. H., LAI, E., GRAYSON, D. R., and DARNELL, J. E. 1988. The cell-specific enhancer of the mouse transthyretin (prealbumin) gene binds a common factor at one site and a liver-specific factor(s) at two other sites. *Mol. Cell. Biol.* **8**: 81.

COUPAR, B. E. H., ANDREW, M. E., BOTH, G. W., and BOYLE, D. B. 1986a. Temporal regulation of influenza hemaglutinin expression in vaccinia virus recombinants and effects on the immune response. *Eur. J. Immunol.* **16**: 1479.

COUPAR, B.E.H., ANDREW, M.E., BOYLE, D.B., and BLANDEN, R.V. 1986b. Immune response to H-2k^d antigen expressd by recombinant vaccinia virus. *Proc. Natl. Acad. Sci. U.S.A.* **83**: 7879.

CRIPE, T. P., HAUGEN, T. H., TURK, J. P., TABATABAI, F., SCHMID, III, P. G., DURST, M., GISSMANN, L., ROMAN, A., and TUREK, L. P. 1987. Transcriptional regulation of the human papilloma virus-16 E6-E7 promoter by a keratinocyte-dependent enhancer, and by viral E2 trans-activator and repressor gene products: implications for cervical carcinogenesis. *EMBO J.* **6**: 3745.

CULOTTA, V. C. and HAMER, D. H. 1989. Fine mapping of a mouse metallothionein gene metal-response element. *Mol. Cell. Biol.* **9**: 1376.

DANDOLO, L., BLANGY, D., and KAMEN, R. 1983. Regulation of polyma virus transcription in murine embryonal carcinoma cells. *J. Virology* **47**: 55.

DANOS, O., ENGEL, L. W., CHEN, E. Y., YANIV, M., and HOWLEY, P. M. 1983. Comparative analysis of the human type 1a and bovine type 1 papilloma virus genomes. *J. Virology* **46**: 557.

DANOS, O. and MULLIGAN, R. C. 1988. Safe and efficient generation of recombinant retroviruses with amphotropic and ecotropic host ranges. *Proc. Natl. Acad. Sci. U.S.A.* **85**: 6460.

DAVIDSON, D. and HASSELL, J. A. 1987. Overproduction of polyoma virus middle T antigen in mammalian cells through the use of an adenovirus vector. *J. Virology* **61**: 1226.

DENNISTON, K. J., YONEYAMA, T., HOYER, B. H., and GERIN, J. L. 1984. Expression of hepatitis B virus surface and e antigen genes cloned in bovine papilloma virus vectors. *Gene* **32**: 357.

DESCHAMPS, J., MEIJLINK, F., and VERMA, I. M. 1985. Identification of a transcriptional enhancer element upstream from the proto-oncogene fos. *Science* **230**: 1174.

DE VILLIERS, J., SCHAFFNER, W., TYNDALL, C., LUPTON, S., and KAMEN, R. 1984. Polyoma virus DNA replication requires an enhancer. *Nature* **312**: 242.

DIMAIO, D., CORBIN, V., SIBLEY, E., and MANIATIS, T. 1984. High-level expression of a cloned HLA heavy-chain gene introduced into mouse cells on a bovine papilloma virus vector. *Mol. Cell. Biol.* **4**: 340.

DOETSCHMAN, T., MAEDA, N., and SMITHIES, O. 1988. Targeted mutation of the Hprt gene in mouse embryonic stem cells. *Proc. Natl. Acad. Sci. U.S.A.* **85**: 8583.

DORNER, A. J. and KAUFMAN, R. J. 1990. Analysis of synthesis, processing and secretion. *Methods Enzymol.* **185**: 577.

DYNAN, W. S. 1989. Modularity in promoters and enhancers. *Cell* **58**: 1.

EDBROOKE, M. R., BURT, D. W., CHESHIRE, J. K., and WOO, P. 1989. Identification of cis-acting sequences responsible for phorbol ester induction of human serum amyloid A gene expression via a nuclear-factor-κB-like transcription factor. *Mol. Cell. Biol.* **9**: 1908.

EDLUND, T., WALKER, M. D., BARR, P. J., and RUTTER, W. J., 1985. Cell-specific expression of the rat insulin gene: evidence for role of two distinct 5' flanking elements. *Science* **230**: 912.

EDWARDS, R. H. and RUTTER, W. J. 1988. Use of vaccinia virus vectors to study protein processing in human disease. *J. Clin. Invest.* **82**: 44.

EIDEN, M., NEWMAN, M., FISHER, A. G., MANN, D. L., HOWLEY, P. M. and REITZ, M. S. 1985. Type 1 human T-cell leukemia virus small envelope protein expressed in mouse cells by using a bovine papilloma-virus-derived shuttle vector. *Mol. Cell. Biol.* **5**: 3320.

ELLIS, J. and BERNSTEIN, A. 1989. Gene targeting with retroviral vectors: recombination by gene conversion into regions of nonhomology. *Mol. Cell. Biol.* **9**: 1621.

EMERMAN, M. and TEMIN, H. M. 1984. Genes with promoters in retrovirus vectors can be independently suppressed by an epigenetic mechanism. *Cell* **39**: 449.

EMERMAN, M. and TEMIN, H. M. 1986a. Comparison of promoter suppression in avian and murine retrovirus vectors. *Nuc. Acids Res.* **14**: 9381.

EMERMAN, M. and TEMIN, H. M. 1986b. Quantitative analysis of gene suppression in integrated retrovirus vectors. *Mol. Cell. Biol.* **6**: 792.

ENGEL, L. W., HEILMAN, C. A., and HOWLEY, P.M. 1983. Transcriptional organization of bovine papillomavirus type 1. *J. Virol.* **47**: 516.

ENVER, T., BREWER, A. C., and PATIENT, R. K. 1987. Role for DNA replication in beta-globin gene activation. *Mol. Cell. Bio.* **8**: 1301.

FALKNER, S. G. and MOSS, B. 1988. *Escherichia coli* gpt gene provides dominant selection of vaccinia virus open-reading-frame expression vectors. *J. Virology* **62**: 1849.

FENG, S. and HOLLAND, E. C. 1988. HIV-I tat trans-activation requires the loop sequence within tar. *Nature* **334**: 6178.

FIRAK, R. A. and SUBRAMANIAN, K. N. 1986. Minimal transcriptional enhancer of simian virus 40 is a 74-base-pair sequence that has interacting domains. *Mol. Cell. Bio.* **6**: 3667.

FLEXNER, C., HÜGIN, A., and MOSS, B. 1987. Prevention of vaccinia virus infection in immunodeficient mice by vector-directed IL-2 expression. *Nature* **330**: 259.

FOECKING, M. K. and HOFSTETTER, H. 1986. Powerful and versatile enhancer-promoter unit for mammalian expression vectors. *Gene* **45**: 101.

FRANZ, T., HILBERG, F., SELIGER, B., STOCKING, C., and OSTERTAG, W. 1986. Retroviral mutants efficiently expressed in embryonal carcinoma cells. *Proc. Natl. Acad. Sci. U.S.A.* **83**: 3292.

FUERST, T. R., EARL, P. L. and MOSS, B. 1987. Use of a hybrid vaccinia virus-T7 RNA polymerase system for expression of target genes. *Mol. Cell. Biol.* **7**: 2538.

FUERST, T. R., FERNANDEZ, M. P., and MOSS, B. 1989. Transfer of the inducible lac repressor-operator system from *Escherichia coli* to a vaccinia virus expression vector. *Proc. Natl. Acad. Sci. U.S.A.* **86**: 2549.

FUJITA, T., SHIBUYA, H., HOTTA, H., YAMANISHI, K., and TANIGUCHI, T. 1987. Interferon-beta gene regulation: tandemly repeated sequences of a synthetic 6-bp oligomer function as a virus-inducible enhancer. *Cell* **49**: 357.

FUKUNAGA, R., SOKAWA, Y., and NAGATA, S. 1984. Constitutive production of human interferons by mouse cells with bovine papilloma virus as a vector. *Proc. Natl. Acad. Sci. U.S.A.* **81**: 5086.

FUKUSHIMA, J., TANI, K., ATSUMI, Y., HAMAJIMA, K., KAWAMOTO, and OKUDA, K. 1988. Trans-activation of class II (I-Aα) gene by I-A-β gene transfection using bovine papilloma virus as a shuttle vector system. *J. Immunol.* **141**: 302.

GELLER, A. J. and BREAKEFIELD, X. O. 1988. A defective HSV-1 vector expresses *E. coli* beta-galactosidase in cultured peripheral neurons. *Science* **241**: 1667.

GILBOA, E., MITRA, S. W., GOFF, S., and BALTIMORE, D. 1979. A detailed model of reverse transcription and tests of crucial aspects. *Cell* **18**: 94.

GILLES, S. D., MORRIS, S. L., OI, V. T., and TONEGAWA, S. 1983. A tissue-specific transcription enhancer element is located in the major intron of a rearranged immunoglobulin heavy-chain gene. *Cell* **33**: 717.

GIUS, D., GROSSMAN, S., BEDELL, M. A., and LAIMINS, L. A. 1988. Inducible and constitutive enhancer domains in the noncoding region of human papilloma virus type 18. *J. Virology* **62**: 665.

GLOSS, B., BERNARD, H. U., SEEDORF, K., and KLOCK, G. 1987. The upstream regulatory region of the human papilloma virus-16 contains an E2 protein-independent enhancer which is specific for cervical carcinoma cells and regulated by glucocorticoid hormones. *EMBO J.* **6**: 3735.

GODBOUT, R., INGRAM, R. S., and TILGHMAN, S. M. 1988. Fine-structure mapping of the three mouse alpha-fetoprotein gene enhancers. *Mol. Cell. Biol.* **8**: 1169.

GOODBOURN, S., BURSTEIN, H., and MANIATIS, T. 1986. The human beta-interferon gene enhancer is under negative control. *Cell* **45**: 601.

GOODBOURN, S. and MANIATIS, T. 1988. Overlapping positive and negative regulatory domains of the human β-interferon gene. *Proc. Natl. Acad. Sci. U.S.A.* **85**: 1447.

GRASS, D. S., READ, D., LEWIS, E. D., and MANLEY, J. L. 1987. Cell- and promoter-specific activation of transcription by DNA replication. *Genes and Dev.* **1**: 1065.

GREEN, L., SCHLAFFER, I., WRIGHT, K., MORENO, M. L., BERAND, D., HAGER, G., STEIN, J., and STEIN, G. 1986. Cell-cycle-dependent expression of a stable episomal human histone gene in a mouse cell. *Proc. Natl. Acad. Sci. U.S.A.* **83**: 2315.

GREENE, W. C., BÖHNLEIN, and BALLARD, D. W. 1989. HIV-1, and normal T-cell growth: transcriptional strategies and surprises. *Immunology Today* **10**: 272.

GROSSCHEDL, R. and BALTIMORE, D. 1985. Cell-type specificity of immunoglobulin gene expression is regulated by at least three DNA sequence elements. *Cell* **41**: 885.

GUILD, B. C., FINER, M. H., HOUSMAN, D. E., and MULLIGAN, R. C. 1988. Development of retrovirus vectors useful for expressing genes in cultured murine embryonal cells and hematopoietic cells in vivo. *J. Virol.* **62**: 3795.

HAJ-AHMAD, Y. and GRAHAM, F. L. 1986. Development of a helper-independent human adenovirus vector and its use in the transfer of the herpes simplex virus thymidine kinase gene. *J. Virology* **57**: 267.

HAMBOR, J. E., HAUER, C. A., SHU, H.-K., GROGER, R. K., KAPLAN, D. R., and TYKOCINSKI, M. L. 1988. Use of an Epstein-Barr virus episomal replicon for anti-sense RNA-mediated gene inhibition in a human cytotoxic T-cell clone. *Proc. Natl. Acad. Sci. U.S.A.* **85**: 4010.

HAMMER, R. E., KRUMLAUF, R., CAMPER, S. A., BRINSTER, R. L., and TILGHMAN, S. M. 1987. Diversity of alpha-fetoprotein gene expression in mice is generated by a combination of separate enhancer elements. *Science* **235**: 53.

HANTZOPOULOS, P. E., SULLENGER, B. A., UNGERS, G., and GILBOA, E., 1989. Improved gene expression upon transfer of the adenosine deaminase minigene outside the transcriptional unit of a retroviral vector. *Proc. Natl. Acad. Sci. U.S.A.* **86**: 3519.

HASLINGER, A. and KARIN, M. 1985. Upstream promoter element of the human metallothionein-II gene can act like an enhancer element. *Proc. Natl. Acad. Sci. U.S.A.* **82**: 8572.

HAUBER, J. and CULLEN, B. R. 1988. Mutational analysis of the trans-activation-responsive region of the human immunodeficiency virus type I long terminal repeat. *J. Virology* **62**: 673.

HEINZEL, S. S., KRYSAN, P. J., CALOS, M. P., and DuBRIDGE, R. B. 1988. Use of simian virus 40 replication to amplify Epstein-Barr virus shuttle vectors in human cells. *J. Virology* **62**: 3738.

HEN, R., BORRELLI, E., FROMENTAL, C., SASSONE-CORSI, P., and CHAMBON, P. 1986. A mutated polyoma virus enhancer which is active in undifferentiated embryonal carcinoma cells is not repressed by adenovirus-2 E1A products. *Nature* **321**: 249.

HENSEL, G., MEICHLE, A., PFIZENMAIER, K., and KRONKE, M. 1989. PMA-responsive 5' flanking sequences of the human TNF gene. *Lymphokine Res.* **8**: 347.

HERR, W. and CLARKE, J. 1986. The SV40 enhancer is composed of multiple functional elements that can compensate for one another. *Cell* **45**: 461.

HIROCHIKA, H., BROWKER, T. R., and CHOW, L. T. 1987. Enhancers and trans-acting E2 transcriptional factors of papilloma viruses. *J. Virol* **61**: 2599.

HIRSCH, M. R., GAUGLER, L., DEAGOSTINI-BAUZIN, H., BALLY-CUIF, L., and GORDIS, C. 1990. Identification of positive and negative regulatory elements governing cell-type-specific expression of the neural-cell-adhesion-molecule gene. *Mol. Cell. Biol.* **10**: 1959.

HIRT, B. 1967. Selective extraction of polyoma DNA from infected mouse cell cultures. *J. Mol. Biol.* **26**: 365.

HOLBROOK, N., GULINO, A., and RUSCETTI, F. 1987. Cis-acting transcriptional regulatory sequences in the gibbon ape leukemia virus (GALV) long terminal repeat. *Virology* **157**: 211.

HORLICK, R. A. and BENFIELD, P. A. 1989. The upstream muscle-specific enhancer of the rat muscle creatine kinase gene is composed of multiple elements. *Mol. Cell. Biol.* **9**: 2396.

HOYOS, B., BALLARD, D. W., BÖHNLEIN, E., SIEKEVITZ, M., and GREENE, W. C. 1989. Kappa B-specific DNA-binding proteins: role in the regulation of human interleukin-2 gene expression. *Science* **244**: 457.

HRUBY, D. E., HODGES, W. M., WILSON, E. M., FRANKE, C. A., and FISCHETTI, V. A., 1988. Expression of streptococcal M. protein in mammalian cells. *Proc. Natl. Acad. Sci. U.S.A.* **85**: 5714.

HSIUNG, N., FITTS, R., WILSON, S., MILNE, A., and HAMER, D. 1984. Efficient production of hepatitis B surface antigen using a bovine-papilloma-virus-metallothionein vector. *J. Mol. Appl. Genet.* **2**: 497.

HUG, H., COSTAS, M., STAEHELI, P., AEBI, M., and WEISSMANN, C. 1988. Organization of the murine Mx gene and characterization of its interferon- and virus-inducible promoter. *Mol. Cell. Biol.* **8**: 3065.

HUANG, A. L., OSTROWSKI, M. C., BERARD, D., and HAGAR, G. L. 1981. Glucocorticoid regulation of the Ha-MuSV p21 gene conferred by sequences from mouse mammary tumor virus. *Cell* **27**: 245.

HWANG, I., LIM, K., and CHAE, C. 1990. Characterization of the S-phase-specific transcription regulatory elements in a DNA-replication-independent testis-specific H2B (TH2B) histone gene. *Mol. Cell. Biol.* **10**: 585.

HWANG, L. H. and GILBOA, E. 1984. Exppression of genes introduced into cells by retroviral infection is more efficient than that of genes introduced into cells by DNA transfection. *J. Virology*, **50**: 417.

IMAGAWA, M., CHIU, R., and KARIN, M. 1987. Transcription factor AP-2 mediates induction by two different signal-transduction pathways: protein kinase C and cAMP. *Cell* **51**: 251.

IMBRA, R. J. and KARIN, M. 1986. Phorbol ester induces the transcriptional stimulatory activity of the SV40 enhancer. *Nature* **323**: 555.

IMLER, J. L., LEMAIRE, C., WASVLYK, C., and WASLYK, B. 1987. Negative regulation contributes to tissue specificity of the immunoglobulin heavy-chain enhancer. *Mol. Cell. Biol.* **7**: 2558.

IMPERIALE, M. J. and NEVINS, J. R. 1984. Adenovirus 5 E2 transcription unit: an E1A-inducible promoter with an essential element that functions independently of position or orientation. *Mol. Cell. Biol.* **4**: 875.

ISRAEL, D. I. and KAUFMAN, R. J. 1989. Highly inducible expression from vectors containing multiple GREs in CHO cells overexpressing the glucocorticoid receptor. *Nuc. Acids Res.* **17**: 4589.

JAHNER, D. and JAENISCH, R. 1985. Chromosomal position and specific demethylation in enhancer sequences of germ-line-transmitted retroviral genomes during mouse development. *Mol. Cell. Biol.* **5**: 2212.

JAKOBOVITS, A., SMITH, D. H., JAKOBOVITS, E. B., and CAPON, D. J. 1988. A discrete element 3' of human immunodeficiency virus 1 (HIV-1) and HIV-2 mRNA initiation sites mediates transcriptional activation by an HIV trans-activator. *Mol. Cell. Biol.* **8**: 2555.

JAMEEL, S. and SIDDIQUI, A. 1986. The human hepatitis B virus enhancer requires transacting cellular factor(s) for activity. *Mol. Cell. Biol.* **6**: 710.

JANG, S. K., KRÄUSSLICH, H., NICKLIN, M. J. H., DUKE, G. M. PALMENBERG, A. C., and WIMMER, E. 1988. A segment of the 5' nontranslated region of encephalomyocarditis virus RNA directs internal entry of ribosomes during in vitro translation. *J. Virology* **62**: 2636.

JAYNES, J. B., JOHNSON, J. E., BUSKIN, J. N., GARTSIDE, C. L., and HAUSCHKA, S. D. 1988. The muscle creatine kinase gene is regulated by multiple upstream elements, including a muscle-specific enhancer. *Mol. Cell. Biol.* **8**: 62.

JOHNSON, J. E., WOLD, B. J., and HAUSCHKA, S. D. 1989a. Muscle creatine kinase sequence elements regulating skeletal and cardiac muscle expression in transgenic mice. *Mol. Cell. Biol.* **9**: 3393.

JOHNSON, R. S., SHENG, M., GREENBERG, M. E., KOLODNER, R. D., PAPAIOANNOU, V. E., and SPIEGELMAN, B. M. 1989b. Targeting of nonexpressed genes in embryonic stem cells via homologous recombination. *Science* **245**: 1234.

KADESCH, T. and BERG, P. 1986. Effects of the position of the simian virus 40 enhancer on expression of multiple transcription units in a single plasmid. *Mol. Cell. Biol.* **6**: 2593.

KARASUYAMA, H. and MELCHERS, F. 1988. Establishment of mouse cell lines which constitutively secrete large quantities of interleukin 2, 3, 4, or 5, using modified cDNA expression vectors. *Eur. J. Immunol.* **18**: 97.

KARIN, M., CATHALA, G., and NGUYEN-HUU, M. C. 1983. Expression and regulation of a human metallothionein gene carried on an autonomously replicating shuttle vector. *Proc. Natl. Acad. Sci. U.S.A.* **80**: 4040.

KARIN, M., HASLINGER, A., HEGUY, A., DIETLIN, T., and COOKE, T. 1987. Metal-responsive elements act as positive modulators of human metallothionein-IIA enhancer activity. *Mol. Cell. Biol.* **7**: 606.

KARLSSON, S., PAPAYANNOPOULOU, T., SCHWEIGER, S. G., STAMATOYANNOPOULOS, G., and NIENHUIS, A. W. 1987. Retroviral-mediated transfer of genomic globin genes leads to regulated production of RNA and protein. *Proc. Natl. Acad. Sci. U.S.A.* **84**: 2411.

KASID, A., MORECKI, S., AEBERSOLD, P., CORNETTA, K., CULVER, K., FREEMAN, S., DIRECTOR, E., LOTZE, M. T., BLAESE, R. M., ANDERSON, W. F., and ROSENBERG, S. A. 1990. Human gene transfer: characterization of human tumor-infiltrating lymphocytes as vehicles for retroviral-mediated gene transfer in man. *Proc. Natl. Acad. Sci. U.S.A.* **87**: 473.

KATINKA, M., VASSEUR, M., MONTREAU, N., YANIV, M., and BLANGY, D. 1981. Polyoma DNA sequences involved in the control of viral gene expression in murine embryonal carcinoma cells. *Nature* **290**: 720.

KATINKA, M., YANIV, M., VASSEUR, M., and BLANGY, D. 1980. Expression of polyoma early functions in mouse embryonal carcinoma cells depends on sequence rearrangements in the beginning of the late region. *Cell* **20**: 393.

KAUFMAN, R. J. 1985. Identification of the components necessary for adenovirus translational control and their utilization in cDNA expression vectors. *Proc. Natl. Acad. Sci. U.S.A.* **82**: 689.

KAUFMAN, R. J. 1987. Translational control mediated by eukaryotic initiation factor 2 in transfected cells. In *Gene Transfer Vectors for Mammalian Cells*, eds. J. Miller and M. Calos, Cold Spring Harbor: Cold Spring Harbor Press.

KAUFMAN, R. J. 1990. Vectors used for expression. *Methods Enzymol.* **185**: 487.

KAUFMAN, R. J., DAVIES, M. V., PATHAK, V. K., and HERSHEY, J. W. 1989. The phosphorylation state of eukaryotic initiation factor 2 alters translational efficiency of specific mRNAs. *Mol. Cell. Biol.* **9**: 946.

KAUFMAN, R. J. and MURTHA, P. 1987. Translational control mediated by eukaryotic initiation factor-2 is restricted to specific mRNAs in transfected cells. *Mol. Cell. Biol.* **7**: 1568.

KAWAMOTO, T., MAKINO, K., NIW, H., SUGIYAMA, H., KIMURA, S., ANEMURA, M., NAKATA, A., and KAKUNAGA, T. 1988. Identification of the human beta-actin enhancer and its binding factor. *Mol. Cell. Biol.* **8**: 267.

KELLEY, D. E., POLLOK, B. A., ATCHISON, M. L, and PERRY, R. P. 1988. The coupling between enhancer activity and hypomethylation of kappa immunoglobulin genes is developmentally regulated. *Mol. Cell. Biol.* **8**: 930.

KERN, F. G. and BASILICO, C. 1986. An inducible eukaryotic host-vector expression system: amplification of genes under the control of the polyoma late promoter in a cell line producing a thermolabile large T antigen. *Gene* **43**: 237.

KIENY, M. P., LATHE, R., SPEHNER, D., DRILLON, R., SKORY, S., SCHMITT, D., WIKTOR, T., KOPROWSKI, H., and LECOCQ, J. P. 1984. Expression of rabies virus glycoprotein from a recombinant vaccinia virus. *Nature* **312**: 163.

KILEDJIAN, M., SU, L. K., and KADESCH, T. 1988. Identification and characterization of two functional domains within the murine heavy-chain enhancer. *Mol. Cell. Biol.* **8**: 145.

KIOUSSIS, D., WILSON, F., DANIELS, C., LEVETON, C., TAVERNE, J., and PLAYFAIR, J. H. L. 1987. Expression and rescuing of a cloned human tumor necrosis factor gene using an EBV-based shuttle cosmid vector. *EMBO J.* **6**: 355.

KITAJEWSKI, J., SCHNEIDER, R. J., SAFER, B., MUNEMITSU, S. M., SAMUEL, C. E., THIMMAPPAYA, B., and SHENK, T. 1986. Adenovirus VA RNA antagonizes the antiviral action of interferon by preventing activation of the interferon-induced eIF-2 alpha kinase. *Cell* **45**: 195.

KLAMUT, H. J., GANGOPADYHAY, S. B., WORTON, R. G., and RAY, P. N. 1990. Molecular and functional analysis of the muscle-specific promoter region of the Duchenne Muscular Dystrophy gene. *Mol. Cell. Biol.* **10**: 193.

KOCH, W., BENOIST, C., and MATHIS D. 1989. Anatomy of a new B-cell-specific enhancer. *Mol. Cell. Biol.* **9**: 303.

KOLLER, B. H., HAGEMANN, L. J., DOETSCHMAN, T., HAGAMAN, J. R., HUANG, S., and WILLIAMS, P. J. 1989. Germ-line transmission of a planned alteration made in a hypoxanthine phosphoribosyltransferase gene by homologous recombination in embryonic stem cells. *Proc. Natl. Acad. Sci. U.S.A.* **86**: 8927.

KOLLER, B. H., MARRACK, P., KAPPLER, J. W., and SMITHIES, O. 1990. Normal development of mice deficient in β_2M MHC class I proteins, and CD8$^+$ T cells. *Science* **248**: 1227.

KOZAK, M. 1986. Point mutations define a sequence flanking the AUG intiator codon that modulates translation by eukaryotic ribosomes. *Cell* **44**: 283.

KRIEGLER, M. and BOTCHAN, M. 1982. A retrovirus LTR contains a new type of eukaryotic regulatory element. *Eukaryotic Viral Vectors* ed. Y. GLUZMAN. Cold Spring Harbor: Cold Spring Harbor Laboratory. NY.

KRIEGLER, M. and BOTCHAN, M. 1983. Enhanced transformation by a simian virus 40 recombinant virus containing a Harvey murine sarcoma virus long terminal repeat. *Mol. Cell. Biol.* **3**: 325.

KRIEGLER, M., PEREZ, C., and BOTCHAN, M. 1983. Promoter substitution and enhancer augmentation increases the penetrance of the SV40 A gene to levels comparable to that of the Harvey murine sarcoma virus ras gene in morphologic transformation. In *Gene Expression* eds. D. Hamer and M. Rosenberg. New York: Alan R. Liss.

KRIEGLER, M., PEREZ, C., DEFAY, K., ALBERT, I., and LIU, S. D. 1988. A novel form of TNF/cachectin is a cell-surface cytotoxic transmembrane protein: ramifications for the complex physiology of TNF. *Cell* **53**: 45.

KRIEGLER, M., PEREZ, C. F., HARDY, C., and BOTCHAN, M. 1984a. Transformation mediated by the SV40 T antigens: separation of the overlapping SV40 early genes with a retroviral vector. *Cell* **38**: 483.

KRIEGLER, M. P., PEREZ, C., HARDY, C., and BOTCHAN, M. 1984b. Viral integration and early gene expression both affect the efficiency of SV40 transformation of murine cells: biochemical and biological characterization of an SV40 retrovirus. In *Cancer Cells 2/Oncogenes and Viral Genes*, G. F. Van de Woude, A. J. Levine, W. C. Topp, and J. D. Watson. Cold Spring Harbor: Cold Spring Harbor Laboratory.

KUCHERLAPATI, R. S., EVES, E. M., SONG, K. Y., MORSE, B. S., and SMITHIES, O. 1984. Homologous recombination between plasmids in mammalian cells can be enhanced by treatment of input DNA. *Proc. Natl. Acad. Sci. U.S.A.* **81**: 3153.

KUCHERLAPATI, R. S., SPENCER, J., and MOORE, P. D. 1985. Homologous recombination catalyzed by human cell extracts. *Mol. Cell. Biol.* **5**: 714.

KUHL, D., DE LA FUENTA, J., CHATURVEDI, M., PARINOOL, S., RYALS, J., MEYER, F., and WEISSMAN, C. 1987. Reversible silencing of enhancers by sequences derived from the human IFN-alpha promoter. *Cell* **50**: 1057.

KUNZ, D., ZIMMERMAN, R., HEISIG, M., and HEINRICH, P. C. 1989. Identification of the promoter sequences involved in the interleukin-6-dependent expression of the rat alpha 2-macroglobulin gene. *Nucl. Acids Res.* **17**: 1121.

KWONG, A. D. and FRENKEL, N. 1984. Herpes simplex virus amplicon: effect of size on replication of constructed defective genomes containing eukaryotic DNA sequences. *J. Virology* **51**: 595.

KWONG, A. D. and FRENKEL, N. 1985. The herpes simplex virus amplicon IV: efficient expression of a chimeric chicken ovalbumin gene amplified within defective virus genomes. *Virology* **142**: 421.

LARSEN, P. R., HARNEY, J. W., and MOORE, D. D. 1986. Repression mediates cell-type-specific expression of the rat growth hormone gene. *Proc. Natl. Acad. Sci. U.S.A.* **83**: 8283.

LASPIA, M. F., RICE, A. P., and MATHEWS, M. B. 1989. HIV-1 tat protein increases transcriptional initiation and stabilizes elongation. *Cell* **59**: 283.

LATIMER, J. J., BERGER, F. G., and BAUMANN, H. 1990. Highly conserved upstream regions of the α_1-antitrypsin gene in two mouse species govern liver-specific expression by different mechanisms. *Mol. Cell. Biol.* **10**: 760.

LEE, F., HALL, C. V., RINGOLD, G. M., DOBSON, D. E., LUH, J., and JACOB, P. E. 1984. Functional analysis of the steroid hormone control region of mouse mammary tumor virus. *Nuc. Acids Res.* **12**: 4191.

LEE, F., MULLIGAN, R., BERG, P., and RINGOLD, G. 1981. Glucocorticoids regulate expression of dihydrofolate reductase cDNA in mouse mammary tumor virus chimaeric plasmids. *Nature* **294**: 228.

LENARDO, M. J., FAN, C., MANIATIS, T., and BALTIMORE, D. 1989. The involvement of NF-kappa-B in interferon gene regulation reveals its role as a widely inducible second messenger. *Cell* **57**: 287.

LENARDO, M. J., PIERCE, J. W., and BALTIMORE, D. 1987. Protein-binding sites in Ig gene enhancers determine transcriptional activity and inducibility. *Science* **236**: 1573.

LEUNG, K. and NABEL, G. J. 1988. HTLV-l trans-activator induces interleukin-2-receptor expression through an NF-kappa-B-like factor. *Nature* **333**: 776.

LEVINSON, B., KHOURY, G., VANDEWOUDE, G., and GRUSS, P. 1982. Activation of SV40 genome by 72-base-pair tandem repeats of Moloney sarcoma virus. *Nature* **295**: 79.

LEWIN, B. L. 1990. Commitment and activation at pol II promoters: A tail of protein-protein interactions: *Cell* **61**: 1161.

LI, Y., GOLEMIS, E., HARTLEY, J. W., and HOPKINS, N. 1987. Disease specificity of nondefective Friend and Moloney murine leukemia viruses is controlled by a small number of nucleotides. *J. Virology*, **61**: 693.

LIBERMAN, T. A. and BALTIMORE, D. 1990. Activation of interleukin-6 gene expression through the NF-kappa-B transcription factor. *Mol. Cell. Biol.* **10**: 2327.

LIM, B., WILLIAMS, D. A., and ORKIN, S. H. 1987. Retrovirus-mediated gene transfer of human adenosine deaminase: expression of functional enzyme in murine hematopoietic stem cells in vivo. *Mol. Cell. Biol.* **7**: 3459.

LIN, B. B., CROSS, S. L., HALDEN, N. F., DRAGOS, G. R., TOLEDANO, M. B., and LEONARD, W. J. 1990. Delineation of an enhancerlike positive regulatory element in the interleukin-2 receptor α-chain gene. *Mol. Cell. Biol.* **10**: 850.

LOWENTHAL, J. W., BALLARD, D. W., BOHNLEIN, E., and GREENE, W. C. 1989. Tumor necrosis factor alpha induces proteins that bind specifically to kappa-B-like enhancer elements and regulate interleukin-2 receptor alpha-chain gene expression in primary human lymphocytes. *Proc. Natl. Acad. Sci. U.S.A.* **86**: 2331.

LOWY, D. R., DVORETZKY, I., SHOBER, R., LAW, M. F., ENGEL, L., and HOWLEY, P. M. 1980. In vitro tumorigenic transformation by a defined sub-genomic fragment of bovine papilloma virus DNA. *Nature* **287**: 72.

LURIA, S., GROSS, G., HOROWITZ, M., and GIVOL, D. 1987. Promoter and enhancer elements in the rearranged alpha-chain gene of the human T-cell receptor. *EMBO J* **6**: 3307.

LUSKY, M., BERG, L., WEIHER, H., and BOTCHAN, M. R. 1983. Bovine papilloma virus contains an activator of gene expression at the distal end of the early transcription unit. *Mol. Cell. Biol.* **3**: 1108.

LUSKY, M. and BOTCHAN, M. 1981. Inhibition of SV40 replication in simian cells by specific pBR322 sequences. *Nature* **283**: 253.

LUSKY, M. and BOTCHAN, M. R. 1986. Transient replication of bovine papilloma virus type 1 plasmids: cis and trans requirements. *Proc. Natl. Acad. Sci. U.S.A.* **83**: 3609.

MACKETT, M., SMITH, G. L., and MOSS, B. 1982. Vaccinia virus: a selectable eukaryotic cloning and expression vector. *Proc. Natl. Acad. Sci. U.S.A.* **79**: 7415.

MAJORS, J. and VARMUS, H. E. 1983. A small region of the mouse mammary tumor virus long terminal repeat confers glucocorticoid hormone regulation on a linked heterologous gene. *Proc. Natl. Acad. Sci. U.S.A.* **80**: 5866.

MANN, R., MULLIGAN, R., and BALTIMORE, D. 1983. Construction of a retrovirus packaging mutant and its use to produce helper-free defective retrovirus. *Cell* **33**: 153.

MANSOUR, S. L., GRODZICKER, T., and TJIAN, R. 1985. An adenovirus vector system used to express polyoma virus tumor antigens. *Proc. Natl. Acad. Sci. U.S.A.* **82**: 1359.

MANSOUR, S. L., THOMAS, K. R., and CAPECCHI, M. R. 1988. Disruption mutagenesis by gene targeting in mouse-embryo-derived stem cells. *Cell* **51**: 503.

MARGOLSKEE, R. F., KAVATHAS, P., and BERG, P. 1988. Epstein-Barr virus shuttle vector for stable episomal replication of cDNA expression libraries in human cells. *Mol. Cell. Biol.* **8**: 2837.

MARKOWITZ, D., GOFF, S., and BANK, A. 1988. A safe packaging line for gene transfer: separating viral genes on two different plasmids. *J. Virology* **62**: 1120.

MAROTEAUX, L., CHEN, L., MITRANI-ROSENBAUM, S., HOWLEY, P. M., and REVEL, M. 1983. Cycloheximide induces expression of the human interferon beta 1 gene in mouse cells transformed by bovine papilloma virus-interferon beta 1 recombinants. *J. Virology*, **47**: 89.

MASSIE, B., GLUZMAN, Y., and HASSELL, J. A. 1986. Construction of a helper-free recombinant adenovirus that expresses polyoma virus large T antigen. *Mol. Cell. Biol.* **6**: 2872.

MATTHIAS, P., BOEGER, U., DANESCH, U., SCHUTZ, G., and BERNARD, H. U. 1986. Physical state, expression, and regulation of two glucocorticoid-controlled genes on bovine papilloma virus vectors. *J. Mol. Biol.* **187**: 557.

MCKNIGHT, S. L. and KINGSBURY, R. 1982. Transcriptional control of a eukaryotic protein-coding gene. *Science* **217**: 316.

MCKNIGHT, S. L., KINGSBURY, R. C., SPENCE, A., and SMITH, M. 1984. The distal transcription signals of the herpes virus tk gene share a common hexanucleotide control sequence. *Cell* **37**: 253.

MCNEALL, J., SANCHEZ, A., GRAY, P. P., CHESTERMAN, C. N., and SLEIGH, M. J. 1989. Hyperinducible gene expression from a metallothionein promoter containing additional metal-responsive elements. *Gene* **76**: 81.

MELLITS, K. H., KOSTURA, M., and MATHEWS, M. B. 1990. Interaction of adenovirus VA RNA1 with the protein kinase DAI: Nonequivalence of binding and function. *Cell* **61**: 843.

MEYER, D. I. 1988. Preprotein confirmation: the year's major theme in translocation studies. *TIBS* **13**: 471.

MIKSICEK, R., HEBER, A., SCHMID, W., DANESCH, U., POSSECKERT, G., BEATO, M., and SCHUTZ, G. 1986. Glucocorticoid responsiveness of the transcriptional enhancer of Moloney murine sarcoma virus. *Cell* **46**: 203.

MILLER, A. D. and BUTTIMORE, C. 1986. Redesign of retrovirus packaging cell lines to avoid recombination leading to helper virus production. *Mol. Cell. Biol.* **6**: 2895.

MILLER, A. D. and ROSMAN, G. J. 1989. Improved retroviral vectors for gene transfer and expression. *BioTechniques* **7**: 980.

MITCHELL, P. J. and TJIAN, R. 1989. Transcriptional regulation in mammalian cells by sequence-specific DNA-binding proteins. *Science* **245**: 371.

MITRANI-ROSENBAUM, S., MAROTEAUX, L., MORY, Y., REVEL, M., and HOWLEY, P. M. 1983. Inducible expression of the human interferon beta 1 gene linked to a bovine papilloma virus DNA vector and maintained extrachromosomally in mouse cells. *Mol. Cell. Biol.* **3**: 233.

MORDACQ, J. C. and LINZER, D. I. 1989. Co-localization of elements required for phorbol ester stimulation and glucocorticoid repression of proliferin gene expression. *Genes and Dev.* **3**: 760.

MOREAU, P., HEN, R., WASYLYK, B., EVERETT, R., GAUB, M. P., and CHAMBON, P. 1981. The SV40 base-repair repeat has a striking effect on gene expression both in SV40 and other chimeric recombinants. *Nucl. Acids Res.* **9**: 6047.

MUESING, M. A., SMITH, D. H., and CAPON, D. J. 1987. Regulation of mRNA accumulation by a human immunodeficiency virus trans-activator protein. *Cell* **48**: 691.

MULLER, H., SOGO, J. M., and SCHAFFNER, W. 1989. An enhancer stimulates transcription in trans when attached to the promoter via a protein bridge. *Cell* **58**: 767.

MÜLLER, W. J., NAUJOKAS, M. A., and HASSELL, J. A. 1984. Isolation of large T-antigen-producing mouse cell lines capable of supporting replication of polyoma virus-plasmid recombinants. *Mol. Cell. Biol.* **4**: 2406.

NANDI, A. K., ROGINSKI, R. S., GREGG, R. G., SMITHIES, O., and SKOULTCHI, A. I. 1988. Regulated expression of genes inserted at the human chromosomal beta-globin locus by homologous recombination. *Proc. Natl. Acad. Sci. U.S.A.* **85**: 3845.

NG, S. Y., GUNNING, P., LIU, S. H., LEAVITT, J., and KEDES, L. 1989. Regulation of the human beta-actin promoter by upstream and intron domains. *Nuc. Acids Res.* **17**: 601.

NEUBERGER, M. S. and WILLIAMS, G. T. 1988. The intron requirement for immunoglobulin gene expression is dependent upon the promoter. *Nuc. Acids Res.* **16**: 6713.

OKAYAMA, H. and BERG, P. 1983. A cDNA-cloning vector that permits expression of cDNA inserts in mammalian cells. *Mol. Cell. Biol.* **3**: 280.

O'MALLEY, R. P., MARIANO, T. M., SIEKIERKA, J., and MATHEWS, M. B. 1986. A mechanism for the control of protein synthesis by adenovirus VA RNA1. *Cell* **44**: 391.

ONDEK, B., SHEPPARD, A., and HERR, W. 1987. Discrete elements within the SV40 enhancer region display different cell-specific enhancer activities. *EMBO J.* **6**: 1017.

ORNITZ, D. M., HAMMER, R. E., DAVISON, B. L., BRINSTER, R. L., and PALMITER, R. D. 1987. Promoter and enhancer elements from the rat elastase I gene function independently of each other and of heterologous enhancers. *Mol. Cell. Biol.* **7**: 3466.

PALELLA, T. D., SILVERMAN, L. J., SCHROLL, C. T., HOMA, F. L., LEVINE, M., and KELLEY, W. N. 1988. Herpes-simplex-virus-mediated human hypoxanthine-guanine phosphoribosyltransferase gene transfer into neuronal cells. *Mol. Cell. Biol.* **8**: 457.

PALMITER, R. D., CHEN, H. Y., and BRINSTER, R. L. 1982. Differential regulation of metallothionein-thymidine kinase fusion genes in transgenic mice and their offspring. *Cell* **29**: 701.

PANICALI, D. and PAOLETTI, E. 1982. Construction of poxviruses as cloning vectors: insertion of thymidine kinase gene from herpes simplex virus into the DNA of infectious vaccinia virus. *Proc. Natl. Acad. Sci. U.S.A.* **79**: 4927.

PAOLETTI, E., LIPINSKAS, B. R., SAMSONOFF, C., MERCER, S., and PANICALI, D. 1984. Construction of live vaccines using genetically engineered poxviruses: biological activity of vaccinia virus recombinants expressing the hepatitis B surface antigen and the herpes simplex virus glycoprotein D. *Proc. Natl. Acad. Sci. U.S.A.* **81**: 193.

PAVLAKIS, G. N. and HAMER, D. H. 1983. Regulation of a metallothionein-growth hormone hybrid gene in bovine papilloma virus. *Proc. Natl. Acad. Sci. U.S.A.* **80**: 397.

PECH, M., RAO, C. D., ROBBINS, K. C., and AARONSON, S. A. 1989. Functional identification of regulatory elements within the promoter region of platelet-derived growth factor 2. *Mol. Cell. Biol.* **9**: 396.

PEREZ, C., ALBERT, I., DEFAY, K., ZACHARIADES, N., GOODING, L., and KRIEGLER, M. 1990. A non-secretable cell surface mutant of tumor necrosis factor (TNF) kills by cell to cell contact. *Cell*, in press.

PEREZ-STABLE, C. and CONSTANTINI, F. 1990. Roles of fetal γ-globin promoter elements and the adult β-globin 3' enhancer in the stage-specific expression of globin genes. *Mol. Cell. Biol.* **10**: 1116.

PETTERSSON, S., COOK, G. P., BRUGGEMANN, WILLIAMS, G. T., and NEUBERGER, M. S. 1990. A second B-cell-specific enhancer 3' of the immunoglobulin heavy-chain locus. *Nature* **344**: 165.

PICARD, D. and SCHAFFNER, W. 1984. A lymphocyte-specific enhancer in the mouse immunoglobulin kappa gene. *Nature* **307**: 83.

PICCINI, A., PERKUS, M. E., and PAOLETTI, E. 1987. Vaccinia virus as an expression vector. *Methods Enzymol.* **153**: 547.

PINKERT, C. A., ORNITZ, D. M., BRINSTER, R. L., and PALMITER, R. D. 1987. An albumin enhancer located 10 kb upstream functions along with its promoter to direct efficient, liver-specific expression in transgenic mice. *Genes and Dev.* **1**: 268.

PONTA, H., KENNEDY, N., SKROCH, P., HYNES, N. E., and GRONER, B. 1985. Hormonal response region in the mouse mammary tumor virus long terminal repeat can be dissociated from the proviral promoter and has enhancer properties. *Proc. Natl. Acad. Sci. U.S.A.* **82**: 1020.

PORTON, B., ZALLER, D. M., LIEBERSON, R., and ECKHARDT, L. A. 1990. Immunoglobulin heavy-chain enhancer is required to maintain transfected γ2A gene expression in a pre-B-cell line. *Mol. Cell. Biol.* **10**: 1076.

PTASHNE, M. and GANN, A. 1990. Activators and targets. *Nature* **346**: 329.

QUEEN, C. and BALTIMORE, D. 1983. Immunoglobulin gene transcription is activated by downstream sequence elements. *Cell* **35**: 741.

QUINN, J. P., FARINA, A. R., GARDNER, K., KRUTZSCH, H., and LEVENS, D. 1989. Multiple components are required for sequence recognition of the AP1 site in the gibbon ape leukemia virus enhancer. *Mol. Cell. Biol.* **9**: 4713.

RAMSHAW, I. A., ANDREW, M. E., PHILLIPS, S. M., BOYLE, D. B., and COUPAR, B. E. H. 1987. Recovery of immunodeficient mice from a vaccinia virus/IL-2 recombinant infection. *Nature* **329**: 545.

REDDY, V. B., GARRAMONE, A. J., SASAK, H., WEI, C., WATKINS, P., GALLI, J., and HSIUNG, N. 1987. Expression of human uterine tissue-type plasminogen activator in mouse cells using BPV vectors. *DNA* **6**: 461.

REDONDO, J. M., HATA, S., BROCKLEHURST, C., and KRANGEL, M. S. 1990. A T-cell-specific transcriptional enhancer within the human T-cell receptor δ locus. *Science* **247**: 1225.

REISMAN, D. and ROTTER, C. 1989. Induced expression from the Moloney murine leukemia virus long terminal repeat during differentiation of human myeloid cells is mediated through its transcriptional enhancer. *Mol. Cell. Biol.* **9**: 3571.

REISMAN, D. and SUGDEN, B. 1986. Trans-activation of an Epstein-barr viral transcriptional enhancer by the Epstein-Barr viral nuclear antigen 1. *Mol. Cell. Biol.* **6**: 3838.

REITZ, B. A., RAMABHADRAN, T. V., and PINTEL, D. 1987. The p39 promoter of minute virus of mice directs high levels of bovine-growth-hormone gene expression in the bovine papilloma virus shuttle vector. *Gene* **56**: 297.

RESENDEZ, JR., E., WOODEN, S. K., and LEE, A. S. 1988. Identification of highly conserved regulatory domains and protein-binding sites in the promoters of the rat and human genes encoding the stress-inducible 78-kilodalton glucose-regulated protein. *Mol. Cell. Biol.* **8**: 4579.

RICHARDS, R. I., HEGUY, A., and KARIN, M. 1984. Structural and functional analysis of the human metallothionein-IA gene: differential induction by metal ions and glucocorticoids. *Cell* **37**: 263.

RIPE, R. A., LORENZEN, S. BRENNER, D. A., and BREINDL, M. 1989. Regulatory elements in the 5' flanking region and the first intron contribute to transcriptional control of the mouse alpha-1-type collagen gene. *Mol. Cell. Biol.* **9**: 2224.

RITTLING, S. R., COUTINHO, L., AMARM, T., and KOLBE, M. 1989. AP-1/jun-binding sites mediate serum inducibility of the human vimentin promoter. *Nuc. Acids Res.* **17**: 1619.

ROSEN, C. A., SODROSKI, J. G., and HASELTINE, W. A. 1988. The location of cis-acting regulatory sequences in the human T-cell lymphotropic virus type III (HTLV-111/LAV) long terminal repeat. *Cell* **41**: 813.

SAKAI, D. D., HELMS, S., CARLSTEDT-DUKE, J., GUSTAFSSON, ROTTMAN, F. M., and YAMAMOTO, K. R. 1988. Hormone-mediated repression: a negative glucocorticoid-response element from the bovine prolactin gene. *Genes and Dev.* **2**: 1144.

SAMBROOK, J., RODGERS, L., WHITE, J., and GETHING, M. J. 1985. Lines of BPV-transformed murine cells that constitutively express influenza virus hemagglutinin. *EMBO J*, **4**: 91.

SARVER, N. MUSCHEL, R., BYRNE, J. C., KHOURY, and HOWLEY, P. M. 1985. Enhancer-dependent expression of the rat preproinsulin gene in bovine papilloma virus type 1 vectors. *Mol. Cell. Biol.* **5**: 3507.

SARVER, N., RICCA, G. A., LINK, J., NATHAN, M. H., NEWMAN, J., and DROHAN, W. N. 1987. Stable expression of recombinant factor VIII molecules using a bovine papilloma virus vector. *DNA* **6**: 553.

SATAKE, M., FURUKAWA, K., and ITO, Y. 1988. Biological activities of oligonucleotides spanning the F9 point mutation within the enhancer region of polyoma virus DNA. *J. Virology.* **62**: 970.

SAUER, B., WHEALY, M., ROBBINS, A., and ENQUIST, L. 1987. Site-specific insertion of DNA into a pseudorabies virus vector. *Proc. Natl. Acad. Sci. U.S.A.* **84**: 9108.

SCHAFFNER, G., SCHIRM, S., MULLER-BADEN, B., WEVER, G., and SCHAFFNER, W. 1988. Redundancy of information in enhancers as a principle of mammalian transcription control. *J. Mol. Biol.* **201**: 81.

SEARLE, P. F., STUART, G. W., and PALMITER, R. D. 1985. Building a metal-responsive promoter with synthetic regulatory elements. *Mol. Cell. Biol.* **5**: 1480.

SEDIVY, J. M. and SHARP, P. A. 1989. Positive genetic selection for gene disruption in mammalian cells by homologous recombination. *Proc. Natl. Acad. Sci. U.S.A.* **86**: 227.

SEN, R. and BALTIMORE, D. 1986. Multiple nuclear factors interact with the immunoglobulin enhancer sequence. *Cell* **46**: 705.

SHARP, P. A. and MARCINIAK, R. A. 1989. HIV tar: an RNA enhancer? *Cell* **59**: 229.

SHAUL, Y. and BEN-LEVY, R. 1987. Multiple nuclear proteins in liver cells are bound to hepatitis B virus enhancer element and its upstream sequences. *EMBO J.* **6**: 1913.

SHAW, G. and KAMEN, R. 1986. A conserved AU sequence from the 3' untranslated region of GM-CSF mRNA mediates selective mRNA degradation. *Cell* **46**: 659.

SHERMAN, P. A., BASTA, P. V., MOORE, T. L., BROWN, A. M., and TING, J. P. 1989. Class II box consensus sequences in the HLA-DRα gene: transcriptional function and interaction with nuclear proteins. *Mol. Cell. Biol.* **9**: 50.

SHIH, M., ARSENAKIS, M., TIOLLAIS, P., and ROIZMAN, B. 1984. Expression of hepatitis B virus S gene by herpes simplex virus type 1 vectors carrying α- and β-regulated gene chimeras. *Proc. Natl. Acad. Sci. U.S.A.* **81**: 5867.

SHIMIZU, H., MITOMO, K., WATANABE, T., OKAMOTO, S., and YAMAMOTO K. 1990. Involvement of a NK-κB-like transcription factor in the activation of the interleukin-6 gene by inflammatory lymphokines. *Mol. Cell. Biol.* **10**: 561.

SHYU, A. B., GREENBERG, M. E., and BELASCO, J. G. 1989. The c-fos transcript is targeted for rapid decay by two distinct mRNA-degradation pathways. *Genes and Dev.* **3**: 60.

SLEIGH, M. J. and LOCKETT, T. J. 1985. SV40 enhancer activation during retinoic-acid-induced differentiation of F9 embryonal carcinoma cells. *J. EMBO* **4**: 3831.

SMITHIES, O., GREGG, R. G., BOGGS, S. S., KORALEWSKI, M. A., and KUCHERLAPATI, R. S. 1985. Insertion of DNA sequences into the human chromosomal beta-globin locus by homologous recombination. *Nature* **317**: 230.

SONG, K. Y., SCHWARTZ, F., MAEDA, N., SMITHIES, O., and KUCHERLAPATI, R. 1987. Accurate modification of a chromosomal plasmid by homologous recombination in human cells. *Proc. Natl. Acad. Sci. U.S.A.* **84**: 6820.

SPAETE, R. R. and FRENKEL, N. 1982. The herpes simplex virus amplicon: a new eukaryotic defective-virus cloning-amplifying vector. *Cell* **30**: 295.

SPALHOLZ, B. A., YANG, Y. C., and HOWLEY, P. M. 1985. Transactivation of a bovine papilloma virus transcriptional regulatory element by the E2 gene product. *Cell* **42**: 183.

SPANDAU, D. F. and LEE, C. H. 1988. Trans-activation of viral enhancers by the hepatitis B virus X protein. *J. Virology* **62**: 427.

SPANDIDOS, D. A. and WILKIE, A. M. 1983. Host-specificities of papilloma virus, Moloney murine sarcoma virus and simian virus 40 enhancer sequences. *EMBO J.* **2**: 1193.

STACEY, A., ARBUTHNOTT, C., KOLLEK, R., COGGINS, L., and OSTERTAG, W. 1984. Comparison of myeloproliferative sarcoma virus with Moloney murine sarcoma virus variants by nucleotide sequencing and heteroduplex analysis. *J Virology* **50**: 725.

STAMATOS, N. M., CHAKRABARTI, S., MOSS, B., and HARE, J. D. 1987. Expression of polyoma virus proteins by a vaccinia virus vector: association of VP1 and VP2 with the nuclear framework. *J. Virology* **61**: 516.

STENLUND, A., LAMY, D., MORENO-LOPEZ, J., AHOLA, H., PETTERSSON, U., and TIOLLAIS, P. 1983. Secretion of the hepatitis B virus surface antigen from mouse cells using an extra-chromosomal eukaryotic vector. *EMBO J.* **2**: 669.

STEPHENS, E. B., COMPANS, R. W., EARL, P., and MOSS, B. 1986. Surface expression of viral glycoproteins is polarized in epithelial cells infected with recombinant vaccinia viral vectors. *EMBO J.* **5**: 237.

STEPHENS, P. E. and HENTSCHEL, C. C. 1987. The bovine papilloma virus genome and its uses as a eukaryotic vector. *Biochem. J.* **248**: 1.

STEWART, C., VANEK, M., and WAGNER, E. F. 1985. Expression of foreign genes from retroviral vectors in mouse teratocarcinoma chimaeras. *EMBO J.* **4**: 3701.

STUART, G. W., SEARLE, P. F., and PALMITER, R. F. 1985. Identification of multiple metal regulatory elements in mouse metallothionein-I promoter by assaying synthetic sequences. *Nature* **317**: 828.

SUGDEN, B., MARSH, K., and YATES, J. 1985. A vector that replicates as a plasmid and can be efficiently selected in B-lymphoblasts transformed by Epstein-Barr virus. *Mol. Cell. Biol.* **5**: 410.

SULLIVAN, K. M. and PETERLIN, B. M. 1987. Transcriptional enhancers in the HLA-DQ subregion. *Mol. Cell. Biol.* **7**: 3315.

SVENSSON, C. and AKUSJARVI, G. 1985. Adenovirus VA RNA1 mediates a translational stimulation which is not restricted to the viral mRNAs. *EMBO J.* **4**: 957.

SWARTZENDRUBER, D. E. and LEHMAN, J. M. 1975. Neoplastic differentiation: interaction of simian virus 40 and polyoma virus with murine teratocarcinoma cells. *J. Cell. Physiology* **85**: 179.

TABIN, C. J., HOFFMAN, J. W., GOFF, S. P., and WEINBERG, R. A. 1982. Adaptation of a retrovirus as a eukaryotic vector in transmitting the herpes simplex virus thymidine kinase gene. *Mol. Cell. Biol.* **2**: 426.

TAKEBE, Y., SEIKI, M., FUJISAWA, J., HOY, P., YOKOTA, K., ARAI, K., YOSHIDA, M., and ARAI, N. 1988. SRα promoter: an efficient and versatile mammalian cDNA expression system composed of the simian virus 40 early promoter and the R-U5 segment of human T-cell leukemia virus type 1 long terminal repeat. *Mol. Cell. Biol.* **8**: 466.

TAVERNIER, J., GHEYSEN, D., DUERINCK, F., CAN DER HEYDEN, J., and FIERS, W. 1983. Deletion mapping of the inducible promoter of human IFN-beta gene. *Nature* **301**: 634.

TAYLOR, I. C., SOLOMON, W., WEINER, B. M., PAUCHA, E., BRADLEY, M., and KINGSTON, R. E. 1989. Stimulation of the human heat-shock protein 70 promoter in vitro by simian virus 40 large T antigen. *J. Biol. Chem.* **264**: 15160.

TAYLOR, I. C. A. and KINGSTON, R. E. 1990a. Factor substitution in a human HSP70 gene promoter: TATA-dependent and TATA-independent interactions. *Mol. Cell. Biol.* **10**: 165.

TAYLOR, I. C. A. and KINGSTON, R. E. 1990b. Ela trans-activation of human HSP70 gene promoter substitution mutants is independent of the composition of upstream and TATA elements. *Mol. Cell. Biol.* **10**: 176.

THIESEN, H. J., BOSZE, Z., HENRY, J., and CHARNAY, P. 1988. A DNA element responsible for the different tissue specificities of Friend and Moloney retroviral enhancers. *J. Virology* **62**: 614.

THOMAS, K. R. and CAPECCHI, M. R. 1986. Introduction of homologous DNA sequences into mammalian cells induces mutations in the cognate gene. *Nature* **324**: 34.

THOMAS, K. R. and CAPECCHI, M. R. 1987. Site-directed mutagenesis by gene targeting in mouse-embryo-derived stem cells. *Cell* **51**: 503.

THOMSEN, D. R., MAROTTI, K. R., PALERMO, D. P., and POST, L. E. 1987. Pseudorabies virus as a live virus vector for expression of foreign genes. *Gene* **57**: 261.

THUMMEL, C., TJIAN, R., HU, S., and GRODZICKER, Y. 1983. Translational control of SV40 T antigen expressed from the adenovirus late promoter. *Cell* **33**: 455.

TOOZE, J. 1981 *DNA Tumor Viruses, Part 2*. Cold Spring Harbor Press, Cold Spring Harbor, NY.

TREISMAN, R. 1985. Transient accumulation of c-fos RNA following serum stimulation requires a conserved 5' element and c-fos 3' sequences. *Cell* **42**: 889.

TRONCHE, F., ROLLIER, A., BACH, I., WEISS, M. C., and YANIV, M. 1989. The rat albumin promoter: cooperation with upstream elements is required when binding of APF/HNF 1 to the proximal element is partially impaired by mutation or bacterial methylation. *Mol. Cell Biol.* **9**: 4759.

TRONCHE, F., ROLLIER, A., HERBOMEL, P., BACH, I., CEREGHINI, S., WEISS, M., and YANIV, M. 1990. Anatomy of the rat albumin promoter. *Mol Biol. Med.* **7**: 173.

TRUDEL, M. and CONSTANTINI, F. 1987. A 3' enhancer contributes to the stage-specific expression of the human beta-globin gene. *Genes and Dev.* **6**: 954.

TYNDALL, C., LA MANTIA, G., THACKER, C. M., FAVALORO, J., and KAMEN, R. 1981. A region of the polyoma virus genome between the replication origin and late protein-coding sequences is required in cis for both early gene expression and viral DNA replication. *Nuc. Acids Res.* **9**: 6231.

VALERIO, D., EINERHAND, M. P. W., WAMSLEY, P. M., BAKX, T. A., LI, C. L., and VERMA, I. M. 1989. Retrovirus-mediated gene transfer into embryonal carcinoma and hemopoietic stem cells: expression from a hybrid long terminal repeat. *Gene* **84**: 419.

VAN DOREN, K., HANAHAN, D., and GLUZMAN, Y. 1984. Infection of eukaryotic cells by helper-independent recombinant adenoviruses: early region 1 is not obligatory for integration of viral DNA. *J. Virology* **50**: 606.

VANNICE, J. L. and LEVINSON, A. D. 1988. Properties of the human hepatitis B virus enhancer: position effects and cell-type nonspecificity. *J. Virology* **62**: 1305.

VASSEUR, M., KRESS, C., MONTREAU, N., and BLANGY, D. 1980. Isolation and characterization of polyoma virus mutants able to develop in multipotential murine embryonal carcinoma cells. *Proc. Natl. Acad. Sci. U.S.A.* **77**: 1068.

VELDMAN, G. M., LUPTON, S., and KAMEN, R. 1985. Polyoma virus enhancer contains multiple redundant sequence elements that activate both DNA replication and gene expression. *Mol. Cell. Biol.* **5**: 649.

VERNER, K. and SCHATZ, G. 1988. Protein translocation across membranes. *Science* **241**: 1307.

WAGNER, E. F., VANEK, M., and VENNSTROM, B. 1985. Transfer of genes into embryonal carcinoma cells by retrovirus infection: efficient expression from an internal promoter. *EMBO J.* **4**: 663.

WANG, X. F. and CALAME, K. 1986. SV40 enhancer-binding factors are required at the establishment but not the maintenance step of enhancer-dependent transcriptional activation. *Cell* **47**: 241.

WANG, Y., STRATOWA, C., SCHAEFER-RIDDER, M., DOEHMER, J., and HOFSCHNEIDER, P. H. 1983. Enhanced production of hepatitis B surface antigen in NIH 3T3 mouse fibroblasts by using extra chromosomally replicating bovine papilloma virus vector. *Mol. Cell. Biol.* **3**: 1032.

WEBER, F., DE VILLIERS, J., and SCHAFFNER, W. 1984. An SV40 "enhancer trap" incorporates exogenous enhancers or generates enhancers from its own sequences. *Cell* **36**: 983.

WEINBERGER, J., JAT, P. S., and SHARP, P. A. 1988. Localization of a repressive sequence contributing to B-cell specificity in the immunoglobulin heavy-chain enhancer. *Mol. Cell. Biol.* **8**: 988.

WHEALY, M. E., BAUMEISTER, K., ROBBINS, A. K., and ENQUIST, L. W. 1988. A herpes virus vector for expression of glycosylated membrane antigens: fusion proteins of pseudorabies virus gIII and human immunodeficiency virus type 1 envelope glycoproteins. *J. Virology* **62**: 4185.

WICKNER, W. 1989. Secretion and membrane assembly. *TIBS* **14**: 280.

WILLIAMS, G. T., MCCLANAHAN, T. K., and MORIMOTO, R. I. 1989. E1a transactivation of the human HSP70 promoter is mediated through the basal transcriptional complex. *Mol. Cell. Biol.* **9**: 2574.

WINGENDER, E. 1988. Compilation of transcription-regulating proteins. *Nuc. Acids Res.* **16**: 1879.

WINOTO, A. and BALTIMORE, D. 1989. $\alpha\beta$-lineage-specific expression of the α T-cell receptor gene by nearby silencers. *Cell* **59**: 649.

YAMADA, M., LEWIS, J. A., and GRODZICKER, T. 1985. Overproduction of the protein product of a nonselected foreign gene carried by an adenovirus vector. *Proc. Natl. Acad. Sci. U.S.A.* **82**: 3567.

YAN, S. C. B., GRINNELL, W., and WOLD, F. 1989. Post-translational modifications of proteins: some problems left to solve. *TIBS* **14**: 264.

YATES, J. W., WAREN, N., and SUGDEN, B. 1985. Stable replication of plasmids derived from Epstein-Barr virus in various mammalian cells. *Mol. Cell. Biol.* **5**: 410.

YEE, J. K., MOORES, J. C., JOLLY, D. J., WOLFF, J. A., RESPESS, J. G., and FRIEDMANN, T. 1987. Gene expression from transcriptionally disabled retroviral vectors. *Proc. Natl. Acad. Sci. U.S.A.* **84**: 5197.

YU, S. F., VON RUDEN, T., KANTOFF, P. W., GARBER, C., SEIBERG, M., RUTHER, U., ANDERSON, W. F., WAGNER, E. F., and GILBOA, E. 1986. Self-inactivating retroviral vectors designed for transfer of whole genes into mammalian cells. *Proc. Natl. Acad. Sci. U.S.A.* **83**: 3194.

YUTZEY, K. E., KLINE, R. L., and KONIECZNY, S. F. 1989. An internal regulatory element controls troponin I gene expression. *Mol. Cell. Biol.* **9**: 1397.

ZHENG, H. and WILSON, J. H. 1990. Gene targeting in normal and amplified cell lines. *Nature* **344**: 170.

ZIJLSTRA, M., LI, E., SAJJADI, F., SUBRAMANI, S., and JAENISCH, R. 1989. Germ-line transmission of a disrupted β_2-microglobulin gene produced by homologous recombination in embryonic stem cells. *Nature* **342**: 435.

ZIJLSTRA, M., BIX, M., SIMISTER, N. E., LORING, J. M., RAULET, D. H., and JAENISCH, R. 1990. β_2-microglobulin-deficient mice lack CD4−8+ cytolytic T cells. *Nature* **344**: 742.

ZINN, K., DIMAIO, D., and MANIATIS, T. 1983. Identification of two distinct regulatory regions adjacent to the human beta-interferon gene. *Cell* **34**: 865.

ZINN, K., MELLON, P., PTASHNE, M., and MANIATIS, T. 1982. Regulated expression of an extrachromosomal human beta-interferon gene in mouse cells. *Proc. Natl. Acad. Sci. U.S.A.* **79**: 4897.

GENE TRANSFER METHODS

This section of the manual contains a collection of proven, thoroughly tested methods that facilitate the transfer of genes into a variety of cell types as well as the subsequent selection, isolation, and characterization of cells containing the transferred DNA. Included is a catalog of biochemical selection and gene-amplification strategies. Key methods for the expression cloning of cytokines and cell-surface molecules are described, including a new, highly useful method for plasmid-cDNA-library amplification. Methods describing the generation of subtracted probes and the generation of subtracted libraries are also included. This section closes with a collection of methods enabling the rescue of high-titer recombinant retroviruses for use in retroviral-mediated gene transfer and a relatively new technique, PCR-based expression.

Each section begins with a brief introduction of the technology, including an explanation of the biology and chemistry involved in the process. This is followed by a list of materials required for the successful execution of the protocol. Next I describe the protocol itself. Last, where appropriate, potential problems with each method and their solutions are described.

4
CELLS AND CELL LINES

BASIC TISSUE-CULTURE TECHNIQUES

The establishment of successful tissue cultures requires several important pieces of equipment: a flow hood, a temperature-, humidity-, and CO_2-controlled incubator, an inverted microscope, a 4°C refrigerator, a −20°C freezer, and a liquid-nitrogen freezer for long-term cell storage. It is important that the flow hood, incubator, and microscope are located such that they are separate from the facilities used for the manipulation of bacterial or yeast cultures and that this area be kept absolutely clean. This is because sterility is the most important concern for the tissue culturist attempting to propagate and manipulate tissue culture cells for long periods of time. In fact, culture contamination is perhaps the greatest cause of wasted time and materials in gene-transfer experiments. Careful execution of the techniques described in this chapter will ensure a high degree of reproducibility and sterility and eliminate the possibility of cross-contamination during culture manipulations. A few minutes a day spent ensuring sterility will save the investigator weeks or months of grief regenerating precious but contaminated cultures.

In addition to the specialized equipment just mentioned, a number of specialized materials are required. You will need sterile media, of which there are many types, each appropriate for the propagation of particular lines of cells. Sterile serum or a serum substitute is also required to provide growth factors not present in media but essential for cell growth. Serum is usually generated in serum lots. Serum performance is usually consistent within a lot; however, it is not unusual to observe variations in performance *between* lots. Because some cell lines exhibit sensitivity, in the form of toxicity, to different lots of presumably identical sera, it is advisable to stick with one serum lot for a given experiment if possible, and to test thoroughly any new lots of serum to be introduced into the laboratory for toxicity relative to an acceptable control. Serum toxicity is characterized by poor cell growth as well as the appearance of numerous vacuoles in the cultured cells. Under phase-contrast microscopy, healthy cells appear translucent and bright, whereas unhealthy cells appear darker, somewhat gray.

Media, which contain salts as well as a variety of combinations of amino acids, are stable when stored at −20°C and are stable at 4°C for a couple of weeks. However, one amino acid, glutamine, is relatively unstable, and so it is best provided fresh, *even if the medium already contains glutamine*. In general, it is best to assume that all the glutamine present in frozen media has degraded, and to replenish the frozen media just prior to their use with sufficient glutamine to reconstitute the intended glutamine concentration of the media.

The addition of the antibiotics penicillin (PEN) and streptomycin (STREP) to media is a common practice. Although it is not absolutely necessary, and in some cases is undesirable, the addition of these substances to media is a convenient safeguard against bacterial infection and should have no discernible effect on the outcome of an experiment.

When dealing with substrate-attached cells, you will use an isotonic solution of trypsin to release the monolayer cultures from their substrate and from each other. This solution, like all of the solutions mentioned previously, is available already prepared from a number of sources.

Cells can be grown in a number of plastic containers such as tissue-culture plates, T-flasks, multiwell plates, and roller bottles. These containers are chemically treated by the manufacturers to facilitate adherence of cells to the plastic surface. Thus, bacterial petri plates are inappropriate substitutes for tissue-culture dishes. Each manufacturer claims advantages for its product. We typically use Falcon or Corning plasticware.

Within the last few years, a number of adherence substrates such as collagen and fibronectin have become available to facilitate the attachment or differentiation of cells in culture. Such substances are plated on the tissue-culture plasticware, and the cells under study are plated on these protein matrices.

We deal with substrate-attached cells and suspension cells separately because there are unique tricks for handling each cell type. First, the propagation of cell lines is discussed, including routine procedures for testing for subtle mycoplasma contamination, freezing and thawing of tissue-culture cells, staining of cell cultures, and viability analysis with trypan-blue staining. Second, the propagation of embryonal stem (ES) cells is described. Last, assays for colony formation, anchorage-independent growth, and focus formation are described.

PROPAGATION OF CELL LINES

Passaging of Cells

Materials

1. Dulbecco's modified Eagle's medium (DMEM high-glucose; Irvine Scientific, cat. # 9024).
2. RPMI medium 1640 (Irvine Scientific, cat. # 9160).
3. L-glutamine (200 mM, 29.2 mg/ml; Irvine Scientific, cat. # 9317).
4. Trypsin-EDTA solution (0.5 g/L trypsin, 0.2 g/L EDTA; Irvine Scientific, cat. # 9341).
5. Penicillin-streptomycin (PEN/STREP) (100X; Irvine Scientific, cat. # 9366).
6. Fetal bovine serum (Sigma, cat. # F4884).
7. Phosphate buffered saline (PBS), Ca^{++} and Mg^{++} free (Sigma, cat. # D5527).
8. 60-mm culture dishes (Falcon, cat. # 3002).
9. 100-mm culture dishes (Corning, cat. # 25020-100).
10. T25 culture flasks (25 cm^2; Corning, cat. # 25100).
11. T75 culture flasks (75 cm^2; Corning, cat. # 25110-75).
12. T150 culture flasks (150 cm^2; Corning, cat. # 25120).

Adherent Cells

1. Rinse the cell monolayers briefly with trypsin-EDTA. This step is necessary to remove any traces of serum, which could inactivate the trypsin.
2. Add 1.0 ml of trypsin-EDTA to a 100-mm dish. Incubate 1–4 minutes at 25°–37°C.
3. Inactivate the trypsin by the addition of 9.0 ml of complete medium (DMEM high-glucose, 5–10% serum, penicillin-streptomycin solution, L-glutamine as required). Pipette the cells up and down several times to ensure complete removal of the cells from the dish.
4. Count the cells in a hemacytometer (optional) and seed a dilution of cells that allows for future cell growth but is not too low to retain viability of the culture by cross-feedings. Incubate in new culture vessels.

Suspension Cells

1. Count the cells if necessary.
2. Dilute the cells into fresh medium and seed a dilution of cells that allows for future cell growth but is not too low to retain viability of the culture by cross-feeding. (If necessary, the cells may be pelleted at 250 × g and rinsed with fresh medium, then seeded.)

Mycoplasma Testing

The protocols for detecting mycoplasma contamination described next are simple and reliable. With these procedures, mycoplasma DNA binds the fluorescent dye and, upon examination with a fluorescence microscope, produces what appears to be a fluorescent cytoplasm. The protocol for suspension cultures requires the transmission of mycoplasma from the media of the suspension-cell culture to be tested to the tester-cell monolayer, which is itself subsequently incubated with dye and analyzed for cytoplasmic fluorescence.

Materials

1. Hoechst 33258 stain (2'-[4-hydroxyphenyl]-5-[4-methyl-1-piperazinyl]-2,5'-bi-1H-benzimidazole; Sigma, cat. # B2883): Prepare a stock solution of 50 ng/ml in PBS.

2. Acetic acid/methanol: Prepare fresh a solution of one part glacial acetic acid and three parts absolute methanol. Cool to −20° C.
3. Mounting solution: 50% glycerol, 44.0 mM citrate, 111 mM sodium phosphate, pH 5.5.
4. Clear fingernail polish.
5. Vero cells (ATCC# CCL 81) for testing non-adherent cells or contaminated media.
6. Ultraviolet microscope equipped for epifluorescence with 330/380-nm excitation filter, and LP 440-nm barrier filter.
7. Microscope cover slips.
8. Phosphate buffered saline (PBS).

Method: Adherent Cultures

1. Seed the test cultures at the regular passage density in 35-mm or 60-mm petri dishes.
2. Incubate the cultures until they reach between 20–50% confluence. (If the cultures reach complete confluence, mycoplasma visualization is less efficient.)
3. Remove the media and rinse the cells twice with PBS.
4. Rinse the cells with a solution of one part PBS and one part acetic acid/methanol.
5. Rinse the cells with pure acetic acid/methanol.
6. Add cold acetic acid/methanol to the cells and incubate in −20°C freezer for 10 minutes.
7. Remove the acetic acid/methanol and rinse the cells twice with deionized water. (Samples may be stored if air-dried after removal of fixative. Just before staining, stored plates should be rinsed twice with deionized water.)
8. Add Hoechst 33258 dye solution to the cells and incubate at room temperature for 10 minutes.
9. Remove stain and rinse cells twice with deionized water.
10. Mount a coverslip in a drop of mounting solution and blot off surplus from the edges of the coverslip with a Kimwipe. Seal the edges with clear fingernail polish.
11. Examine by epifluorescence. Mycoplasma will produce bright particulate or filamentous extranuclear fluorescence over the cytoplasm and sometimes in intercellular spaces. If the contamination is low-level, not all of the cells will show extranuclear fluorescence, so scan the entire culture before declaring a culture negative.

Method: Suspension Cultures and Infected Media

1. Grow Vero cells to 20 percent confluence. Replace the media with test media or whole-cell suspension culture. Incubate for 72 hours.
2. Fix and stain the monolayer as above (steps 3–11). A control monolayer of Vero cells should also be processed and examined in parallel.

Note:

1. Although rigorous experimental design demands positive controls, the experimenter must decide if it is wise to carry contaminated cultures in the main cell-culture facility. It may be smarter to maintain a separate incubator for mycoplasma-positive lines and quarantined lines to be tested.
2. Freshly thawed or lysed cells may show some extranuclear fluorescence. If in doubt, retest suspected cells.

Cell Freezing

Materials

1. Dimethylsulfoxide (DMSO; Sigma, cat. # D2650).
2. Freezing vials (Nunc, cat. # 3-68632).
3. Freezing chamber: Fill a polystyrene foam box (with a wall thickness of 15 mm) with layers of cotton wool (Baxter, cat. # C8345).
4. Freezing medium: This consists of culture medium with serum in which the cells are normally cultured, supplemented with 5–10% DMSO. Sterilize the freezing medium by passing through a 0.2-μm filter.

Method

1. Prepare a cell suspension and pellet the cells by centrifugation.
2. Resuspend the cells in freezing medium at a concentration of 10^6–10^7 cells/ml.
3. Dispense 1 ml of cell suspension into each freezing vial.
4. Place the vials in the polystyrene box between layers of cotton wool.
5. Place the box into a $-70°C$ freezer and freeze overnight.
6. Store vials under liquid nitrogen.

Thawing Frozen Cells

1. Remove the vial from the liquid-nitrogen freezer and thaw it in a 37°C water bath.
2. Transfer the cells to a T75 flask containing 35 ml of media. Incubate the culture at 37°C.
3. After 8–12 hours, replace the culture medium with fresh medium.

Staining Cell Cultures

Giemsa Staining

Materials

1. Hanks' balanced salt solution (HBSS; Sigma, cat. # H2513).
2. Methanol (Sigma, cat. # M3641).
3. Giemsa stain (0.4% w/v solution in buffered methanol; Sigma, cat. # GS-1L).

Method

1. Remove the medium from the cell-culture dish.
2. Rinse the cells with HBSS.
3. Add a 50% HBSS/50% methanol mixture to the cells and let stand for 30 seconds. Remove the solution.
4. Add methanol to the cells and let stand for 10 minutes. Remove the solution.
5. Rinse the cells with methanol, remove the solution.
6. Cover the cells entirely with the Giemsa stain and let stand for 2 minutes.
7. Pour off the stain and wash the culture dish vigorously in a beaker filled with tap water, then rinse the dish in a beaker filled with deionized water. Let the dish air-dry.

Methylene-Blue Staining

Materials

1. Hanks' balanced salt solution (HBSS; Sigma, cat. # H2513).
2. Glutaraldehyde (25% aqueous; Sigma, cat. # G6257).

3. 2% methylene-blue solution: prepare a 2% (w/v) methylene-blue solution (Sigma, cat. # MB-1) in deionized water. Stir the solution on a magnetic stirrer for 2–3 hours, then filter the solution through a Whatman No. 1 filter. Store the dye solution at room temperature.
4. Staining/fixing solution: 1.25% glutaraldehyde, 0.06% methylene blue in HBSS. Prepare this reagent just prior to use.

Method

1. In a chemical hood, remove the medium from the cell-culture dish and replace it with the staining/fixing solution. Let the dish stand in the chemical hood for 1 hour.
2. Pour off the staining/fixing solution, then wash the dish vigorously in a beaker of tap water and rinse the dish in a beaker of deionized water. Let the dish air-dry.

Trypan-Blue Staining for Cell Viability

Materials

1. Hanks' balanced salt solution (HBSS; Sigma, cat. # H2513).
2. Trypan-blue solution (Sigma, cat. # T8154): 0.4% (w/v) trypan blue in 0.81% (w/v) NaCl and 0.06% (w/v) K_2HPO_4.
3. Hemacytometer (Baxter, cat. # B3178-1).

Method

1. Prepare a cell suspension in HBSS at a concentration of approximately 5×10^5 cells/ml.
2. Mix 0.5 ml of the trypan-blue solution with 0.5 ml of the cell suspension. Allow to stand for 5–15 minutes.
3. Transfer a drop of the trypan-blue cell suspension to the edge of the hemacytometer, and allow the chamber to be filled by capillary action.
4. Count the number of unstained cells and the number of stained cells in five 1-mm squares. The percentage of viable cells is the ratio of *unstained cells* to the total number of cells (stained + unstained) counted.

PROPAGATION OF EMBRYONAL STEM CELLS

A specific discussion of the propagation of embryonal stem (ES) cells is included because ES cells have proven to be of tremendous utility in a wide variety of applications, many discussed in Part I of this manual, and because the maintenance of *undifferentiated* ES cells, a requirement to maintain totipotency, requires special attention. Execution of this protocol will serve to ensure the preservation of healthy undifferentiated ES cells.

The classical method, described next, for the maintenance of ES cells requires the preparation of STO feeder monolayers. A time-consuming task, it now appears that a far simpler alternative is commercially available. The AMRAD Corporation Ltd. (Level 2, 17-27 Cotham Road, Kew, Victoria 3101, Australia. Tel: 61+3 853 0022. Fax: 61+3 853 0202) is producing and selling recombinant murine leukemia inhibitory factor (LIF) under the name of ESGRO. LIF is a simple, suitable alternative to STO monolayers and is highly efficient at inhibiting the spontaneous differentiation of ES cells in culture.

Materials

1. Medium: Dulbecco's modified Eagle's medium (DMEM high-glucose; Irvine Scientific, cat. # 9024)
2. L-glutamine (200 mM; 29.2 mg/ml; Irvine Scientific, cat. # 9317). Add to DMEM for a final concentration of 2 mM.
3. Phosphate buffered saline (PBS), Ca^{++}- and Mg^{++}-free (Sigma, cat. # D5527).
4. Trypsin-EDTA solution (0.5 g/L Trypsin, 0.2 g/L EDTA; Irvine Scientific, cat. # 9341).
5. Serum: This is by far the most critical reagent in the propagation of stem cells. The feeder cells, STO (thioguanine-resistant, ouabain-resistant SIM mouse fibroblasts; Martin and Evans, 1975), require newborn calf serum (NCS; Gibco, cat. # 200-6010AJ; Flow, cat. # 29-121-54), and the ES cells require fetal calf serum (FCS) (Hyclone cat. # A-1115-L).
6. Non-essential amino acids (NAA; 100X, Gibco, cat. # 320-1140AG; 100X, Flow, cat. # 16-810-49).
7. 2-mercaptoethanol (β-mercaptoethanol, BME; Sigma, cat. # M7522). This stock is at 14.4 M. Working stock is made fresh by adding 35 μl to 50.0 ml of PBS and sterilizing by filtration.
8. Gelatin: 0.1% (w/v) in water (Sigma, Swine skin, type I; Sigma, cat. # G 1890). Sterilize by autoclaving.
9. Mitomycin C: Dissolve 2 mg mitomycin C (Sigma, cat. # M0503) by adding 2 ml of sterile water or sterile PBS to the injection vial. The drug should be stored in the dark at 4°C and used for no more than one week. This is a 100X stock.
10. Penicillin/streptomycin (PEN/STREP) (100X; Irvine Scientific, cat. # 9366)

Media Formulations

STO feeder cells

DMEM + 10% NCS
450 ml DMEM
50 ml NCS
5 ml L-glutamine
5 ml PEN/STREP

	DMEM + 15% FCS
ES cells	425 ml DMEM
	75 ml FCS
	5 ml L-glutamine
	5 ml PEN/STREP
	5 ml NAA
	5 ml BME working solution (10 mm)

Method

Preparation of STO Feeder Layers for ES Cells

Propagation of STO Cells

STO fibroblasts (Ware and Axelrad, 1972) that have been mitotically inactivated by treatment with mitomycin C are used as feeder cells for ES cells.

A good feeder layer is made from STO cultures that retain their ability to be contact-inhibited; thus any culture that reaches a cell density greater than 8×10^6/100-mm dish should be discarded and replaced with cells that were frozen earlier.

1. When STO cells reach confluence ($6-8 \times 10^6$ cells/100-mm dish), aspirate the media, and wash the monolayer once with 5 ml of PBS.
2. Add 1.0 ml of trypsin-EDTA to the monolayer and incubate at 37°C for about 5 minutes.
3. Inactivate the trypsin by adding 9.0 ml DMEM, 10% NCS to the monolayer. Repeatedly pipette forcibly to produce a single-cell suspension.
4. Seed 1.0 ml of the cell suspension per 100-mm dish with 10.0 ml media.
5. Incubate at 37° C from 4–5 days until cells reach confluence and passage again.

Preparation of Feeder Cell Layers

1. Treat freshly confluent 100-mm dishes of STO cells with inactivation media (10.0 ml of DMEM, 10% NCS, supplemented with mitomycin C to a final concentration of 10.0 μg/ml) for 2 hours at 37°C. Use a mitomycin C stock that is less than one week old and prepare the inactivation media fresh.
2. While the STO cells are being treated, incubate the appropriate number of new 60-mm petri dishes with enough sterile 0.1% gelatin to cover the bottoms of the dishes, and incubate for one hour at room temperature. Aspirate the gelatin solution.
3. Aspirate the mitomycin-C-supplemented media and wash the monolayers three times with 5–10 ml of PBS.
4. Incubate the monolayers with 1.0 ml of trypsin for 5 minutes at 37°C.
5. Suspend the trypsinized cells in 14 ml of media and centrifuge at $250 \times g$ for 5 minutes. Aspirate the media, resuspend the cells in 10 ml of media, and count in a hemacytometer.
6. Seed 9×10^5 cells in 60-mm dish that has been gelatin-treated. The final volume of the dish should be 4.0 ml.
7. Incubate the feeder plates at 37°C. If the monolayer is not confluent the next day, more mitomycin-C-treated cells can be added.
8. These feeder layers are useful for about 7–10 days.

Propagation of Embryonal Stem Cells ES cells should be passaged every 3–4 days on feeder layers. Once an ES line has been established, many frozen vials should be generated to ensure that low-passage-number cultures can be used for experimentation. ES cell cultures should be split just as they reach confluence, and they should be fed with fresh media a few hours before subculture.

Thawing of Frozen ES Cells

1. Remove an ampule of ES cells from the freezer and thaw at 37°C until ice crystals disappear.
2. Sterilize the outside of the tube with ethanol.
3. Pipette the contents of the ampule into a 15-ml centrifuge tube and add 5 ml of fresh complete media with constant, gentle agitation.
4. Pellet the cells at $250 \times g$.
5. Aspirate off the media, resuspend the pellet in fresh media, and add the ES cells to an STO monolayer.
6. One day after plating, remove media and replace with fresh media. If the thawing procedure has been carried out properly, expect approximately 90 percent viability.

Continuous Culture of ES Cells

1. Aspirate off the media, and rinse the ES/STO culture with 5.0 ml PBS.
2. Add 0.5 ml of trypsin to the 60-mm culture dish and incubate at 37°C for 4–5 minutes.
3. Add 1.0 ml of DMEM, 15% FCS, and repeatedly pipette the cells with a cotton-plugged 9.0-inch sterile Pasteur pipette until the cell suspension is free of clumps.
4. Resuspend the cells in a total volume of 15 ml of complete media and centrifuge for 5 minutes at 1,000 rpm.
5. Aspirate the media, resuspend the cells in 10 ml of complete media and count. A typical 60-mm dish will contain about 2×10^7 ES cells and 9×10^5 STO cells. Therefore, the contribution of the STO cells to the total cell suspension is negligible.
6. Seed 1.0×10^6 cells in 4 ml of complete media into a 60-mm dish with an STO feeder layer. Incubate.
7. Subculture again in 3–4 days.

ASSAYS FOR COLONY FORMATION, ANCHORAGE-INDEPENDENT GROWTH, AND FOCUS FORMATION

Upon transfer into cells in culture of a given recombinant molecule, such as that encoding a cytokine, an oncogene, a selectable marker, or any expression library containing such genes, you may intend to select or screen for either colony formation or anchorage-independent growth of the transfected cells. Procedures for both methocel (Risser and Pollack, 1974) and soft-agar (MacPhearson and Montagnier, 1964) suspension are described here. In our laboratory, we prefer the soft-agar method. Alternative, less stringent assays for oncogenic transformation are selection for growth in low serum and focus formation, both of which are described here. The methods described here are appropriate for virtually all types of media.

Soft-Agar Suspension

Materials

1. 2% agar: Prepare 2% agar (Noble, Difco, cat. # 0142-01-8) in distilled H_2O. Autoclave. Let cool. Store at room temperature. Melt by heating or by microwave. Let cool to 60°C, and place in water bath at 45°C.
2. 2X complete medium: Prepare from 10X concentrate to half of the recommended volume. Filter-sterilize. Add 80.0 ml of 2X concentrate to 20.0 ml serum and place in 45°C water bath.
3. 0.5% agar underlay: Melt the 2% agar stock. Dilute 1:2 with sterile distilled H_2O. Place in 45°C water bath. Place 2X complete medium in 45°C water bath. Mix 2X complete medium and 1% agar 1:1 and plate 5.0 ml/60-mm dish. Allow to gel at room temperature for 20–30 minutes.
4. 0.33% plating agar: Dilute 2% agar 1:3 with sterile distilled H_2O. Place in 45°C water bath. Mix 0.67% agar with 2X complete medium 1:1, and maintain at 45°C.

Method

1. Trypsinize the cells and resuspend them in medium at 10X the desired plating concentration. If you are unsure of the correct concentration, plate serial dilutions.
2. Add 200 μl of the cell suspension to a 5-ml test tube. Add 2.0 ml of plating agar. Plate 1.5 ml of agar/cell suspension over the agar underlay. Allow the gel to harden at room temperature for 30 minutes.
3. Incubate at 37°C.
4. Feed culture 5–7 days after plating by overlaying with 1.5 ml of plating agar. Repeat as necessary. Depending on the cell type, colonies should be visible to the unaided eye 10–14 days after plating.

Plating in Methylcellulose (Methocel)

Materials

1. Methylcellulose stock (1.6% in medium): Bring 250 ml of distilled H_2O to 80–100°C in a 500-ml pyrex Erlenmeyer flask containing a magnetic stirring bar. Add 8.0 g methylcellulose (4000 centipoise; Sigma, cat. # M0512) and wet by swirling. Place the flask on a magnetic stirrer and stir until cooled to room temperature. Stir overnight at 4°C. Autoclave (methocel will form an opaque solid), allow the stock to cool to room temperature, and then

cool to 4°C. Add 250 ml of sterile 2X medium. Stir overnight at 4°C. Dispense into sterile bottles, and store at −20°C. Thawed methocel is good for 3–4 weeks at 4°C.

2. 2% agar: Prepare 2% agar (Noble, Difco, cat. # 0142-01-8) in distilled H_2O. Autoclave. Let the solution cool. Store at room temperature. Melt the agar by heating or in a microwave oven. Let cool to 60°C, and place in water bath at 45°C.

3. 2X Complete medium: Prepare the 2X solution from 10X media concentrate. Filter-sterilize. Add 80 ml of the 2X concentrate to 20 ml of serum and place medium in a 45°C water bath.

Method

1. Combine 1:1 2% agar and 2X complete medium and immediately plate 5 ml/60-mm dish. Allow to gel at room temperature for 20–30 minutes.
2. Prepare the cell suspension and dilute it to 2X the desired final concentration.
3. Dilute the cell suspension with one volume of methocel and plate out 2.0 ml of the mix over the agar underlay.
4. Incubate at 37°C until colonies form.
5. After one week, feed the culture by adding 3.0 ml methocel (0.8%) without disturbing the colonies at the agar/methocel interface.
6. After an additional week, carefully remove 4.0 ml of methocel from the top of the culture and replace with 4.0 ml of fresh methocel (0.8%).

Two alternative methods of selection for morphologic transformation of tissue-culture cells are selection for growth in low serum (Risser and Pollack, 1974) and selection for the ability to overgrow a monolayer (Aaronson et al., 1970) both abnormal phenotypes for normal cells in culture. A variety of tissue-culture cells transformed with a variety of viral and cellular oncogenes display these altered phenotypes. These selection procedures are outlined here:

Growth in Medium Containing 1% Serum

Plate 2×10^4 cells/35-mm dish in 10% (control) and 1% serum. Change media every third day and trypsinize and count sample dishes daily. *Saturation densities* are the densities that the cells maintain for the last 3–4 days of the experiment. *Doubling times* are determined by calculating the initial slope of the logarithmic growth curve before saturation occurs.

Growth on Normal Monolayers

Plate 100–1,000 cells on confluent monolayers of density-dependent contact-inhibited cells (e.g., NIH 3T3, Rat2). Change media every 3–4 days, and after 10–14 days, fix and stain in methylene blue; see above. Score colonies that are sufficiently dense to see without the aid of a microscope.

5

DNA TRANSFER

The experimental introduction of DNA into cells is accomplished by methods that (1) form DNA precipitates that can be internalized by the target cell; (2) create DNA-containing complexes whose charge characteristics are compatible with DNA uptake by the target cell; or (3) result in the transient formation of pores in the plasma membrane of a cell exposed to an electric pulse, pores of sufficient size to allow DNA to enter the target cell.

The factors that determine which method one selects are (1) the duration of expression required—that is, transient vs. stable expression, and (2) the type of cell to be transfected. The specific details of each procedure are described here.

CALCIUM PHOSPHATE TRANSFECTION METHOD

The most commonly used method to transfer DNA into recipient cells is the co-precipitation of the DNA of interest with calcium phosphate, after which the precipitate is added to the cells. With this technique, DNA entering the cell is taken up into phagocytic vesicles (Graham and van der Eb, 1973). Nevertheless, sufficient quantities of DNA enter the nucleus of the treated cells to allow relatively high frequencies of genetic transformation. This technique was originally employed by Graham and van der Eb to increase the infectivity of adenoviral DNA, but it was popularized by Wigler and colleagues (1977) and Maitland and McDougall (1977).

This procedure has been shown to be appropriate for the transfer into a variety of cell lines of single-copy genes present in the total genomic DNA derived from a donor cell or tissue sample. It has been the method of choice for identifying a number of cellular oncogenes as well as cellular genes for which a biochemical selection strategy exists. Using a variety of cell types, transfection efficiencies of up to 10^{-3} have been obtained. This is the method of choice for the generation of stable transfectants.

Over time, a number of variations of the basic technique have been developed. If the experiment involves the transfer of plasmid DNA, one may choose to include high-molecular-weight genomic DNA isolated from a defined cell or tissue source. The addition of such DNA, called *carrier DNA*, often serves to increase the efficiency of transformation by the plasmid DNA. Upon arrival of the plasmid DNA/carrier DNA calcium phosphate co-precipitate to the nucleus of the treated cell, the plasmid DNA appears to integrate into the carrier DNA, often in a tandem array, and this assembly of plasmid and carrier DNA, called a *transgenome*, subsequently integrates into the chromosome of the host cell.

Another procedural option available to you is the addition of a chemical shock step to the transfection protocol. Either dimethylsulfoxide or glycerol are appropriate. The optimal concentrations and lengths of treatment vary from cell line to cell line. The use of these agents can dramatically affect cell viability. We do not include a chemical shock in our protocol and prefer to allow transfected cells to sit overnight in the presence of the co-precipitate.

Chen and Okayama (1987) carefully optimized this transfer technique. They, like us, do not employ a chemical shock. They reported that incubation of the cells and the co-precipitate is optimal at 35°C in 2–4% CO_2 for 15–24 hours; that circular DNA is far more active than linear DNA, and that an optimal, finer precipitate is obtained when the DNA concentration is between 20–30 μg/ml in the precipitation mix.

If the protocol presented here does not yield the desired result, it may be wise to alter the incubator temperature, CO_2 concentration, and DNA concentration to those just cited. However, those temperature and CO_2 concentrations are *not* optimal for cell growth and should be maintained only temporarily.

Materials

1. 2X HBS (HEPES buffered saline)
   salt solution
   KCl 3.7 g
   D(+)glucose 10.0 g
   Na_2HPO_4 1.0 g
   Bring up to 50.0 ml with gd H_2O.
   In a separate beaker, add
   salt solution 1.0 ml
   HEPES 1.0 g (N-2-hydroxyethylpiperazine-N'-2-ethane sulfonic acid)
   NaCl 1.0 g
   Bring up to 85 ml with gd H_2O.
   pH to 7.05–7.10 with 1 N NaOH, bring up to 100 ml with gd H_2O.
2. 2.5 M $CaCl_2$
   $CaCl_2 \cdot 2H_2O$ 36.8 g
   Bring up to 100 ml with gd H_2O.
3. Distilled H_2O.
   pH of water must be in the range of 7.05–7.10.
4. TE (Tris-EDTA), pH 7.05
   0.01 M Tris base
   0.001 M EDTA

Filter-sterilize all reagents through a 0.2-μm filter prior to use. Store in sterile 50.0-ml conical polypropylene tubes.

Method

Day 1:

Seed 1.3×10^6 cells per 100-mm dish. Cells should be about 75% confluent when used to seed the dishes. Do not use "old" or very dense cell passages.

Day 2:

Generally, we prepare a large calcium phosphate cocktail mixture to transfect many plates simultaneously. Described here is the protocol for 1 ml (or

1 × 100-mm dish equivalent) of solution. Scale up the amounts as necessary, and allow for an appropriate amount of sample-transfer errors.

Adherence to sterile technique is critical. Use sterile reagents, tips, and tubes.

1. Add 1–20 µg DNA (1 mg/ml in sterile TE, pH 7.05) to 0.45 ml sterile H_2O.

   Note: First "sterilize" DNA by ethanol precipitation with NaCl (0.1 M final aqueous concentration) and 2X volume 100% ethanol.
2. Add 0.5 ml 2X HBS. Mix well.
3. Add 50 µl of 2.5 M $CaCl_2$ and vortex immediately.
4. Allow the DNA mixture to sit undisturbed for 15–30 minutes at room temperature. If a precipitate has not formed, use a sterile 1-ml cotton-plugged pipette to bubble air into the mixture; the precipitate will usually form. If it doesn't, it is likely that one of the reagents was improperly made or an incorrect volume was added.
5. Add 1 ml of the DNA transfection cocktail directly to the medium in the 100-mm dish (plated with cells on day 1).
6. Incubate the dishes containing the DNA precipitate for 16 hours at 37°C. Remove the media containing the precipitate and add fresh complete growth media.
7. Allow the cells to incubate for 24 hours. Post incubation, the cultures may be split into selective media. We generally split our cultures 1:5; however, if you want to isolate individual colonies for further analysis, split your cultures 1:10 and 1:100 as well. (For experimental details on drug selection, see the section on selection and amplification.)

DEAE DEXTRAN TRANSFECTION METHOD

DEAE dextran transfection is the method of choice for transient transfection of cells in culture. This is due to the relative ease of preparation of the DNA/DEAE dextran mixture for transfection as well as the high efficiency of transient gene transfer and expression that can be accomplished with this technique.

With this procedure, a DEAE dextran mixture is prepared, and the DNA sample of interest is added, mixed, and then transferred to the cells in culture. The ease of sample preparation facilitates the preparation of a large number of samples, as many as 1,000 or more. Thus, this method is the preferred method for analyzing the numerous samples of plasmid DNA required in the expression cloning of either secreted molecules or cell-surface molecules. The chemical events that lead to the cellular uptake of DNA are not clearly understood. Nevertheless, the transfection efficiency of this method can be dramatically improved by including chloroquine in the treatment of the transfected cells. It is felt that chloroquine serves to neutralize the pH of the lysosomes of the transfected cells, thus inhibiting the degradation of the DNA internalized by the cell as the DNA makes its way to the nucleus.

This method has been reported to yield transfection efficiencies of as high as 80 percent. DNA introduced into cells with this method appears to undergo mutations at a higher rate than that observed with calcium-phosphate-mediated transfection (Calos et al., 1983; Razzaque et al., 1983; Ashman and Davidson, 1985). This method is not useful for isolating stable transfectants.

Materials

1. DEAE dextran
   Sigma, cat. # D-9885 (M.W. 5×10^5)
   Stock solution: 0.1% (1 mg/ml) in TBS
   Filter-sterilize. Store at 4°C.
2. Chloroquine
   Sigma, cat. # C-6628 (M.W. 319.89)
   Stock solution: 100 mM in H_2O
   Filter-sterilize. Aliquot in 0.5 ml amounts.
   Store frozen at −20°C. Protect from light.
3. Phosphate-buffered saline (PBS) with Ca^{++} and Mg^{++}
4. Tris-EDTA (TE), pH 8.0
   0.01 M Tris base
   0.001 M EDTA
5. 10X "A"
NaCl	8.00 g
KCl	0.38 g
Na_2HPO_4	0.20 g
Tris base	3.00 g

   Adjust pH to 7.5 with HCl.
   Bring up to 100.0 ml with gd H_2O.
   Filter-sterilize. Store at room temperature.
6. 100X "B"
$CaCl_2 \cdot 2H_2O$	1.5 g
$MgCl_2 \cdot 6H_2O$	1.0 g

   Bring up to 100.0 ml with gd H_2O.
   Filter-sterilize. Store at room temperature.

7. Tris-buffered saline (TBS)— prepare fresh
   - 10X "A"   10 ml
   - 100X "B"   1 ml
   - H_2O 89 ml

   Filter-sterilize. Store at 4°C.

Method

The following transfection protocol is adapted for COS cells, but it can be used for a variety of cell types. We use Falcon 3046 six-well cluster dishes with a surface area of 9.6 cm^2 and maximum volume capacity of 3 ml.

Day 1:

1. Seed cells at a concentration of 2×10^4 cells/cm^2 in a total volume of 2 ml/well (1.92×10^5 cells/well of a six-well cluster dish). Cells should be about 75% confluent when used to seed the dishes. Do not use old or very dense cell passages.

Day 2:

1. Using sterile technique, reagents, tubes, and tips, resuspend 0.5 μg DNA in 10–25 μl TE. Add 50 μl TBS and 50 μl DEAE Dextran (in TBS). Final DEAE Dextran concentration should be about 0.04%.
2. Prior to proceeding with transfection, observe cell monolayers microscopically. Cells should appear about 60–70% confluent and well distributed.

   Bring all reagents to room temperature.

3. Aspirate off growth media and wash monolayer once with 3 ml of PBS followed by one wash with 3 ml of TBS.
4. Aspirate off TBS solution and add 100–125 μl of the appropriate DNA/DEAE-Dextran/TBS mixture to the wells.
5. Incubate dishes at room temperature inside a laminar flow hood. Rock the dishes every 5 minutes for 1 hour, making sure the DNA solution covers the cells. This is critical and must be done to avoid dehydration of the cells.
6. After the 1-hour incubation period, aspirate off the DNA solution and wash once with 3 ml of TBS followed by 3 ml of PBS.
7. Remove the PBS solution by aspiration and replace with 2 ml of complete growth media containing 100 μM chloroquine. Incubate the dishes in an incubator set at 37°C + 5% CO_2 for 4 hours.
8. Remove the media containing chloroquine and replace with 2–3 ml of complete growth media (no chloroquine).
9. Incubate the transfected cells for 1–3 days, after which the cells will be ready for experimental analysis. The exact incubation period will depend on the intent of the experiment. We find that optimal expression of transfected cytokine genes and cell-surface proteins occurs at 3 days post transfection.

ELECTROPORATION

Electroporation is a process whereby cells in suspension are mixed with the DNA to be transferred. This cell/DNA mixture is subsequently exposed to a high-voltage electric field. This creates pores in the membranes of treated cells that are large enough to allow the passage of macromolecules such as DNA into the cells. Such DNA molecules are ultimately transported to the nucleus, and a subset of these molecules are stably integrated into the host chromosome. The reclosing of the membrane pores is both time- and temperature-dependent and thus is delayed by incubation at 0°C, thereby increasing the probability that the molecule of interest will enter the cell.

Although calcium-phosphate- and DEAE-dextran-mediated DNA transfer have been the preferred modalities for transferring DNA into mammalian cells, the process of electroporation has gained ever-widening acceptance. This is due, in large part, to a key technical advantage that electroporation offers: you can successfully electroporate cells totally resistant to calcium-phosphate- or DEAE-dextran-mediated gene transfer. In fact, electroporation appears to work on virtually every cell type. With this technique, the efficiency of gene transfer is high for both transient transfection and stable transformation. One important technical difference between electroporation and other competing technologies is that the number of input cells required for electroporation is considerably higher.

Unlike calcium-phosphate-mediated and DEAE-dextran-mediated gene transfer, with electroporation the conformation of the input DNA appears to affect the outcome of the gene-transfer experiment. Linearized DNA appears more recombinagenic than supercoiled or nicked circular DNA and thus leads to a higher percentage of stably transformed cells. Supercoiled DNA serves as a superior transcriptional template in transient transfection experiments.

It is generally recognized that there are six key parameters for successful gene transfer with electroporation: the field strength, which is a function of the electrode gap size and the voltage applied; the duration of the current pulse; the pulse shape; the conductivity of the electroporation buffer; the DNA concentration; and the cell concentration. The control of the first three parameters are, of course, a function of the electroporation machine you select, and everyone seems to have his or her favorite. We have a variety of machines and all seem to work to some degree. New machines are introduced regularly into this highly competitive market. The cost/feature ratio of these machines continues to shrink as well. The Bio-Rad Gene Pulser® has been a popular choice; however, another company that produces premium, potentially superior machines is BTX, Biotechnologies and Experimental Research, Inc. They currently offer a new machine, the Electrocell Manipulator® 600, that is housed in a single unit, is highly flexible, and represents a good value. You can't go wrong with either of these companies' machines.

For many cell types, optimal transfection occurs at a field strength that results in cell death of 50 percent or more. Cell death can be rapidly monitored by trypan blue staining post electroporation and the 5-minute incubation on ice described here (see section on propagation of cell lines for details of trypan-blue staining).

There exists a substantial amount of technical literature that describes both the theory behind this technique as well as cell-type-specific electroporation protocols. Unfortunately, information derived for one cell type does not enable the investigator to make predictions for another cell type. Thus, we provide no specific information on the setting of these parameters. However, this information is readily available at no cost to the investigator, by a toll-free phone call. One company, BTX, maintains a current literature database on electroporation and cell types that have been successfully electroporated. If you call 1-800-289-2465 and request information on specific cell or tissue types, they will provide relevant literature references. Only the basic electroporation procedure is described here. For more specific information refer to the owner's manual for your particular power device.

Materials

1. Electroporation power device
2. Electroporation cuvettes
3. Electroporation buffer
   21 mM HEPES (pH 7.05)
   137 mM NaCl
   5 mM KCl
   0.7 mM Na_2HPO_2
   6 mM glucose

Method

1. Harvest, by trypsinization, and pellet exponentially growing cells, then wash twice with electroporation buffer.
2. Resuspend cells in electroporation buffer at a concentration of 2–20×10^6 cells/ml in an electroporation cuvette.
3. Add 5–25 µg of DNA that has been linearized to the cell suspension.
4. Insert or connect the electroporation electrode according to the manufacturer's instructions, and subject cell/DNA mixture to an electric field (pulse).
5. Return cell/DNA mixture to ice and incubate for 5 minutes.
6. Plate cells in non-selective medium. Biochemical selection may be carried out 24–48 hours later.

6

SELECTION AND AMPLIFICATION

POSITIVE SELECTION

Methods for gene transfer vary widely in their relative efficiencies. At an appropriate multiplicity of infection, virtually all the cells in a population can be infected with a recombinant retrovirus. All other means of stable transformation are less efficient. To facilitate the isolation of cells stably transformed with the DNA of interest, a gene encoding a dominant selectable marker is usually included in the transfection protocol. Such a gene encoding a dominant selectable marker can be provided either in *cis* or in *trans* to the other transferred gene. Both approaches work well. When the selectable marker is provided in *cis*, the rate of phenotypic co-transformation—the rate at which both the selectable marker and non-selectable gene are expressed in the same transfected cell—is virtually 100 percent. When the selectable marker and non-selectable gene are provided in *trans*, the phenotypic co-transformation rate is approximately 80 percent and if three different DNAs are transfected, including one selectable marker, the phenotypic co-transformation rate for all three is approximately 60 percent (Kriegler et al., 1983).

The variety of dominant selectable markers is quite broad and continues to grow. Protocols for all the currently available biochemical selections are included here.

Typically, the DNA or DNA mixture containing the resistance-conferring gene is transfected into the recipient cell by any of the methods described previously. Post transfection, the treated cells are allowed to grow for a period of time—24–48 hours to allow for efficient expression of the selectable marker. After the appropriate expression time, transfected cells are treated with media containing the concentration of drug appropriate for the selective survival and expansion of the transfected and now drug-resistant cells. Such colonies may be examined en masse or cloned individually for further analysis.

It is often insufficient to merely select the transformed drug-resistant cells and remove the drug selection. This is because transfected DNA can be relatively unstable in the host genome, and therefore the reversion rates of transfected cells can be quite high. Over several generations without selection, a substantial percentage of the transfectants can revert to the untransfected phenotype. Therefore, to maintain maximal levels of expression of the transferred gene, it is a good idea to maintain selection throughout the time course of the experiment, removing selection only when it is necessary to remove the selective drug for experimental purposes. The importance of continuous drug selection is clearly illustrated with the amphotropic retrovirus packaging cell line PA317.

This line, which was generated with HAT selection, readily loses its selectable marker and co-transfected DNA if selection is not maintained, and if the line is passaged for several months in the absence of drug, it will, when replated in HAT, demonstrate perhaps 50 percent viability. If one assumes that the loss of the dominant selectable marker occurs concomitantly with the loss of the non-selectable gene (an assumption that appears to be borne out by experimentation), then production of the desired genetic products from the cell line might drop by 50 percent. In this case, the packaging materials and perhaps recombinant retroviral titer might drop 50 percent as well.

Selectable Markers

General Selection Protocol

1. Sixteen hours after either addition of the calcium-phosphate precipitate, electroporation, or infection with a recombinant retrovirus, feed the transfected/infected cells with fresh, non-selective media.
2. Twenty-four to forty-eight hours later, split cultures to a 1:5 or greater dilution and plate in drug-containing media. Note: cells are not placed in drug-containing media immediately after gene transfer to provide a sufficient amount of time for the drug-resistance gene to express itself and manifest a biochemical phenotype.
3. Re-feed cell cultures with drug-containing media every three days, at which time cultures are examined under a microscope to determine the efficiency of drug selection.

The details of virtually all currently available drug-selection strategies are listed here:

A. Thymidine kinase
Thymidine kinase (Wigler et al., 1977; Colbere-Garapin et al., 1979) is encoded by the cellular, HSV, or vaccinia virus tk genes. When transferred to TK^- cell lines, these genes confer resistance to HAT medium—complete medium supplemented with 100 μM hypoxanthine (Sigma, cat. # H9636), 0.4 μM aminopterin (Sigma, cat. # A3411), and 16 μM thymidine (Sigma, cat. # T1895). After selection of TK^+ cells in HAT medium, it is important to feed the cells with HT medium (HAT medium minus aminopterin) before feeding the cells with non-selective medium. This procedure allows the TK^+ cells to recover from the HAT drug selection gradually.
HAT (50X) lyopilized (Sigma, cat. # H0262): Each vial contains sufficient concentrate to supplement 500 ml of media.
HAT (100X).
1. Add 136.1 mg of hypoxanthine (Sigma, cat. # H9636), 1.76 mg of aminopterin (Sigma, cat. # A3411), and 37.8 mg of thymidine (Sigma, cat. # T1895) to 99.2 ml of H_2O.
2. Add 0.8 ml of 10 N NaOH. When the chemicals have dissolved, sterilize the solution by passage through a 0.2 μM filter. HAT medium is light-sensitive. Store the concentrate at 4°C in a bottle covered with foil.

B. Adenine phosphoribosyltransferase
Adenine phosphoribosyltransferase (Murray et al., 1984; Lowy et al., 1980; Stambrook et al., 1984) is encoded by the APRT gene. When transferred to $APRT^-$ cells (CHO aa^{r}7, Ltk$^-$ aprt$^-$, and LS-24-b), this gene confers resistance to complete medium supplemented with 0.05 mM azaserine (Sigma, cat. # A1164), 0.1 mM adenine (Sigma, cat. # A9795), and 4 μg/ml of alanosine (M.W. 149.1, NSC # 153353, Drug Synthesis Branch, Div. Cancer Treatment, N.C.I.).

C. Hypoxanthine-guanine phosphoribosyltransferase
Hypoxanthine-guanine phosphoribosyltransferase (Jolly et al., 1983) is encoded by the hgprt gene. When transferred to LA9 cells (hgprt$^-$, aprt$^-$), this gene confers resistance to HAT medium (see procedures for thymidine kinase selection).

D. Aspartate transcarbamylase
Aspartate transcarbamylase (Ruiz and Wahl, 1986) is encoded by pyrB. When transferred to CHO D2O (UrdA mutant, deficient in the first three enzymatic activities of *de novo* uridine biosynthesis: carbamyl phosphate synthetase, aspartate transcarbamylase, and dihydroorotase), this gene confers resistance to Ham F-12 medium (minus uridine).

E. Ornithine decarboxylase
Ornithine decarboxylase (Chiang and McConlogue, 1988; Kaufman, 1990) is encoded by the odc gene. When transferred to CHO C55.7 cells (ODC−; maintained in DMEM with putrescine), this gene confers resistance to DMEM minus putrescine, with 10% FCS and 0.1 mM DFMO (α-difluoromethylornithine; obtain from Dr. P. McCann, Merril Dow Research Center, Cincinnati, Ohio).

F. Aminoglycoside phosphotransferase
Aminoglycoside phosphotransferase (Southern and Berg, 1982) is encoded by the aph gene. When transferred to virtually any cell type, this dominant selectable marker confers resistance to 100–800 μg/ml of G418 (Geneticin; Sigma, cat. # G9516) in complete media.

G. Hygromycin-B-phosphotransferase
Hygromycin-B-phosphotransferase (Gritz and Davies, 1983; Sugden et al., 1985; Palmer et al,. 1987) is encoded by the hph gene. When transferred to virtually any cell type, this dominant selectable marker confers resistance to complete media containing 10–400 μg/ml of hygromycin-B (Sigma, cat. # H2638).

H. Xanthine-guanine phosphoribosyltransferase
Xanthine-guanine phosphoribosyltransferase (Mulligan and Berg, 1981) is encoded by the gpt gene. When transferred to virtually any cell type, this dominant selectable marker confers resistance to XMAT medium containing dialyzed serum, 250 μg/ml of xanthine (Sigma, cat. # X2001); 15 μg/ml of hypoxanthine (Sigma, cat. # H9636); 10 μg/ml thymidine (Sigma, cat. # T1895); 2 μg/ml of aminopterin (Sigma, cat. # A3411); 25 μg/ml of mycophenolic acid (Sigma, cat. # M5255); and 150 μg/ml of L-glutamine (Sigma, cat. # G5763).

I. Tryptophan synthetase
Tryptophan synthetase (Hartman and Mulligan, 1988) is encoded by the trpB gene. When transferred to virtually any cell type, this dominant selectable marker confers resistance to tryptophan-minus medium (Irvine Scientific custom preparation) supplemented with 9 μg/ml of L-alanine (Sigma, cat. # A3534); 15 μg/ml of L-aspartic acid (Sigma, cat. # A4534); 14.7 μg/ml L-glutamic acid (Sigma, cat. # G5889); 34.5 μg/ml of L-proline (Sigma, cat. # P4655); 42 μg/ml of L-serine (Sigma, cat. # S5511); 4.1 μg/ml of hypoxanthine (Sigma, cat. # H9636); 0.73 μg/ml of thymidine (Sigma, cat. # T1895); 0.8 μg/ml of $ZnSO_4$ (Sigma, cat. # Z0251); 0.0025 μg/ml of $CuSO_4 \cdot 5H_2O$ (Sigma, cat. # C8027); 80 ng/ml of sodium linoleate (Sigma, cat. # L8134); 200 pg/ml of lipoic acid (Sigma, cat. # T1395); 160 ng/ml of putrescine dihydrochloride (Sigma, cat. # P5780); 8 ng/ml of biotin (Sigma, cat. # B4639); 1.4 ng/ml of vitamin B_{12} (Sigma, cat. # V6629); 10 ng/ml of epidermal growth factor (Sigma, cat. # E1257; Collaborative Research, cat. # 40001); 1:1000 dilution of stock

of insulin, transferrin, and selenous acid (Collaborative Research, cat. # 40351); 1% dialyzed calf serum (Sigma, cat. # C5542); and 300 µM indole (Sigma, cat. # I0750).

J. Histidinol dehydrogenase

Histidinol dehydrogenase (Hartman and Mulligan, 1988) is encoded by the hisD gene. When transferred to virtually any cell type, this dominant selectable marker confers resistance to DMEM containing 10% serum and 2.5 mM histidinol (Sigma, cat. # H8750).

K. Multiple drug resistance

The multiple drug resistance biochemical marker (Kane et al., 1988; Choi et al., 1988) is encoded by the mdr1 gene. When transferred to virtually any cell type, this dominant selectable marker confers resistance to complete medium supplemented with 0.06 µg/ml of colchicine (Sigma, cat. # C9754).

L. Dihydrofolate reductase

Dihydrofolate reductase (Subramani et al., 1981; Kaufman and Sharp, 1982; Simonsen and Levenson; 1983) is encoded by the dhfr gene and a dhfr mutant can be employed as a dominant selectable marker. When the wild-type allele is transferred to DHFR$^-$ cells, or the dhfr mutant is transferred to virtually any cell type, this marker confers resistance to α^- MEM (Flow, cat. # 12-313-54; Irvine Scientific, cat. # 9472; Gibco, cat. # 320-2561AJ) containing dialyzed 10% FCS (Sigma, cat. # F0392). When using the mutant dhfr gene as a dominant selectable marker, add 0.250–0.500 µM methotrexate (Sigma, cat. # A6770: dissolve to 5 mM in α^- MEM medium, filter-sterilize, and store at $-20°C$ in a foil-wrapped container. Potency may vary between commercial preparations, so use the same stock for an entire experiment) to the medium.

M. CAD gene

CAD (de Saint Vincent et al., 1981; Patterson and Carnwright, 1977), encoded by the cad gene, is a single protein that possesses the first three enzymatic activities of *de novo* uridine biosynthesis: carbamyl phosphate synthetase, aspartate transcarbamylase, and dihydroorotase. When transferred to CHO D20 cells (UrdA mutant, deficient in the first three enzymatic acitvities of *de novo* uridine biosynthesis), this gene confers resistance to HAM F-12 medium (Flow, cat. # 12-423-54; Irvine Scientific, cat. # 9436; normally deficient in uridine).

N. Adenosine deaminase

Adenosine deaminase (Bonthron et al., 1986; Kaufman et al., 1986; Yeung et al., 1983) is encoded by the ada gene. When transferred to CHO DUKX B11 cells, this gene, in the appropriate setting is a dominant selectable marker (see below), confers resistance to α^- MEM (Flow, cat. # 12-313-54; Irvine Scientific, cat. # 9472; Gibco, cat. # 320-2561AJ) supplemented with 10 µg/ml of thymidine (Sigma, cat. # T1895), 15 µg/ml of hypoxanthine (Sigma, cat. # H9636), 4 µM of Xyla-A (9-β-D-xylofuranosyl, M.W. 267.0; NSC # 7539, Drug Synthesis and Chemistry Branch, Division of Cancer Treatment, N.C.I.: prepare a stock solution of 0.4 mM in H_2O, filter-sterilize, and store in aliquots at $-20°C$), 0.01 µM dCF (2'-deoxycoformycin; Sigma, cat. # D6032: prepare a stock solution of 20 mg/ml (75 mM) in H_2O, filter-sterilize, and store in aliquots at $-20°C$), and 10% dialyzed FCS (Sigma, cat. # F0392). FCS contains low levels of adenosine deaminase; therefore, add dialyzed FCS to media just prior to use to minimize detoxification of the cytotoxic selective agents. To use ada as a dominant selectable marker in murine fibroblasts, supplement the medium with 0.05 mM alanosine (M.W. 149.1; NSC # 153353, Drug Synthesis and Chemistry Branch, Division of Cancer Treamtment, N.C.I.: prepare a fresh 5.0 mM stock solution in H$_2$O and filter-sterilize).

O. Asparagine synthetase

Asparagine synthetase (Cartier et al., 1987) is encoded by the as gene. When transferred to CHO N3 (AS$^-$) cells, this gene confers resistance to α^- MEM minus asparagine (Irvine Scientific, custom preparation) supplemented with 100 μg/ml of L-glutamine, 10% dialyzed FCS (Sigma, cat. # F0392), and 0.1 mM β-AH (β-aspartyl hydroxamate; Sigma, cat. # A6508): prepare a stock solution at 500 mg/ml in H$_2$O, filter-sterilize, and store in aliquots at $-20°$C). For use as a dominant selectable marker with AS$^+$ cells, add Albizziin (Sigma, cat. # A5398: prepare a stock at 50 mg/ml in H$_2$O, filter-sterilize, and store in aliquots at $-20°$C.)

P. Glutamine synthetase

Glutamine synthetase (Bebbington and Hentschel, 1987; Wilson and Bebbington, 1987) is encoded by the gs gene. When transferred to CHO K-1 cells, this gene confers resistance to GMEM-S: Glasgow MEM without glutamine (Irvine Scientific, cat. # 9427) and with 10% dialyzed FCS (Sigma, cat. # F0392); sodium pyruvate (Sigma, cat. # S8636); the non-essential amino acids alanine (Sigma, cat. # A3534), aspartate (Sigma, cat. # A4534), glycine (Sigma, cat. # G6388), proline (Sigma, cat. # P4655), and serine (Sigma, cat. # S5511) at 100 μM each; 500 μM glutamate (Sigma, cat. # G5889); 500 μM asparagine (Sigma, cat. # A4159); 10 μM thymidine (Sigma, cat. # T1895); adenosine (Sigma, cat. # A4036), guanosine (Sigma, cat. # G6264), and uridine (Sigma, cat. # U3003) at 30 μM each; and 25 μM MSX (methionine sulphoximine; Sigma, cat. # M5379)).

AMPLIFICATION

Incorporation of the appropriate *cis*-acting elements into a recombinant cistron can serve to increase the amount of expression only to a point. The ability of a promoter to initiate transcription, or of a transcript to elongate post initiation, is probably rate-limiting in this situation. A logical and proven method for further enhancing the expression of an exogenous gene is based on a naturally occurring phenomenon, gene amplification. Gene amplification occurs in a variety of organisms. The phenomenon drew wide attention when it was suggested that cells comprising drug-resistant tumors contained multiple copies of transcriptionally active alleles encoding gene products that could by some unknown mechanism protect the tumor cells from chemical attack. The archetype of this type of gene is dihydrofolate reductase, or DHFR. This enzyme serves to detoxify the drug methotrexate (MTX) by cleaving MTX into biologically inactive subunits. Thus, cells containing multiple copies of the DHFR gene will be resistant to higher levels of MTX (Schimke, 1984, 1988).

Gene amplification appears to be a phenomenon that occurs somewhat randomly in somatic cells. It is postulated that expansion of a single allele of a given gene into multiple copies occurs relatively frequently; however, in the absence of selection for this amplified genotype, the amplified structure resolves itself and soon disappears.

In the presence of selection, the cell containing the amplified gene survives whereas its siblings die. Presumably, the spontaneous amplification of cistrons continues, and the drug-resistance gene is further amplified. Thus, if one continues to increase the selective pressure on the cell population, one will derive a cell line that carries a tremendously amplified array of the drug-resistance gene. If the selection pressure is raised too quickly, allowing insufficient time for amplified variants to occur, the entire population may be lost.

Transfection of cells normally sensitive to the drug for which the drug-resistance gene confers protection, followed by drug selection, can ultimately result in the stable integration and subsequent amplification of the transferred drug-resistance gene.

Amplification of a given gene does not begin just 5' to the promoter of the resistance gene and end just 3' to the 3' end of the sequence encoding an expressible transcript. Amplified regions can be quite large and often include copies of genes flanking the selectable marker. In the case of amplification of an endogenous selectable marker, the flanking genes will also be amplified, a process termed *co-amplification*. In the case of an exogenous selectable marker, the DNA (including any non-selectable genes) flanking the selectable marker will also be amplified. It is this observation that one exploits in attempts to maximize gene expression in a transfected cell. Essentially, one assembles a recombinant molecule containing both an expressible, drug-selectable marker and an expressible, non-selectable marker. Post transfection, one selects for the drug-resistant phenotype and gradually increases the selective pressure. As selective pressure is increased, one should simultaneously monitor the levels of expression of the non-selectable gene product of interest to determine if co-amplification is occurring. In all cases, selection should be maintained to ensure that the co-amplified cistron is not lost.

General Amplification Protocol

1. Transfect cells with plasmids constructed with both the target gene and the amplifiable gene, or co-transfect cells with target-gene plasmids and amplifiable-gene plasmids at a ratio of 10:1. Transfect for 16 hours, then change media to *maintenance media*. Incubate for 24 hours.

2. Plate transfected cells into maintenance media at a density that would equal a 1:15 split of a confluent culture.
3. After a 24-hour incubation, change media to *selection media*, and subsequently change media every 3–4 days until stable colonies form.
4. Amplification: Plate either pooled stable colonies or single isolated clones at a split ratio of 1:6 to 1:15 in *amplification media*, which contain the cytotoxic reagent that selects for the amplifiable cistron. Continue sub-culturing until resistant populations (or single clones) form. Each selection step may take 2–3 weeks. Continue amplification by plating resistant populations or single clones in amplification media with the cytotoxic reagent at the next higher concentration.
5. During the course of amplification, resistant populations or clones should be tested and frozen at each amplification step.

The details of virtually all currently available drug-amplification strategies are listed here:

A. Ornithine decarboxylase
The biochemical marker ornithine decarboxylase (Chiang and McConlogue, 1988; Kaufman, 1990) is encoded by the odc gene. When transferred to CHO C55.7 cells (ODC$^-$), this gene confers resistance to DMEM minus putrescine, with 10% FCS and 0.1 mM DFMO (α-difluoromethylornithine; obtain from Dr. P. McCann, Merril Dow Research Center, Cincinnati, Ohio).

To amplify this biochemical marker, first incubate the cells in selection medium, DMEM, 10% FCS, 0.1 mM DFMO for 24 hours, then feed with amplification medium, DMEM, 10% FCS, and DFMO at concentrations of 0.16, 0.60, 1.00, 3.00, 9.00 and 15.00 mM at the respective amplification steps.

B. Multiple drug resistance
The multiple drug resistance biochemical marker (Kane et al., 1988; Choi et al., 1988) is encoded by the mdrl gene. When transferred to virtually any cell type, this dominant selectable marker confers resistance to complete medium supplemented with 0.06 μg/ml of colchicine (Sigma, cat. # C9754).

To amplify this biochemical marker, first incubate the cells in selection medium, DMEM, 10% FCS, 0.06 μg/ml colchicine for 24 hours, then feed the cells with amplification medium, DMEM, 10% FCS, and colchicine at concentrations of 0.12, 0.24, 0.48, 1.00, and 1.50 μg/ml at the respective amplification steps.

C. Dihydrofolate reductase
The biochemical marker dihydrofolate reductase (Subramani et al., 1981; Kaufman and Sharp, 1982; Simonsen and Levenson, 1983) is encoded by the dhfr gene, and a dhfr mutant can be employed as a dominant selectable marker. When the wild-type allele is transferred to DHFR$^-$ cells, or the dhfr mutant is transferred to virtually any cell type, this marker confers resistance to α^- MEM (Flow, cat. # 12-313-54; Irvine Scientific, cat. # 9472; Gibco, cat. # 320-256-1AJ) containing dialyzed 10% FCS (Sigma, cat. # F0392). When using the mutant dhfr gene as a dominant selectable marker, add 0.250–0.500 μM methotrexate (Sigma, cat. # A6770: dissolve to 5 mM in α^- MEM medium, filter-sterilize, and store at $-20°$C in foil-wrapped container. Potency may vary between commercial preparations, so use the same stock for the entire amplification experiment) to the medium.

Amplification is normally performed with DHFR-deficient CHO cells. To amplify this biochemical marker, first incubate cells in selection medium, α^- MEM (Flow, cat. #12-313-54; Irvine Scientific, cat. # 9472; Gibco, cat. # 320-256-1AJ) supplemented with 10 μg/ml of adenosine (Sigma, cat. # A4036), 10μg/ml of deoxyadenosine (Sigma, cat. # D0651), 10 μg/ml of thymidine (Sigma, cat. # T1895), and 10% FCS for 24 hours, then feed with amplification medium, α^- MEM, dialyzed 10% FCS (Sigma, cat. # F0392), and methotrexate (Sigma, cat. # A6770) at concentrations of 0.005, 0.002, 0.08, 0.32, 1.28, 5.0, 12, 20, 50, and 80 μM at the respective amplification steps.

D. CAD gene

The biochemical marker CAD (de Saint Vincent et al., 1981; Patterson and Carnwright, 1977), encoded by the cad gene, is a single protein that possesses the first three enzymatic activities of *de novo* uridine biosynthesis: carbamyl phosphate synthetase, aspartate transcarbamylase, and dihydroorotase. When transferred to CHO D2O cells (UrdA mutant, deficient in the first three enzymatic activities of *de novo* uridine biosynthesis), this gene confers resistance to HAM F-12 medium.

To amplify this biochemical marker, first incubate the cells in selection medium, HAM F-12 medium, 10% FCS for 24 hours, then feed with amplification medium, HAM F-12 medium, 10% dialyzed FCS (Sigma, cat. # F0392) and PALA (N-phosphonacetyl-L-aspartate, M.W. 299.1, NSC # 224131, Drug Synthesis and Chemistry Branch, Division of Cancer Treatment, N.C.I.: prepare a stock solution at 250 mM in H_2O, filter-sterilize, and store at $-20°C$) at concentrations of 0.05, 0.20, 1.0, 5.0, 20.0, 60.0, 100.0, and 250.0 μM at the respective amplification steps.

E. Adenosine deaminase

The biochemical marker adenosine deaminase (Bonthron et al., 1986; Kaufman et al., 1986; Yeung et al., 1983) is encoded by the ada gene. When transferred to CHO DUKX B11 cells, this gene, which in the appropriate setting is a dominant selectable marker (see below), confers resistance to α^- MEM (Flow, cat. # 12-313-54; Irvine Scientific, cat. # 9472; Gibco, cat. # 320-2561AJ) supplemented with 10 μg/ml of thymidine (Sigma, cat. # T1895), 15 μg/ml of hypoxanthine (Sigma, cat. # H9636), 4μM Xyla-A (9-β-D-xylofuranosyl, M.W. 267.0, NSC # 7539, Drug Synthesis and Chemistry Branch, Division of Cancer Treatment, N.C.I.: prepare a stock solution of 0.4 mM in H_2O, filter-sterilize, and store in aliquots at $-20°C$), 0.01 μM dCF (2-deoxycoformycin, Sigma, cat. # D6032: prepare a stock solution of 20 mg/ml (75 mM) in H_2O, filter-sterilize, and store in aliquots at $-20°C$), and 10% dialyzed FCS. FCS contains low levels of adenosine deaminase; therefore, add dialyzed FCS (Sigma, cat. # F0392) to media just prior to use to minimize detoxification of the cytotoxic selective agents. To use ada as a dominant selectable marker in murine fibroblasts, supplement the medium with 0.05 mM alanosine (M.W. 149.1, NSC # 153353, Drug Synthesis and Chemistry Branch, Division of Cancer Treatment, N.C.I.: prepare a fresh 5.0 mM stock solution in H_2O and filter-sterilize.)

To amplify this biochemical marker, first incubate cells in selection medium, α^- MEM supplemented with 15 μg/ml of hypoxanthine (Sigma, cat. # H9636), 10 μg/ml of thymidine (Sigma, cat. # T1895), 4 μM Xyla-A, 0.01 μM dCF (for murine fibroblasts supplemented with 0.05 μM alanosine) and 10% dialyzed FCS for 24 hours, then feed with amplification medium, α^- MEM supplemented with 10 μg/ml of thymidine (Sigma, cat. # T1985), 10 μg/ml of deoxyadenosine (Sigma, cat. # D0651), 1 mM uridine (Sigma, cat. # U3003), 1.1 mM adenosine (Sigma, cat. # A4036), 10% dialyzed FCS (added just prior to use, 0.05 mM alanosine for murine fibroblasts) and dCF at concentrations of 0.03, 0.10, 0.50, 20.0, and 100.0 μM at the respective amplification steps.

F. Asparagine synthetase

The biochemical marker asparagine synthetase (Cartier et al., 1987) is encoded by the as gene. When transferred to CHO N3 (AS$^-$) cells, this gene confers resistance to α^- MEM minus asparagine (Irvine Scientific, custom preparation) supplemented with 100 μg/ml of L-glutamine (Sigma, cat. # G5763), 10% dialyzed FCS (Sigma, cat. # F0392), 0.1 mM β-AH (β-aspartyl hydroxamate; Sigma, cat. # A6508: prepare a stock solution at 500 mg/ml in H$_2$O, filter-sterilize, and store in aliquots at $-20°C$).

To amplify this biochemical marker, first incubate the cells in selection medium, α^- MEM, 10% dialyzed FCS (Sigma, cat. # F0392) for 24 hours, then feed with amplification medium, α^- MEM minus asparagine (Irvine Scientific, custom preparation), supplemented with 100 μg/ml of L-glutamine (Sigma, cat. # G5763), 10% dialyzed FCS, and β-AH at concentrations of 0.2, 0.6, 0.8, 1.0, 1.2, 1.4, 2.0, 3.0, 4.0, and 5.0 mM at the respective amplification steps. To amplify this biochemical marker with AS$^+$ cells, add Albizziin (Sigma, cat. # A5398: prepare a stock at 50 mg/ml in H$_2$O, filter-sterilize, and store in aliquots at $-20°C$).

G. Glutamine synthetase

The biochemical marker glutamine synthetase (Bebbington and Hentschel, 1987; Wilson and Bebbington, 1987) is encoded by the gs gene. When transferred to CHO K-1 cells, this gene confers resistance to GMEM-S: Glasgow MEM without glutamine (Irvine Scientific, cat. # 9427), and with 10% dialyzed FCS (Sigma, cat. # F0392); sodium pyruvate (Sigma, cat. # S8636); the non-essential amino acids alanine (Sigma, cat. # A3534), aspartate (Sigma, cat. # A4534), glycine (Sigma, cat. # G6388), proline (Sigma, cat. # P4655), and serine (Sigma, cat. # S5511) at 100 μM each; 500 μM glutamate (Sigma, cat. # G5889); 500 μM asparagine (Sigma, cat. # A4159); 10 μM thymidine (Sigma, cat. # T1895); adenosine (Sigma, cat. # A4036), guanosine (Sigma, cat. # G6264), and uridine (Sigma cat. # U3003) at 30 μM each; 25 μM MSX (methionine sulphoximine, Sigma, cat. # M5379).

To amplify this biochemical marker, first incubate the cells in selection medium, GMEM-S, for 24 hours, then feed with amplification medium, GMEM-S, with MSX at concentrations of 0.10, 0.5, 2.5, 7.5, and 10 mM at the respective amplification steps.

H. Metallothionein

The biochemical marker metallothionein (Beach and Palmiter, 1981) is encoded by the mt gene. When the mt gene is co-transfected with a dominant selectable marker, such as neo^r, CHO K-1 cells are first subjected to selection of the dominant selectable marker (see previous section). To amplify the mt marker, first incubate cells in selection medium, HAM F-12, 10% FCS, then feed with amplification medium; the selection medium used to select for the dominant selectable marker supplemented with CdCl$_2$ (Sigma, cat. # B4639) at concentrations between 1–5 μM. Cd^{++}-resistant clones may be subjected to another round of amplification with Cd^{++} at 15–60 μM.

NEGATIVE SELECTION

Negative selection is a procedure whereby one selects against cells transfected with a given marker gene. Negative-selective protocols are utilized in two particular technologies: the generation of cells that have been transfected with the intent of selecting for homologous recombination of the transferred DNA (Mansour et al., 1988); and the generation of so-called *toxigenic* cells and mice (Palmiter et al., 1987; Breitman et al., 1987; Landel et al., 1988; Behringer et al., 1988; Borelli et al., 1988; Heyman et al., 1989). In both cases, cells transfected with particular genes and carrying and expressing those genes are killed either through the mere expression of that gene or through the ability of the transfected gene to metabolize a drug into a toxic substance that ultimately kills the transfected cell.

It is not immediately obvious why one might want to destroy all transfected cells. A brief explanation of the applications should make the reasons clear.

Negative selection for the occurrence of homologous recombination is a method that was developed by Mansour et al. (1988). The genetic mechanics of positive/negative selection have been described previously in the section on homologous recombination and gene replacement vectors in Part I. One objective of these experiments was to knock out, via homologous recombination, a gene or genes of interest in embryonal stem cells. Such embryonal stem cells, now hemizygous for a particular genotype, can be used to generate germ-line chimeric mice also hemizygous for the gene of interest. Such mice can then be crossed to generate so-called *nulls*, mice entirely lacking a functional gene at a particular locus. Such animals are of tremendous importance in understanding the physiologic rules of a variety of genes, as well as functioning as murine models of human diseases, where they are potentially invaluable in the development of treatments and cures for such maladies.

The second strategy for negative selection involves the attachment of a cell-type-specific or inducible enhancer/promoter assembly to a gene encoding a cellular toxin such as diphtheria toxin or the ricin A chain, or a biochemical marker that in the presence of a drug kills the host cell. The classic example is the herpes simplex virus thymidine kinase gene in combination with the drugs acyclovir and ganciclovir (Furman et al., 1980; Evans, 1989). This recombinant DNA molecule is introduced into a fertilized mouse egg and subsequently implanted into a foster mother to generate a transgenic, or in this case toxigenic, mouse. During the development of such a mouse, cells expressing the toxic protein are killed and thus removed from the developmental program of the mouse at specific places in space and time. Such animals may lack particular tissues, a subset of cells from a particular tissue, or a specific lineage of hematologic cells. Indeed, the possibilities are endless. Nevertheless, the rules for constructing recombinant toxigenic cistrons are virtually identical to those in the construction of virtually any engineered cistron. So, although toxigenic technology bears attention, the focus of this section will be techniques for negative selection, with a description of the methods for generating homologous recombinants to be discussed in the section on homologous recombination.

The protocol for negative selection of cells expressing HSVtk with ganciclovir is described here:

Materials

1. Ganciclovir (obtain from Julian Verheyden, Syntex Research): Prepare a 2 mM stock solution in water. Filter-sterilize. Aliquot. Store at $-20°C$.

2. Complete media, supplemented with ganciclovir to a final concentration of 2 μM.

Method

1. Plate cells in complete media with 2 μM ganciclovir.
2. Every three days, refeed cells with fresh media.

7
EXPRESSION CLONING

Expression cloning is a technique whereby you isolate a molecular clone encoding a protein that can be either selected or assayed for. Molecules of this type include selectable markers and oncogenes, as well as cytokines (Wong et al., 1985) and membrane proteins including cell-surface antigens (Aruffo and Seed, 1987) and cellular receptors (D'Andrea et al., 1989).

Typically, a cDNA library is constructed from a source of mRNA, one of which encodes the factor of interest. This cDNA library is constructed in an appropriate expression vector, usually SV40-based. Such vectors are described in the section on vectors in Chapter 2.

The resultant cDNA bearing plasmid is used to transform *E. coli* to generate the cDNA library. The transformed *E. coli* are amplified, and the amplified library is broken down into pools composed of subpopulations of the library. In theory, at least one of these pools contains an expressible cDNA clone of interest. This process is known as *SIB selection*. The transformed bacteria within each plasmid pool are propagated, plasmid DNA is prepared, and the DNA is transfected into the appropriate cell. Twenty-four to seventy-two hours post transfection, the cells can be subjected to various forms of analysis (Figure 7-1).

Analysis for the transfer of a dominant selectable marker or a gene conferring an oncogenic phenotype has been discussed previously in this text. One can, in addition analyze for the production of a cytokine or a membrane protein such as an antigen or a cell surface receptor (Wong et al., 1985; Aruffo and Seed, 1987; D'Andrea et al., 1989).

To analyze for the presence of a cytokine, one removes the media from the expressing cells and subjects the media, potentially containing the cytokine of interest, to a sensitive, accurate bioassay. Plasmid pools that score positive are re-transfected and tested again, after which the positive pools are subjected to the SIB-selection protocol detailed in this section. The final step in the SIB-selection protocol yields a single molecular clone that scores positive in the bioassay of interest.

Identification and isolation of cDNAs encoding cell-surface receptors are carried out in an analogous fashion. Once again, plasmid pools are prepared and transfected into the host cell. In this case, however, a radiolabeled- or fluorescent-receptor ligand is added and allowed to bind. After the cells are washed, they are monitored for the retention of the radioactive ligand or for the presence of the fluorescent tag. Positive pools are reassayed and subjected to SIB selection, re-transfection, and assay, ultimately resulting in the identification and isolation of a single cDNA clone encoding the receptor or receptor component of interest.

EXPRESSION CLONING

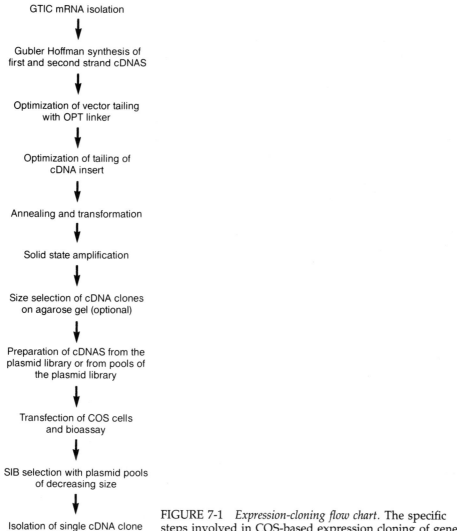

FIGURE 7-1 *Expression-cloning flow chart*. The specific steps involved in COS-based expression cloning of genes described in the text are ordered here.

The strategy employed in isolating a cDNA encoding a cell-surface antigen of interest differs somewhat from that just described. In this case, cells are transfected with the plasmid pools of the cDNA library, and, after allowing for sufficient expression time, the transfected cells are placed in a tissue-culture dish coated with an antibody directed against the antigen of interest. The experimenter subsequently "pans" for those cells expressing the antigen that binds to the plated antibody. Cells that lack the antigen fail to stick to the plate and can be washed away. Cells that express the antigen of interest remain. These cells also carry the expression plasmid, capable of replicating in both mammalian and bacterial cells and containing the cDNA encoding the antigen of interest. These remaining cells are lysed (Hirt, 1967) and the plasmid DNA contained within is purified and used to transform *E. coli*. This population of plasmid DNAs will be highly enriched for the cDNA encoding the binding antigen. Enrichment and ultimate purification of the correct cDNA clone is achieved by further rounds of transfection, panning, bacterial transformation, plasmid purification, and re-transfection.

In attempts to expression-clone a cytokine or cell-surface protein, numerical analysis is a key consideration. The two most important numerical factors to be considered are (1) the relative abundance of the mRNA to be expression-cloned and (2) the sensitivity of the bioassay employed in the expression/detection system. Any information one can acquire about these two variables prior to the attempt to expression-clone will greatly assist in the design of a successful experiment.

The relative abundance of the mRNA encoding the protein of interest is difficult to determine for an unknown protein, as is the sensitivity of a given bioassay for an unknown protein. Several preliminary experiments can serve to indicate the likelihood of success in expression-cloning experiments. One absolute requirement is that you identify and are working with the genetic source of the protein. You should be able to readily detect the cytokine of interest in supernatants derived from the cell source. A dilution series of the supernatant derived from the genetic source and tested in the bioassay can provide you with a "feel" for the experiment. The greater the dilution detectable in the bioassay, the higher the probability that an expression cDNA clone will be detected in a larger plasmid pool.

The ability to detect a transfected receptor will depend in part on the affinity of the ligand for the receptor, the specificity of the ligand/receptor interaction, and the specific activity of the labeled ligand.

Detection of a transfected cell-surface antigen will depend, in part, on the quality of the antibody employed. In any case, transfection of cells with the entire cDNA library and subsequent detection in a given assay virtually guarantees success in further rounds of transfection and analysis.

If the activity of the protein in the transfection supernatant or on the cell surface is too low to be detected in the library, one can segregate the library into pools and then transfect and assay the pools for activity. The relative abundance of the mRNAs encoding the protein of interest will be reflected in the number of cDNA clones present in the library, and this will, in turn, determine the number of plasmid pools scoring positive in the bioassay. Absolute detectability is more a function of apparent specific activity of the protein of interest in the assay rather than the relative abundance of the cDNA clone. If, upon segregation of the pools, one fails to detect the protein, one may fractionate the library into even smaller pools and, post fractionation, transfect and assay once again. If this procedure is unsuccessful there are several possible explanations. It is possible the bioassay is insufficiently sensitive to detect the expressed protein. It is also possible the cDNA library does not contain a functional, expressible copy of the cDNA encoding the protein of interest.

Although the sensitivity of the detection assay is a variable I will not address in this manual, I have provided a cDNA-synthesis protocol we have found to be superior to any other we have tested. This procedure is optimized for the synthesis of full-length cDNAs, up to 7.5 kb in length (greater than 90 percent full length to the mRNA cap). It has been optimized for the maximum yield of cDNAs from mRNAs, thus enabling the investigator to immortalize mRNAs present in very small amounts (≤ 1 in 10^6) in a *plasmid* cDNA library with efficiencies equivalent to the best λ-phage cloning techniques.

The cDNA/vector-joining technology has also been optimized to allow for the maximal expression of the inserted cDNA upon transfection of COS cells. Further, we provide a novel library-amplification technique that we term *solid-state amplification* that permits accurate, highly reproducible amplification of the plasmid cDNA library from generation to generation and appears to function in a manner superior to λ-phage library-amplification methods. The specifics of these procedures are described next.

PREPARATION AND TECHNICAL REQUIREMENTS

The successful execution of this methodology requires a highly reproducible assay or screen for the protein of interest as well as a source of mRNA encoding that protein. We have been most successful employing an mRNA-isolation protocol in which guanidinium isothiocyanate is used as an RNAse-inactivating agent. The specific details of this protocol are described in Part III in the section on analysis of RNA. This method of mRNA isolation should be employed when constructing cDNA expression libraries with the method we describe herein.

LIBRARY CONSTRUCTION AND AMPLIFICATION

Comparison of Construction Methodologies

During the development of this expression-cloning technology, we evaluated a number of cDNA-insertion methodologies. We evaluated tailing, linker addition,

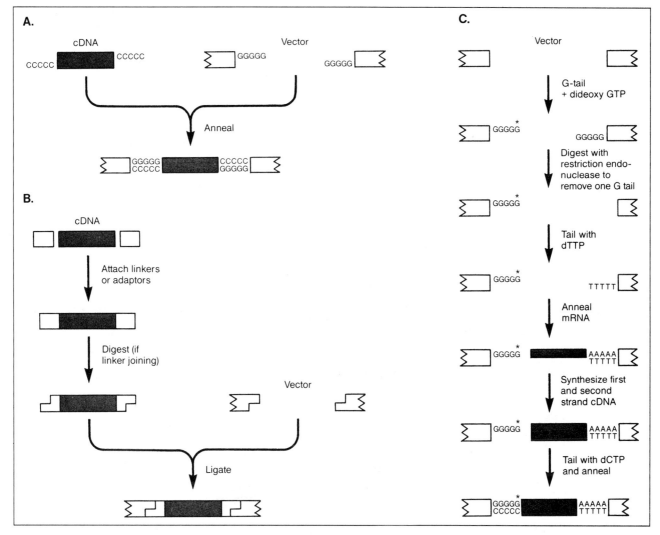

FIGURE 7-2 *cDNA-insertion methods.* The various methods for cDNA insertion described in the text are depicted here. **Panel A,** tailing; **Panel B,** linker/adaptor insertion; **Panel C,** vector priming. See the text for details.

and vector priming. We were attempting to determine under what conditions we could obtain the greatest number of full-length cDNA clones per microgram of mRNA.

We also had to consider the effects the joining procedure might have on the expression of the cDNAs from the promoter in the expression vector. The effects of joining sequences on expression are not well understood. However, in the case of G/C tailing, it is clear that as the G/C tail increases from 10 to 30 nucleotides, expression plummets.

Our next concern was library complexity. If the factor we intended to expression-clone was of low abundance, we wanted to be sure it was represented in the library. We found that we obtain cDNA libraries of the greatest complexity when we use G/C tailing to join the cDNA to the vector. In relative terms, linker addition yields one-tenth the number of plasmid clones per unit of cDNA and vector. Vector-priming—employing a tailed vector to drive the synthesis of the cDNA—presents its own problems, the worst of which is an unacceptably high background of plasmid that contains no insert at all. These methods of insertion are diagrammed in Figure 7-2 and a comparison of our analysis of these joining technologies is represented in Table 7-1.

During our analysis of various joining technologies, we sought to optimize the vector- and cDNA-tailing reaction to obtain the highest number of bacterial transformants per unit of cDNA. We did this empirically by limiting the nucleotide substrate and, after tailing for various periods of time and annealing the vector to the insert, we then transformed *E. coli* (Figure 7-3). We observed that, as the tails grew, there was an initial increase in the number of bacterial transformants, after which the number of transformants reproducibly dropped (Table 7-2).

After sequencing numerous cDNA-containing plasmids from a number of our libraries, we observed that the optimal number of tailing nucleotides for generation of bacterial transformants is 12–15, well below the number that inhibits expression of the gene from the eukaryotic promoter within the expression vector upon transfection of COS cells. Thus, we had arrived at our cDNA-cloning methodology.

We then sought to develop a simple technique that would allow us to amplify a plasmid library in a strictly representative fashion, decreasing the possibility that less-abundant clones would vanish during amplification. Disproportionate amplification is a common problem when working with plasmid libraries, and it is the reason the majority of investigators prefer to work with a λ-phage library when possible. Our amplification method successfully addresses the deficit.

TABLE 7-1 Comparison of Library-Generation Methodologies

Library Generation Method	Orientation Specific?	Colony-Forming Units /μg RNA	Compatible with Solid-State Amplification?	Volume of Competent Bacteria Needed to Produce 1×10^6 cfu
Vector Priming[1,2]	Yes	3.4×10^5	No	200 ml
Linker[3]	Yes	9.34×10^4	No	22.7 ml
Tailing	No	9.68×10^6	Yes	0.7 ml

[1] CFU reported by Okayama and Berg are 1×10^5 cfu/μg RNA
[2] Modification of Gubler/Hoffman; no adapter fragment
[3] XbaI-dT$_{15}$; EcoRI adapter, no methylation

LIBRARY CONSTRUCTION AND AMPLIFICATION

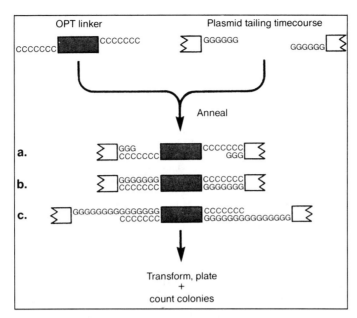

FIGURE 7-3 *Tailing time course.* The method of optimization of vector tailing with the optimization (OPT) linker is depicted here. (*a*) Initially the G tails are insufficiently long to permit optimal hybrid formation of polyG/polyC tails. (*b*) As the incubation time of the tailing reaction increases, the tails grow first to the point where the number of tailing residues is optimal for interaction with the optimization linker and then beyond that point, where the tails grow too long for optimal interaction with the optimization linker (*c*) and, thus, with the tailed cDNA insert.

We developed a solid-state amplification technique wherein bacterial transformants are *suspended*, not plated, in low-melting-temperature agarose in a 50-ml conical tube. We have repeatedly tested this procedure and have found that, from generation to generation of amplification, the relative number of bacterial harboring plasmids encoding selected hematopoietic factors varies less than two-fold (Table 7-3). We can, in fact, propagate 30,000 colonies in each tube, easily allowing for representative amplification of our libraries, some of which are as large as 2,500,000 independent transformants with complexities (a reflection of the size of the population in which you can find a single cDNA copy of an mRNA) approaching 500,000 or more.

Thus, you can generate a large number of cDNA clones in virtually any plasmid vector. Vector selection is an important consideration in expression-cloning experiments. We engaged in an evaluation of two COS-based vectors to drive

TABLE 7-2   dGTP Tailing of pcDL-SRα-296

Time dGTP Tailing	Total # Colonies[a]	Total Colonies[b] Background	Percent Background
0"	4.19×10^4	4.97×10^4	100.00
15"	4.84×10^5	9.02×10^2	0.19
30"	3.20×10^5	9.18×10^2	0.29
45"	2.67×10^5	10.8×10^2	0.40
60"	2.27×10^5	10.6×10^2	0.47
75"	2.03×10^5	9.75×10^2	0.48
90"	1.97×10^5	10.2×10^2	0.52

[a] 7.37 ng SR-α vector (~ 2.6×10^{-3} pmoles) annealed to 1.44×10^{-3} pmoles of OPT-linker (annealing linker) and transformed with 100 μl of DH5α (BRL). This number represents the number of colonies from the entire transformation reaction.
[b] Same as (a), but no annealing linker.

TABLE 7-3 Reproducibility of Solid-State Amplification from Generation to Generation[a]

	Colony Count	Ratio
First Generation	IL-3 $25/1.12 \times 10^5$	
		IL-3/GMCSF = 0.184
	GMCSF $126/1.04 \times 10^5$	
Second Generation	IL-3 $43/1.22 \times 10^5$	
		IL-3/GMCSF = 0.422
	GMCSF $102/1.22 \times 10^5$	

[a] An induced PBL cDNA library was subjected to solid-state amplification and pelleted, plated, and analyzed for IL-3 and GMCSF cDNA by colony hybridization. The second generation involved resuspending a portion of the pellet from the first generation and subjecting those bacteria to a second round of solid-state amplification, after which the bacteria were again pelleted, plated, and analyzed for IL-3 and GMCSF cDNAs by colony hybridization. GMCSF (Granulocyte-Macrophage Colony Stimulating Factor).

the expression of foreign genes, after subcloning the human IL-2 gene and human γ-interferon gene into these vectors. We evaluated the original Okayama-and-Berg vector and selected an Okayama and Berg derivative named pcDL-SRα-296 (SR-α), reportedly vastly superior to the original Okayama-and-Berg vector (see Part I in the section on SV40-based vectors).

Our results support this contention. We have found that this derivative vector serves to express 100X as much γ-interferon and 40X as much IL-2 as the wild-type Okayama-and-Berg equivalent (Table 7-4). Furthermore, based on published information and personal communication, this vector appears at least as potent as the well-known, commonly used SV40-based vectors p91023(B) and its derivatives and πH3M (see Part I in the section on SV40-based vectors).

We found that SR-α IL-2 is sufficiently powerful to generate an IL-2 signal approximately 10X greater than that required for the reliable bioassay of the gene product when, in a reconstruction experiment designed to detect a secreted factor with a specific activity of IL-2, the plasmid DNA encoding that factor is present as 1 percent of the DNA to be transfected.

TABLE 7-4 Comparative Analysis of COS-Based Expression Vectors Driving the Expression of IL-2 and γ-Interferon

COS-Type Vector	units/ml
pCDB-IL-2	36.4
SR-α-IL-2	4.21×10^3
pCDB–γINF	9.0×10^2
SR-α-γ-INF	1.26×10^5

cDNA LIBRARY CONSTRUCTION

The procedure we employ to generate cDNA libraries is diagrammed in Figures 7-1 and 7-3 and described here:

Materials

Enzymes

1. Pst I (New England Biolabs, cat. # 140)
2. DNA Polymerase I (New England Biolabs, cat. # 209)
3. *E. coli* DNA Ligase (New England Biolabs, cat. # 205)
4. RNasin (Promega, cat. # N2112)
5. Ribonuclease H (Promega, cat. # M4281)
6. Terminal transferase (Ratliff Biochemicals, TdT)
7. MoMLV reverse transcriptase (BRL, cat. # 8025SA)

Chemicals

1. 100.0 mM dATP (Pharmacia, cat. # 27-2050-01)
2. 100.0 mM dCTP (Pharmacia, cat. # 27-2060-01)
3. 100.0 mM dGTP (Pharmacia, cat. # 27-2070-01)
4. 100.0 mM dTTP (Pharmacia, cat. # 27-2080-01)
5. $p(dT)_{12-18}$ (Pharmacia, cat. # 27-7858-01)
6. β-NAD, Grade V (Sigma, cat. # N1636)
7. Diethylpyrocarbonate (DEPC; Sigma, cat. # D5758)
8. S.O.C. medium (BRL, cat. # 5544SA)
9. Cacodylic acid, free acid (Sigma, cat. # C0125)
10. Aquasol 2 (New England Nuclear, cat. # NEF-952)
11. Ampicillin, sodium salt (Sigma, cat. # A9518)
12. BSA, DNase-free (Pharmacia, cat. # 27-8915-01)
13. Bromphenol blue (Sigma, cat. # B5525)
14. Dimethyldichlorosilane (silane; Sigma, cat. # D3879)
15. Dithiothreitol (DTT; Sigma, cat. # D9779)
16. 8-hydroxyquinoline (Sigma, cat. # H6878)
17. Xylene cyanol FF (Sigma, cat. # X0377)
18. D-glucose

Isotopes

1. [α-^{32}P]dCTP (in tricine), 10 mCi/ml, 800 Ci/mmole (New England Nuclear, cat # NEG-013A)
2. [8,5-^{13}H] dGTP (in tricine), 2.5 mCi/ml, 20–50 Ci/mmole (New England Nuclear, cat. # NET-448A)

Additional Materials

3. Max Efficiency, DH5α-competent cells (BRL, cat. # 8258SA, 5 × 0.2 ml)
4. SeaKem GTG agarose (FMC Bioproducts, cat. # 50072)
5. SeaPrep agarose (FMC Bioproducts, cat. # 50302)
6. Bio-Gel P-4 Gel (Bio-Rad, cat. # 150-0440)
7. Quik-Sep columns (Isolab, cat. # QS-Q)
8. Micro tubes (Sarstedt, cat. # 72.692)
9. Gene Screen Plus (New England Nuclear, cat. # NEF-976)
10. GENECLEAN (glass beads; BIO-101)
11. JA-10 rotor & 500-ml polypropylene centrifuge bottles (Beckman)
12. 5-ml polypropylene tubes (Falcon, # 2063)
13. 15-ml polypropylene tubes (Falcon, # 2059)
14. 50-ml polypropylene tubes (Falcon, # 2098)
15. Bacto Agar (Difco, cat. # 0140-02, 0.25 lb)

16. Tryptone (Difco, cat. # 0123-02, 0.25 lb)
17. Yeast extract (Difco, cat. # 0127-02, 0.25 lb)
18. Glass fiber filters (GF/C; Whatman)
19. Filter washing manifold (Millipore)
20. Vertical-gel-electrophoresis apparatus (Stratagene)
21. Horizontal-mini–gel-electrophoresis apparatus (Bio-Rad)
22. Petri dishes (Falcon, cat. # 1029)
23. Cryotubes (Nunc, cat. # 3-68632)

Nucleic-Acid Reagents

1. Poly(A)$^+$ RNA, 33 μg
2. DNA vector plasmid, 200 μg
3. OPT.A Oligo: 5'-ATCGATAGGCCTGACTGGCCTCGAGATATC$_{15}$-3'
4. OPT.B Oligo: 5'-ATATCTCGAGGCCAGTCAGGCCTATCGATC$_{15}$-3'
5. Salmon sperm DNA (Sigma, cat. # D1626), 10 mg/ml (sheared by passing 10 times through an 18-gauge needle)
6. pBR322 DNA-Msp I Digest (New England Biolabs, cat. # 303-2)
7. tRNA (yeast—Sigma, cat. # R0128)

Bacterial Growth Media

1. LB-amp^{50} plates
   a. Add 15.0 g Bacto Agar to 500 ml double-distilled H$_2$O. If necessary, pH to 7.0 with 1 N NaOH. Autoclave.
   b. Add 10.0 g Bacto-Tryptone, 5.0 g yeast extract, and 5.0 g NaCl to 500 ml double-distilled H$_2$O. If necessary, pH to 7.0 with 1 N NaOH. Autoclave.
   c. When containers have cooled to about 50°C, combine them and add 4.0 ml of filter-sterilized 50% D-Glucose and 1.0 ml of filter-sterilized 50 mg/ml ampicillin.
   d. Pour plates.
2. 2X LB: Add 20 g Bacto-tryptone, 10 g Bacto-yeast extract, and 10 g NaCl to 1 liter double-distilled H$_2$O. If necessary, pH to 7.0 with 1 N NaOH. Autoclave.
3. 2X LB + 0.3% SeaPrep agarose: Prepare this solution on the day of library amplification. Add 3 g of SeaPrep agarose to 1.0 liter of 2X LB preparation. Autoclave. Cool to around 40°C. Store in 37°C water bath. Just before use, add 1 ml of 50 mg/ml ampicillin.

Solutions and Buffers

1. 10X Pst I Buffer: prepare a solution containing 1.0 M NaCl, 0.5 M Tris-HCl, pH 7.5, 0.5 M MgCl$_2$, and 1 mg/ml DNase-free BSA.
2. 10X annealing buffer (10X AB): prepare a solution containing 0.1 M Tris-HCl, pH 7.6, 1.0 M NaCl, and 10 mM EDTA.
3. 10X terminal transferase buffer (10X TdT)
   a. Prepare a solution containing 13.8 g cacodylic acid, 3.0 g Tris-base, and 60.0 ml H$_2$O
   b. pH to 7.6 by slow addition of solid KOH with constant stirring.
   c. Bring to final volume of 88 ml with H$_2$O. Chill to 0°C, then add 2.0 ml of 0.1 M DTT.
   d. Add 10 ml of 0.1 M MnCl$_2$ drop-wise with constant stirring.
   e. Store at 4°C in small, tightly capped aliquots.

4. 5X first strand buffer (5X FSB): prepare a solution containing 250 mM Tris-HCl, pH 8.3, 375 mM KCl, 15 mM $MgCl_2$, and 50 mM DTT. Make this buffer fresh.
5. 5X second strand buffer (5X SSB): prepare a solution containing 94 mM Tris-HCl, pH 8.3, 453 mM KCl, 23.3 mM $MgCl_2$, and 18.7 mM DTT.
6. 5X RNase H buffer (5X RHB): prepare a solution containing 100 mM Tris-HCl, pH 7.5, 100 mM KCl, 50 mM $MgCl_2$, 0.7 mM DTT, and 0.5 mM EDTA.
7. DEPC-treated H_2O: add diethylpyrocarbonate to sterile H_2O to a final concentration of 0.1% (v/v). Mix vigorously for 10 minutes and allow to stand overnight at room temperature. Autoclave to dissipate the DEPC.
8. 80% glycerol, autoclave.
9. 6X loading buffer (6X LB): prepare a solution containing 0.15% bromophenol blue, 0.25% xylene cyanol, and 15% Ficoll (type 400) in H_2O.
10. 4X alkaline loading buffer (4X ALB): prepare a solution containing 40 mM NaOH, 40% glycerol, and 0.04% bromophenol blue. Make fresh.
11. 10X alkaline gel buffer: prepare a solution containing 0.3 M NaCl, and 20 mM EDTA, pH 7.5.
12. 10X alkaline electrophoresis buffer: prepare a solution containing 0.3 M NaOH and 20 mM EDTA.
13. 10X Tris-acetate electrophoresis buffer (10X TEA): prepare a solution containing 400 mM Tris-acetate, 20 mM EDTA, pH 8.0. Add 242 g of Tris-base, 57.1 ml glacial acetic acid, 100 ml 0.5 M EDTA, and water to a final volume of 5.0 liters.
14. 4% silane in pure carbon tetrachloride.
15. 7.5 M ammonium acetate.
16. Phenol: Add 8-hydroxyquinoline (Sigma, cat. # H6878) to liquified phenol (Sigma, cat. # P1037) to a concentration of 0.1% (w/v). Extract the phenol twice with an equal volume of 1.0 M Tris (pH 8.0). After the final extraction, mix the phenol with chloroform (Aldrich, cat. # 27,063-6) and isoamyl alcohol (Sigma, cat. # I1885) at a ratio of 50:48:2, respectively. Equilibrate the final phenol mixture with an equal volume of 100 mM Tris-HCl, pH 8.0, and 0.2% (v/v) 2-mercaptoethanol (Sigma, cat. # M3148), and store at 4°C in a brown jar.
17. 10% (w/v) TCA + 2% (w/v) sodium pyrophosphate.
18. 5% (w/v) TCA + 1% (w/v) sodium pyrophosphate.
19. TE: prepare a solution containing 10.0 mM Tris-HCl, pH 8.0, 1 mM EDTA.

Preparation of Vector

The first step in the generation of a cDNA expression library is the preparation of the vector to accept the cDNA insert. This involves digestion of the vector followed by optimal tailing.

1. Digestion of vector
   a. Digest 200 μg of plasmid with 700 units of Pst I in 500 μl total volume of 1X Pst I buffer for 60 minutes at 37°C.
   b. Phenol extract once, ethanol precipitate twice with ammonium acetate. (Add 0.5 aqueous volume of 7.5 M ammonium acetate and 2 total volumes ethanol.)
   c. Resuspend the pellet in 200 μl of H_2O (no EDTA!). Determine the concentration by spectrophotometry.

2. G-tailing of digested vector
   Prepare the tailing reaction as described here:
   a. 30.0 μl   10X TdT buffer
       30.0 μl   [^{3}H] dGTP (~70 pmole/μl)
        5.5 μl   1 mM dGTP
          X μl   DNA [112.6 pmole 3' ends ~136 μg of digested SR-α vector]
          Y μl   H$_2$O
       297.0 μl   Total
   b. Remove 30 μl of the tailing reaction and add to 359 μl of 17 mM EDTA ≡ 0" time point.
   c. Prewarm tailing cocktail 15 minutes at 37°C.
   d. Prepare tubes labeled 15", 30", 45", 60", 75", 90", and 105", and add 359 μl of 17 mM EDTA (stop solution) to each.
   e. Add 360 units of terminal transferase (~3 μl) to reaction cocktail. Remove 30 μl of the tailing reaction at each time point and pipet directly into stop solution.
   f. Determine incorporation of label via TCA precipitation method for [^{3}H].
   g. Store tailed vector at −20°C.
3. TCA-precipitation method for [^{3}H]
   a. To a 5.0-ml polypropylene tube add:

          5 μl   sample
        100 μl   TE
          5 μl   sheared salmon sperm DNA (10 mg/ml)
       2500 μl   10% TCA + 2% sodium pyrophosphate.

   b. Place this mixture on ice for at least 30 minutes.
   c. During incubation, mount labeled GF/C filters (mark with #2 pencil on perimeter of filter) on the washing manifold.
   d. Wet filters with 5 ml of cold 10% TCA + 2% sodium pyrophosphate. Adjust flow (via vacuum valves) to drain 5 ml in about 1 minute.
   e. Pour contents of tubes onto appropriate filters. Fill tubes with 5% TCA + 1% sodium pyrophosphate and pour onto corresponding filters. Wash with 20 ml of 5% TCA + 1% sodium pyrophosphate. (Have 5% TCA + 1% sodium pyrophosphate in a polypropylene squirt bottle kept on ice.)
   f. Wash with 10 ml of 95% ethanol.
   g. Dry filters under a heat lamp.
   h. Count filters in 5.0 ml Aquasol 2.
   i. For "Total Available Counts" control, add 5 μl of any time point directly to a scintillation vial and add 5.0 ml of Aquasol 2.
   j. Count vials.
   k. Determine the tail length of the vector as a function of tailing reaction time.
4. Determination of tail length
   To determine the length of the polynucleotide tails, perform the following calculation:
   a. The amount of dGTP available in a total reaction

   $$= [(30\ \mu l) \times (70\ \text{pmoles}/\mu l)\ [^3H]\ dGTP] + [(5.5\ \mu l) \times (10^3\ \text{pmoles}/\mu l)\ dGTP]$$
   $$= 7.6 \times 10^3\ \text{pmoles dGTP}$$

   b. Number of available 3' ends for tailing = 112.6 pmoles.
   c. Tail length

   $$= \text{fraction dGTP incorporated} \times \frac{7.60 \times 10^3\ \text{pmoles dGTP}}{112.6\ \text{pmoles 3' ends}}$$

d. As an example, for a 30-second time point, the average number of incorporated counts was 32,000 cpm and the total number of cpms available was 170,000 cpm.

$$\therefore \text{Tail length (30")} = \frac{32,000\text{cpm}}{170,000\text{cpm}} \times \frac{7.6 \times 10^3}{112.6} = 12.7 \text{ dGTP's/3' end}$$

5. Determination of annealing efficiency

   To determine the annealing efficiency of the tailed vector to tailed cDNAs, we have created an oligonucleotide optimization linker (OPT-linker) that functions as a synthetic cDNA insert. The OPT-linker has 15 unpaired cytosine residues at each end.

   a. Preparation of optimization linker (OPT-linker)
      i. 100 μl   10X AB
          10 μl   OPT A (100 pmole/μl)
          10 μl   OPT B (100 pmole/μl)
          880 μl H_2O
          ─────
          1000 μl   Total (1 pmole/μl)
      ii. Boil 5 min. Cool slowly to room temperature in a beaker partially filled with boiling H_2O.
      iii. Dilute the annealing reaction 1:100 with H_2O. Stock concentration is 0.01 pmole/μl.

   b. Prepare two cocktails.

   OPT-linker cocktail
   150.0 μl   10X AB
   21.6 μl   OPT-linker (0.01 pmole/μl)
   1298.4 μl H_2O
   ─────
   1470.0 μl   Total

   control cocktail
   150.0 μl   10X AB
   1320.0 μl H_2O
   ─────
   1470.0 μl   Total

   c. For each tailing time point, add 2 μl of tailed vector to 98 μl of OPT-linker cocktail.
   d. Anneal by incubating at 65°C for 5 min, then at 57°C for 120 min.

6. Transformation into DH5-α

   The transformation protocol described here results in the highest number of bacterial transformants per μg of plasmid DNA.

   a. Add 10 μl of annealing mix to a Falcon 2059 tube on ice.
   b. Add 100 μl of thawed competent DH5-α cells. Incubate for 30 minutes on ice.
   c. Heat-shock for 45 seconds at 42°C, then incubate for 2 minutes on ice.
   d. Add 900 μl of S.O.C. medium. Incubate for 60 minutes at 37°C at 225 rpm.
   e. Plate 100 μl of 1:10 and 1:100 serial dilutions on LB-amp^{50} plates. Incubate at 37°C overnight.
   f. For cDNA library construction, select the vector-tailing timepoint that generated the greatest number of colonies/ml of transformation mixture.

cDNA Synthesis

The cDNA-synthesis protocol presented here is a modification of Gubler and Hoffman (1983). In it, we substitute MuLV reverse transcriptase for AMV reverse transcriptase.

1. Siliconizing micro tubes
   a. Dispense 4% silane in carbon tetrachloride directly into tubes. Fill tubes entirely.
   b. Decant and remove residual solution with a Pasteur pipet.
   c. Air-dry for 5 minutes. Repeat steps (a)-(c).

d. Rinse tubes in DEPC-treated H_2O twice and 95% ethanol once, then air-dry.
e. Autoclave.
2. First-strand synthesis
   a. Place 33 µg poly (A)$^+$ RNA (for instructions on preparation, see section in Part III on analysis of RNA) in a microcentrifuge tube. Incubate at 68°C, for 5 minutes, then place immediately on ice for 5 minutes.
   b. Add, in this order, to the tube on ice:

X µl	DEPC-H_2O to give 33.0 µg RNA/187 µl H_2O
66.0 µl	5X FSB
9.0 µl	RNasin (40 U/µl)
16.5 µl	10 mM dNTPs
35.0 µl	p(dT)$_{12-18}$ (100 pmole/µl)
16.5 µl	MoMLV-RT 200 U/µl)
330.0 µl	Total

   A second, smaller reaction containing labeled dCTP is set up, in parallel, to enable you to monitor the efficiency of first-strand synthesis.
   c. To a second tube on ice add:

2 µl	[α-^{32}P] dCTP (10.0 µCi/µl)
20 µl	First-strand reaction (from step b).
22 µl	Total

   d. Incubate both tubes at 37°C for 60 minutes, then place on ice.
   e. To the tube with [α-^{32}P] dCTP, add 5 µl of 0.5 M EDTA and 23 µl of H_2O. Store at −20°C for TCA and alkaline gel analyses.
3. Second-strand synthesis
   a. Dispense 37.5 µl of first-strand reaction into each of eight tubes on ice; add 6.25 µl to a ninth tube on ice.
   b. Prepare the second-strand synthesis reaction:

600.0 µl	5X SSB
1802.8 µl	H_2O
56.3 µl	100 mM dNTPs
12.5 µl	[α-^{32}P] dCTP
75.0 µl	6.0 mM β-NAD
75.0 µl	DNA Polymerase I (10 U/µl)
3.4 µl	*E. coli* ligase (9 U/µl)
2625.0 µl	Total

   c. Dispense 262.5 µl of the second-strand reaction into each of the eight tubes. Place on ice. Dispense 43.75 µl of the second-strand reaction and 5 µl of [α-^{32}p] dCATP into the ninth tube. Place on ice.
   d. Incubate all nine tubes at 16°C for 2 hours.
   e. To the ninth tube, add 5 µl of 0.5 M EDTA and 65 µl H_2O. Store at −20°C for TCA and alkaline gel analyses.
   f. Phenol-extract then ethanol-precipitate the eight reactions with 7.5 M ammonium acetate.
   g. Ethanol-precipitate a second time, then combine all pellets in 128 µl TE.
4. RNase H digestion
   a. Prepare this reaction:

128.0 µl	cDNA
128.0 µl	5X RHB
377.3 µl	H_2O
6.7 µl	RNase H (1.9 U/µl)
640.0 µl	Total

b. Incubate at 37°C for 20 minutes, then add 5 µl of 0.5 M EDTA to terminate reaction.
c. Phenol-extract, then ethanol-precipitate twice with ammonium acetate, then resuspend pellet in 12 µl TE.
5. TCA precipitation
   a. Spot 5 µl of [^{32}P]-labeled sample from steps 2e and 3e on a GF/C filter (label filter by using scissors to notch perimeter).
   b. Dry under heat lamp.
   c. Place dried filter into a scintillation vial (with the filter flat at the bottom of the vial) *without* any scintillation fluor. These counts serve as the unwashed control.
   d. Count on the tritium channel, and save filters for the washing steps and for recounting.
   e. Place a 2-liter beaker into an ice bucket and pack ice around it, then place the ice bucket on a rotary shaker.
   f. Place all the filters in the iced beaker, then add cold 10% TCA + 2% sodium pyrophosphate (at least 30 ml/filter). Rotate for 5 minutes.
   g. Pour off 10% TCA + 2% sodium pyrophosphate, and wash three times for 5 minutes each with cold 5% TCA + 1% sodium pyrophosphate.
   h. Wash filters with 95% ethanol for 5 minutes at room temperature.
   i. Dry filters under a heat lamp.
   j. Count dried filters as before.
   k. Determine fraction of label incorporated (unincorporated counts are washed away).
6. First-strand yield calculation
   a. ssDNA synthesized =

   $$\frac{(\text{Frac. Incor. } [\alpha\text{-}^{32}\text{P}] \text{ dCTP}) \times (\text{Available dCTP}) \times 4 \text{ dNTPs/dCTP}}{(\text{amount of dNTPs/}\mu\text{g ssDNA synthesized})}$$

   b. As an example, in a 30-µl first-strand reaction, 30 µg of mRNA are being used as a template. Fifteen µl of 10 mM dCTP are available. The average number of unwashed counts is 2.4×10^6 cpm and the average number of washed counts is 8.7×10^4 cpm.

   ssDNA synthesized =

   $$\frac{\begin{array}{l}[(8.7 \times 10^4 \text{ cpm}/2.4 \times 10^6 \text{ cpm}) \\ \quad \times [15 \, \mu\text{l} \times 10 \, \text{mM dCTP} \\ \quad\quad \times (10^4 \text{ pmoles}/(\mu\text{l} \times 10 \, \text{mM}))] \times 4 \text{ dNTPs/dCTP}]\end{array}}{3080 \text{ pmoles dNTP/}\mu\text{g ssDNA}} = 7.1 \, \mu\text{g ssDNA}$$

   c. $$\text{YIELD} = \frac{7.1 \, \mu\text{g ssDNA synthesized}}{30.0 \, \mu\text{g mRNA template}} \times 100 = 23.7\%$$

7. Second-strand yield calculation
   a. A similar calculation is made, accounting for the second addition of dNTPs in the second-strand reaction.
   b. For example: number of unwashed counts was 2.2×10^6 cpm and the number of washed counts was 2.1×10^4 cpm.

   ssDNA synthesized =

   $$\frac{\begin{array}{l}[(2.1 \times 10^4 \text{ cpm}/2.2 \times 10^6 \text{ cpm}) \\ \quad \times [(15 + 45) \times \mu\text{l } 10 \, \text{mM dCTP} \\ \quad\quad \times (10^4 \text{ pmoles}/(\mu\text{l} \times 10 \, \text{mM}))] \times 4 \text{ dNTPs/dCTP}\end{array}}{3080 \text{ pmoles dNTP/}\mu\text{g ssDNA}} = 7.44 \, \mu\text{g}$$

c. Efficiency of second-strand synthesis is

$$\frac{7.4\ \mu g\ (\text{second strand})}{7.1\ \mu g\ (\text{first strand})} \times 100 = 104\%$$

Assume that the molar yield of ds cDNA is only equal to the molar amount of the available first-strand cDNA template when the efficiency of copying is ≥ 100%. Continue with the production of the library when the efficiency of second-strand synthesis is greater than 80%.

∴ In our example, the amount of ds cDNA synthesized was 2 × 7.1 μg (first-strand template available) = 14.2 μg.

8. Calculating the amount of ds cDNA/cpm
   a. The second-strand labeling reaction consists of two contributions of [α-^{32}P] dCTP: the tracer carried from the second-strand reaction and the 5 μl added to the labeling reaction.
   b. Using our example, the yield of ds cDNA from the 2400 μl second-strand reaction is 14.2 μg.
   c. The yield of ds cDNA in the 50 μl labeling reaction is

   $$14.2\ \mu g \times \frac{50\ \mu l}{2400\ \mu l} = 0.294\ \mu g$$

   d. The amount of tracer label added to labeling reaction is

   $$\frac{12.5\ \mu l\ [\alpha\text{-}^{32}P]\ dCTP}{2625\ \mu l\ \text{cocktail}} \times 43.75\ \mu l = 0.208\ \mu l$$

   e. Total amount of label is 5.0 μl + 0.208 μl = 5.208 μl.
   f. The relative amount of label contributed by the tracer in the labeling reaction is $\frac{0.208}{5.208} = 0.04$.
   g. 5 μl of the labeling reaction gives 21,000 cpm of incorporated counts.
   h. The total amount of incorporated counts is

   $$\frac{21,000\ \text{cpm}}{5\ \mu l} \times 125\ \mu l = 5.25 \times 10^5\ \text{cpm}$$

   i. The amount of incorporated counts contributed by tracer is $5.25 \times 10^5 \times 0.04 = 2.10 \times 10^4$.
   j. And 2.10×10^4 cpm incorporated into 0.294 μg of ds cDNA.
   ∴ 2.10×10^4 cpm/0.294 μg = 7.2×10^4 cpm/μg ds cDNA.

Fractionation and Purification of cDNA

The double-stranded cDNA we have just synthesized represents a heterogeneous population of molecules, some quite long, some quite short. The short cDNAs are often derived from small amounts of degraded mRNA and probably do not contain the entire coding sequence of a gene. The presence of such cDNA inserts in an expression library can only dilute the signals generated by the full-length clones. With this procedure, you remove these smaller double-stranded cDNAs from the population prior to insertion into the expression vector. We generally use 500 bp, as our cut-off point in this fractionation process. First, I describe the conditions for running the gel and refer to a glass-bead purification procedure for the isolation of your fractionated cDNA. I then describe a procedure for purifying DNA from agarose and a variety of solutions. You will use this procedure to purify your fractionated cDNA from the gel.

1. Pour a 35-ml 2% SeaKem agarose gel, 10 cm × 6.5 cm, with an eight tooth, 1-mm-thick comb.
2. Add 2.5 μl of 6X LB to 12 μl of the ds cDNA.
3. Load the cDNA sample into the first lane, and load the size marker (pBR322–MspI digest) in the third lane.

4. Electrophorese, until the bromophenol-blue dye front migrates 2.5 cm.
5. Slice the gel through lane 2, and stain the marker lane with ethidium bromide. *Do not stain cDNA.*
6. Slice the marker lane at the desired low-molecular-weight cut-off for the cDNA to be inserted into the vector.
7. Align the cut agarose-marker-gel slice next to the cDNA containing gel slice and make the same cut in the cDNA-containing lane.
8. Trim excess agarose. Glass-bead purify as below.
9. Elute the ds cDNA, in 107.5 µl gd H_2O.
10. Glass-bead purification of DNA
    With this procedure, we exploit the observation that glass beads treated with nitric acid have a high affinity for DNA. Execution of this procedure results in the generation of very clean DNA that can function as substrate in a number of sensitive biochemical reactions. This procedure is a modification of Vogelstein and Gillespie (1979).
    a. Glass beads:
       1. Suspend 250 ml of silica-325 mesh (available from most ceramic stores) in water to produce a 450-ml slurry.
       2. Stir the slurry with a magnetic stirrer at room temperature for 1 hour.
       3. Allow the coarse silica to settle for one hour, then remove the fine silica suspension, transferring it to 40-ml Beckman polypropylene centrifuge bottles. Centrifuge the suspension in Beckman JA-20 rotor (or equilavent) for 10 minutes at 5000 × g.
       4. Resuspend the pellets in a total volume of 100 ml of H_2O.
       5. Add 100 ml of 11 N nitric acid, and bring close to boiling on a magnetic stirrer/heating plate. Peform this step in a chemical fume hood.
       6. Centrifuge the suspension at 5000 × g for 10 minutes.
       7. Resuspend the pelleted beads in 80 ml of water, then centrifuge at 5000 × g for 10 minutes. Repeat 4 more times.
       8. Store the glass beads as a 50% slurry in H_2O at 4°C. This procedure should yield about 20 g of glass beads.
    b. Sodium iodide solution: Add 90.8 g of NaI (Sigma, cat # S8379) and 1.5 g. of Na_2SO_3 (Sigma, cat. # S8018) to 100 ml of H_2O. Pass the solution through a 0.45-µm filter. Transfer the solution to a brown glass bottle, then add 0.5 g of Na_2SO_3. Store at −20°C. Just before use, allow the stock solution to equilibrate to room temperature. Pass an appropriate volume of solution through a 0.45-µm filter to remove any Na_2SO_3 crystals.
    c. Ethanol wash solution: 10 mM Tris-HCl, 100 mM NaCl, 50% (v/v) ethanol in H_2O, pH 7.5. Store at −20°C.
    Alternatively, glass beads and NaI and ethanol solutions can be obtained as a kit (GENECLEAN, BIO-101).

Method

1. Isolation of DNA from an agarose gel
   a. Fractionate the DNA sample in an agarose gel using Tris-acetate electrophoresis buffer (40 mM Tris-acetate, 2 mM EDTA, pH 7.8)
   b. Stain the gel in ethidium bromide, and excise a gel slice containing the desired DNA.
   c. Weigh the gel slice and grind it by passing it through a disposable 3-ml syringe (without a needle).

d. Add 2 ml of the sodium iodide solution to each gram of agarose.
e. Incubate the tube at 45–55°C until the agarose has dissolved.
f. Add 5–20 µl of the glass-bead slurry. (Twenty µl of the slurry is sufficient to isolate 15 µg of DNA.)
g. Allow the DNA to adsorb to the beads by incubating at 0–25°C for 10 minutes. Occasionally invert the tube to keep the beads in suspension.
h. Transfer the suspension to a microfuge tube and centrifuge at 14 Krpm in a microfuge for 1 minute to pellet the beads.
i. Remove the NaI solution.
j. Resuspend the pellet in 200 µl of cold (−20° to 0°C) ethanol wash solution. Centrifuge at 14 Krpm for 15 seconds. Repeat 3 more times.
k. Remove all of the ethanol wash solution and allow the beads to air-dry for 10–20 minutes at room temperature.
l. Resuspend the glass beads in 25–50 µl of H_2O or TE (10 mM Tris-HCl, 1 mM EDTA, pH 8.0).
m. Incubate the suspension at 45–55°C for 5 minutes.
n. Centrifuge at 14 Krpm for 2 minutes.
o. Recover the supernatant containing the eluted DNA.

2. Isolation of DNA from aqueous solutions
   a. In a microfuge tube, add up to 0.5 ml of a DNA solution (e.g., restriction-endonuclease reactions and ligations) and two volumes of the NaI solution.
   b. Add glass beads and proceed with steps (g–o) described above.

dCTP Tailing of Fractionated cDNA

Both the vector and cDNA are ready to be joined. All that remains is the C-tailing of the cDNA inserts. Previously, you optimized the tailing of the vector relative to the OPT-linker, which contains fifteen C's. Now you will optimize the tailing of the cDNA inserts relative to the tailed vector to obtain the ultimate combination of both, resulting in the highest number of bacterial transformants per microgram of annealed vector and insert. The C-tailing of the cDNAs is described here.

1. Spot 2.5 µl of fractionated cDNA on each of 3 GF/C filters. Dry. Count.
2. Determine the volume of the cDNA solution that contains 0.6 µg ds cDNA.
3. Prepare tailing reaction:

38.3 µl	10X TdT
5.7 µl	1.0 mM dCTP
X µl	cDNA (0.6 µg)
Y µl	H_2O
380.5 µl	Total

4. Prewarm the reaction to 37°C for 15 minutes. Add 360 units of TdT (around 3.0 µl)
5. At 45″, 60″, 75″, 90″, 105″, and 120″ time points, remove 60 µl and add to 468 µl of 1 mm EDTA.
6. For each time point, prepare an annealing reaction

88 µl	C-tailed cDNA time point
2 µl	G-tailed vector
10 µl	10X AB
100 µl	Total

7. Anneal and transform DH5α as before. Include a "no-cDNA" control to ascertain vector background, and a pUC-19 control to ascertain DH5α competence.
8. Scale up the bacterial transformation to employ solid-state amplification (see below), selecting the cDNA-tailing time point that gave rise to the greatest number of colonies.

Alkaline-Gel Analysis

Alkaline-gel analysis of your first- and second-strand reactions facilitates the analysis of the quality of your cDNAs. In this section, the radiolabeled secondary reactions you carried out in parallel with your first-strand syntheses and second-strand syntheses are examined. Electrophoresis of double-stranded DNAs on an alkaline gel denatures the molecules so that you can determine the lengths and relative abundances of both the first and second strands. In one reaction, you labeled the first strand; in another, you labeled the second. Now you will run both, next to each other, on an alkaline gel relative to radiolabeled single-stranded molecules of known molecular weight. Autoradiography of the resulting gel will reveal a smear of radioactivity. The molecular weights of both the first- and second-strand reactions will reflect the quality of that particular reaction. Under the best circumstances, the molecular sizes of both will be similar and both will be high, similar to the size of the mRNA with which the reactions were initiated.

1. Preparation of samples
   a. Pass first- and second-strand reactions through Bio-Gel P4 spin columns separately. Columns should be constructed with Isolab disposable columns and Bio-Gel P4 hydrated in TE.
   b. Ethanol precipitate the effluent using tRNA as a carrier.
   c. Radiolabel appropriate DNA size markers.
2. Pour a vertical 1.4% agarose gel in 1X alkaline gel buffer.
3. Prepare samples, 10,000–30,000 cpm, in 1X alkaline loading buffer.
4. Electrophorese in 1X alkaline electrophoresis buffer, with buffer circulation and with the power supply set for 75 mA, constant, until the bromophenol-blue dye front migrates 9 cm.
5. Dip pre-cut Gene Screen Plus membrane in H_2O, then immerse in alkaline electrophoresis buffer for 15 minutes.
6. Vacuum-dry at 40°C the agarose gel directly onto the membrane. Expose dried gel to Kodak XAR-5 film, with an intensifying screen, at −70°C.

Solid-State Amplification

The plasmid cDNA library-amplification procedure discussed earlier is described here. Bacterial colonies, suspended *in* a gel matrix, grow to a common size, apparently independent of their growth rate *on* agar or *in* solution. Within certain limits, the average size of the colonies suspended in the low-melting-temperature agarose medium depends on the total number of bacteria inoculated—the more bacteria, the smaller the colonies. The simplest interpretation of our observations is that the colonies become nutrient-limited in 3-dimensional space and thus the slower-growing bacterial colonies catch up with the faster-growing colonies, after which the foodstuffs of both types are depleted. This procedure is carried out as follows.

1. Prepare 2 liters of 2X LB + 0.3% SeaPrep agarose. Maintain at 37°C in waterbath.
2. Scale up the appropriate cDNA/vector-annealing reaction to generate up to 2.5×10^6 cfu/2000 ml.

3. Perform the appropriate number of standard bacterial transformations, then pool all the transformations after the 37°C incubation.
4. To the agarose solution, add ampicillin to a final concentration of 100 µg/ml, then add cells ($\leq 1.25 \times 10^6$ cfu), swirling gently to avoid foaming.
5. Aliquot the transformation mix into 50-ml polypropylene centrifuge tubes (Falcon 2098), pouring 25 ml into each tube. Place the tubes in an ice-water bath for 20–60 minutes. Incubate overnight at 37°C undisturbed.
6. Plate 100 µl of the cell/agarose suspension directly on LB-amp^{50} plates for titer determination.
7. Storage of library
    a. Post incubation, pour the colony-containing gels into a sterile 500-ml centrifuge bottle. Pellet the cells at 8 Krpm for 20 minutes at room temperature. The cells will pellet through the agarose.
    b. Resuspend cell pellets in a total volume of 100 ml of 12.5% glycerol in 2X LB. (Filter-sterilize the glycerol solution before adding to the cell pellets.)
    c. Aliquot into sterile Nunc tubes and store at −70°C.

TRANSFECTION AND SIB SELECTION

Transfection

Transfection is carried out by the DEAE dextran method of DNA transfer described in Part II in the section on DEAE Dextran transfection.

SIB Selection

Once a plasmid pool has been identified and verified as positive in an expression-cloning assay, one must execute a series of experimental manipulations that lead to the isolation of a single clone encoding the bioactivity of interest. This process is known as SIB selection.

We have experimentally tested a number of approaches, employing an IL-2 positive plasmid pool derived from one of our expression screens. We SIB-selected IL-2 from this pool by two methodologies, hybridization and, more important for the purposes of this report, bioactivity. The details of these experiments are discussed here.

SIB Selection Method Comparison

The following experiment was designed to determine the SIB selection process that would be used to identify bioactive clones from a library derived from induced peripheral blood leukocytes (PBLs). Two methods were examined. Method 1 involves direct single picking of clones to form pools for screening. This method requires three generations of pools/picks in order to obtain positive single-colony clones. Method 2 involves plating colonies onto filters, cutting the filters into 10 equal slices, each of which constituted a plasmid pool, and subsequently screening the pools. This method requires two generations of filter-slice screening, followed by two generations of colony picks, in order to obtain positive single-colony clones. The specific steps involved are depicted in the flow chart in Figure 7-4.

Method 1

Although this method is quicker, we noted that in the first generation (i.e., 15 pools single picks), only two positive pools were isolated. Subsequent SIB selection from one of the two positive pools yielded 6 positive SIB pools from 20 pools of 4 colonies, and from one of these pools, 7 positive single-pick clones out of 18 single-pick clones were obtained (see Figure 7-4).

Method 2

This involves a larger initial SIB pool size in the first two generations (i.e., 10 filter slices, with 40–60 colonies per filter slice in the first generation and around 20 colonies per slice in the second generation), followed by direct picking of 12 pools of 5 and then 18 single picks. Because of the initial larger pool size, more positive pools were obtained (6 with the first generation, 10 with the second generation, 8 positive pools from 12 pools of 5, and 4 positive single colony picks out of 18). With Method 2, the risk of losing the initial positive signal is reduced. Method 2 is the method we use routinely (see Figure 7-4). See Table 7-5 for IL-2 bioassay results from transfection of COS with plasmid pools scoring positive by hybridization.

SIB Selection for Method 2

1. Obtain the original frozen bacterial stock. Keep frozen. Scrape out a small portion of the frozen cells (around 2 mm^3) and elute in 2X LB. Return the frozen vial **immediately** to −70°C.

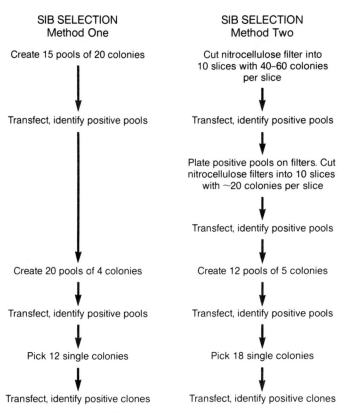

FIGURE 7-4 *Expression-cloning SIB selection flow chart.* The specific steps involved in SIB selection during expression cloning of genes described in the text are ordered here.

2. Perform around 10 1:10 dilutions in 5 ml of 2X LB + amp^{50}. Plate onto filters/LB-amp^{50} plates and incubate overnight. (Because of the variability of sample size from scraping the plate, plate all dilutions to avoid missing the correct/desired plating numbers. For instructions on plating see the section on hybridization analysis of plasmid cDNA libraries.) Grow overnight at 37°C.
3. Select a filter with about *300–600 colonies/filter*. Cut filter into 10 slices. Inoculate 5 ml of 2X LB + amp^{50} with the entire filter slice by stuffing the entire filter slice into the media with a sterile applicator. Grow overnight at 37°C with shaking at 225 rpm.

TABLE 7-5 IL-2 Positive Pool SIB-Selection Protocol Analysis (Method 2) Via Bioassay for IL-2 Positive Clones
Summary of IL-2 Biological Assay Results:

Plasmid name or Pool # Transfected	Pool Size	IL-2 Bioassay Titer U/ml
SR-α 296[a]	N/A	0
SR-α IL-2[b]	N/A	14,500
2	100	54
2A	40	326
2B	40	225
2A-4	20	723
2B-2	20	1041
2B-2(5)	5	4280
2B-2(5)-7	1	8870

[a] SR-α-296 is the expression vector lacking a cDNA insert.
[b] SR-α-IL-2 is the expression vector containing an IL-2 insert.

4. Perform minilysates on 1.5 ml of culture according to standard protocol. Freeze 1.5 ml of O/N culture supplemented with glycerol to 40% (v/v). Store at $-70°C$.
5. Perform COS transfection according to standard protocol. Assay transfection media or cells for signal in assay. Always include the **original** pool (DNA) in each transfection experiment.
6. If assay results are positive, repeat plating as before with frozen SIB pool. Plate in such a manner as to obtain approximately *200 colonies per filter*. Cut filters as before. Grow overnight. Process minilysate DNA. Freeze cells in glycerol at $-70°C$. Transfect second-generation pools *along with* the original pool and the *first-generation positive pool*.
7. Subject samples to assay. Dilute positive pools as before, but plate directly onto LB-amp^{50} plates. Grow overnight at $37°C$. Pick individual colonies and create 12 pools of 5. Grow overnight in 2X LB + amp^{50}. Freeze glycerol stocks. Process the minilysate DNA and transfect according to standard protocols. Include original pool, first-generation SIB pool, and second-generation SIB pool in transfection. Subject samples to assay.
8. Dilute positive pools. Plate directly onto LB-amp^{50} plates. Grow overnight at $37°C$.
9. Pick *18* individual clones. Grow overnight in 2X LB + amp^{50}.
10. Process minilysate DNA. Freeze aliquots at $-70°C$. Transfect according to standard protocols. Include previously positive pools (*original* and *SIBs*) in the transfection. Subject samples to assay.

8
SUBTRACTIVE HYBRIDIZATION

Subtractive hybridization has been successfully used to identify genes that are differentially expressed in different cell types. Subtractive-hybridization analysis can be broken down into two discrete methodologies, the generation of *subtracted probes* and the generation of *subtracted libraries.*

Subtracted probes are radiolabeled cDNAs derived from mRNAs from a cell type that expresses the genes of interest. These cDNAs are mixed with either mRNAs or a cDNA library (plasmid or phage) derived from a cell type that does *not* express the genes of interest. The mRNAs or cDNA-library DNAs are present in vast excess to the labeled cDNAs in the hybridization reaction, and as such are known as the hybridization "drivers" because they drive the kinetics of the hybridization reaction. The driver and the radiolabeled cDNAs are denatured by heat and, under carefully controlled conditions, allowed to renature together, in many cases forming hybrid, double-stranded nucleic-acid molecules. Under these conditions, nucleic-acid sequences present in the mRNAs common to both cell types will form double-stranded structures, and those not common to both cell types will remain single-stranded. These single-stranded molecules can be separated from the double-stranded molecules by a variety of methods, the most common of which is affinity chromatography to hydroxylapatite (HAP). After binding of the reannealed hybridization reaction to HAP, generally in a column, the single-stranded fraction can be eluted by increasing the salt concentration of the column buffer.

The radiolabeled component of this single-stranded eluate is representative of the sequences uniquely expressed in the cells from which it was derived. As such, this subtracted probe can be used to probe a cDNA library derived from its parental cell line. After colony hybridization and autoradiography of the hybridized filters, those colonies scoring positive in the autoradiogram can be picked from replica plates and represent cDNA clones of genes differentially expressed in the cell line of interest. These cDNA clones can be expanded and subjected to further analysis.

Subtracted libraries are generated by a similar methodology but with an additional technical twist. The techniques are virtually identical up to the point where the differentially expressed single-stranded fraction is eluted from the HAP (or a functionally similar) column. At this point, cDNA second strands are synthesized, and the resultant double-stranded cDNAs are inserted into either λ phages or plasmid vectors to create a library. This library will be composed of cDNA sequences that are differentially expressed in the cell type of interest.

At first glance, it may appear that the net results of probing a cDNA library with a subtracted probe and generating a subtracted library from subtracted cDNA are the same. However, there can be subtle but important differences in the properties of these two types of libraries. These differences are due to differences in the method of producing the cDNA first strand for the library. When a subtracted probe is used to probe a preexisting library, the quality of the inserted cDNAs found in the differentially expressed cDNA clones derived from that library will reflect the technology employed in the generation of the library in the first place. If most of the clones in the library are full-length, the differentially expressed clones will probably be full-length as well. However, with the generation of subtracted libraries, one pays a price for generating the differentially expressed cDNA clones directly from the single-stranded eluate of the HAP column: The process of HAP chromatography can be rather brutal to nucleic acids, and the quantitative recovery of high-molecular-weight cDNAs after HAP chromatography is quite difficult. Simply put, if the single-

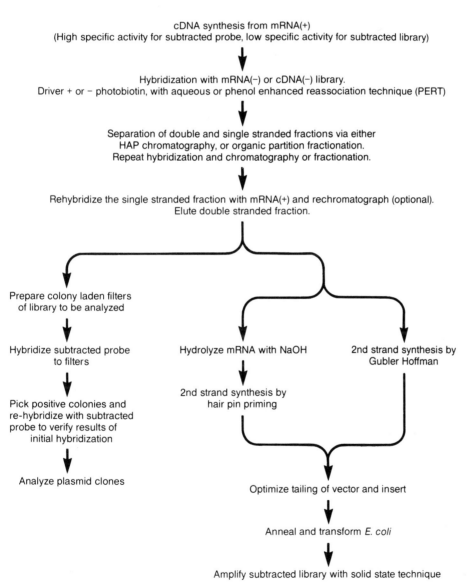

FIGURE 8-1 *Subtracted probe/subtracted library flow chart.* The specific steps involved in the generation of subtracted probes and subtracted libraries described in the text are compared and ordered here.

stranded eluate is used to drive the synthesis of the cDNA second strand, many higher-molecular-weight cDNAs will be truncated. The introduction of alternative chromatographic columns, such as those employing streptavidin/agarose, have not had a significant impact on this problem.

Included next are three different protocols for generating subtracted probes. All three have been used with a good deal of success, and each has advantages and disadvantages. First, I describe the classical approach, aqueous hybridization followed by chromatography on HAP (Davis et al., 1984; Davis, 1986). Next, I describe a fresh approach, aqueous hybridization with a photobiotinylated driver followed by organic extraction of the streptavidin/photobiotinylated driver/cDNA complexes (Sive and St. John, 1988). Third, I describe the classical approach with a twist, the phenol emulsion reassociation technique (PERT) of hybridization followed by chromatography on HAP (Kohne et al., 1977; Travis and Sutcliffe, 1988).

The advantage of aqueous hybridization with HAP chromatography is that the method is tried and true. The disadvantage is that aqueous hybridization to high R_0t values takes several days, and HAP chromatography can be a nuisance. Aqueous hybridization with a photobiotinylated driver takes just as long, but the substitution of an organic extraction step for the HAP column is a major plus. PERT hybridization occurs *three orders of magnitude faster* than its aqueous counterpart but you must still run a HAP column. In that all of these procedures work equally well, the selection of which to use is left to the reader.

A diagram of the steps involved in the generation of subtracted probes and subtracted libraries is shown in Figure 8-1.

GENERATION OF SUBTRACTED PROBES

The first step in the generation of either a subtracted probe or a subtracted library is the generation of a first-strand cDNA copy of the mRNA derived from the cell or tissue encoding the factors of interest. This mRNA is referred to as mRNA(+) or poly(A)$^+$RNA(+), and the corresponding cDNA is referred to as cDNA(+). mRNA, RNA, and cDNA used to subtract the (+) products are termed mRNA(−), poly(A)$^+$RNA(−), and cDNA(−). High-specific-activity cDNA is synthesized for the generation of a subtracted probe, whereas low-specific-activity cDNA is synthesized for the generation of subtracted libraries or as a template for future high-specific-activity labeling by the random-primer method. Although useless as a hybridization probe, low-specific-activity cDNA does not undergo significant radioautolysis, making it ideal for generating a subtracted library and as a template that can be stored for long periods of time and then used for the generation of randomly primed, high-specific-activity hybridization probes.

The cleanest subtracted probes are generated through a three-step process, with the third step optional. First, the cDNA(+) is subtracted with mRNA(−). Then the single-stranded cDNA(+) fraction of that hybridization is rehybridized to mRNA(−). The single-stranded cDNA(+) fraction of that hybridization may optionally be hybridized to mRNA(+), and the resultant double-stranded fraction is retained for use as a probe or to generate a library.

Alternatively, you can use PERT hybridization. First the cDNA(+) is subtracted with cDNA(−), and then the single-stranded fraction is used as a probe. Although the PERT technique lacks the multiple selection steps of the prior procedure, the tremendous acceleration of the hybridization reaction ensures that the hybridization reaction goes to completion. This difference can compensate for the lack of multiple selection steps. All of these procedures are described here.

Production of First-Strand cDNA, High- or Low-Specific-Activity

Materials

1. poly(A)$^+$ RNA,(+) 20–120 µg. (See Part III, in the section on analysis of RNA).
2. MoMLV-RT reverse transcriptase (BRL, cat. # 8025SA).
3. RNasin (Promega, cat. # N2111).
4. p(dT)$_{12-18}$ (Pharmacia, cat. # 27-7858-01).
5. p(dN)$_6$, random primers (Pharmacia, cat. # 27-2166-01).
6. [α-^{32}P]dATP (6000 Ci/mmole, New England Nuclear).
7. dATP, dCTP, dGTP, and dTTP, at 100 mM each (Pharmacia, cat. # 27-2050-01, 27-2060-01, 27-2070-01, and 27-080-01, respectively).
8. Siliconized micro tubes (see the section on library construction and amplification).
9. 5X reverse transcriptase buffer (5X RT): Prepare a solution containing 250 mM Tris-HCl, pH 8.3, 375 mM KCl, 15 mM MgCl$_2$, 10 mM DTT. Make this buffer fresh.
10. tRNA (yeast, Sigma, cat. # R0128).
11. Sephadex G-50 spin columns (Worthington Biochemicals, cat. # LS4404).
12. 7.5 M ammonium acetate
13. DEPC-treated H$_2$O (see the section on library construction and amplification).

14. Bromophenol blue (Sigma, cat. # B5525).
15. Dimethyldichlorosilane (silane, Sigma, cat. # D3879).
16. Dithiothreitol (DTT, Sigma, cat. # D9779).
17. 8-hydroxyquinoline (Sigma, cat. # H6878).
18. Xylene cyanole FF (Sigma, cat. # X0377).

Method: cDNA Synthesis

1. Synthesize high- or low-specific-activity cDNA
   a. Synthesis of high-specific-activity cDNA: Prepare this reaction on ice:

20.0 μl	mRNA(+) (1.0 mg/ml; heat denatured: 68°C, 5 minutes, 0°C, 5 minutes)
40.0 μl	5X RT buffer
5.0 μl	RNasin (40 U/μl)
10.0 μl	dNTPs (10 mM dCTP, dGTP, and dTTP; 0.4 mM dATP)
30.0 μl	[α-^{32}P]dATP (6000 Ci/mmole)
20.0 μl	p(dT)$_{12-18}$ (100 pmole/μl)
20.0 μl	MoMLV-RT (200 U/μl)
55.0 μl	H$_2$O, DEPC-treated
200.0 μl	Total

   Note: Random primers (2.5 mg/ml) can be substituted for oligo dT. Use 50 μl of random primers and 25 μl of H$_2$O.

   b. Synthesis of low-specific-activity cDNA: prepare this reaction on ice:

20.0 μl	mRNA(+) (1.0 mg/ml; heat-denatured: 68°C, 5 minutes, 0°C, 5 minutes)
40.0 μl	5X RT buffer
5.0 μl	RNasin (40 U/μl)
10.0 μl	dNTPs (10 mM each of dATP, dCTP, cGTP, and dTTP)
10.0 μl	[α-^{32}P]dATP (6000 Ci/mmole)
20.0 μl	p(dT)$_{12-18}$ (100 pmole/μl),
20.0 μl	MoMLV-RT (200 U/μl)
75.0 μl	H$_2$O, DEPC-treated
200.0 μl	Total

2. Incubate the reaction at 37°C for 1 hour to synthesize first strand.
3. Add 1 μl of the cDNA-synthesis reaction to 39 μl of water. Remove 5 μl aliquots for TCA-precipitation analysis.
4. Add 20 μl of 1 N NaOH to the 199 μl of the cDNA-synthesis reaction remaining. Incubate for 20 minutes at 68°C to hydrolyze the mRNA.
5. Add 20 μl of 1 M Tris, pH 7.4 to the cDNA-synthesis reaction.
6. Add 6 μl of the cDNA-synthesis reaction to 20 μg of tRNA for an unsubtracted-probe control. Store at -20°C

7. Perform a TCA-precipitation analysis on 5 μl aliquots from step 3 to assess incorporation. (See the section on library construction and amplification; cDNA synthesis. Remember that the concentration of dATP in the *high*-specific-activity-cDNA-synthesis reaction is 20 μM, and remember to set the counter to the ^{3}H channel for Cerenkov analysis.)
8. Removal of unincorporated dNTPs.
   a. Add 117 μl of 7.5 M ammonium acetate to the cDNA-synthesis reaction.
   b. Add 700 μl of 95% ethanol to the cDNA-synthesis reaction.
   c. Centrifuge at 4°C and 14 Krpm in a microfuge for 20 minutes.
   d. Resuspend the pellets in a total volume of 300 μl of TE. Add 150 μl of 7.5 M ammonium acetate and 1.0 ml of 95% ethanol.

e. Centrifuge at 4°C and 14 Krpm in a microfuge for 20 minutes.
   f. Repeat steps (d) and (e) twice.
   g. Resuspend the pellet in 100 µl of H_2O.

Subtracted cDNA Probe: Aqueous Hybridization and Hydroxylapatite (HAP) Chromatography

Materials

1. High-specific-activity cDNA made from 20 µg of poly(A)$^+$ RNA(+) (using p(dT)$_{12-18}$ or random primers).
2. Optional: 60 µg of the same poly(A)$^+$ RNA(+) used to synthesize the cDNA, for positive enrichment of cDNA-probe sequences.
3. 320 µg of poly(A)$^+$ RNA(−) (see Part III, the section on analysis of RNA) used for subtraction.
4. Mineral oil (Sigma, cat. # M5904).
5. Hydroxylapatite (Bio-Rad HTP, cat. # 130-0520).
6. Phosphate buffers (PBs), pH 6.8. (See text for concentration to use). Treat stock buffers with Chelex to remove metal cations.
7. Water-jacketed column; 15 cm × 1 cm I.D. (Bio-Rad, cat. # 737-6200).
8. Circulating water bath (Haake or Lauda).
9. Chelex 100 resin (Bio-Rad, cat. # 143-2832).
10. Sephadex G-50 spin columns (Worthington Biochemicals, cat. # LS4404).
11. Glass beads (GENECLEAN, BIO-101).
12. Sodium iodide solution: Add 90.8 g of NaI (Sigma, cat. # S8379) and 1.5 g of Na_2SO_3 (Sigma, cat. # S8018) to 100 ml of H_2O. Pass the solution through a 0.45-µm filter. Transfer the solution to a brown glass bottle, then add 0.5g of Na_2SO_3. Store at −20°C. Just before use, allow the stock solution to equilibrate to room temperature. Pass an appropriate volume of solution through a 0.45-µm filter to remove any Na_2SO_3 crystals.
13. Ethanol wash solution: 10 mM Tris-HCl, 100 mM NaCl, 50% (v/v) ethanol in H_2O, pH 7.5. Store at −20°C.
14. Sec-Butanol (Sigma, cat. # B1888).

Synthesis

Synthesize high-specific-activity cDNA from 20 µg of mRNA (see Part III, the section on analysis of RNA). Hydrolyze RNA. Recover cDNA by ethanol precipitation and resuspend in H_2O.

Hybridization

1. Add 250 µg of mRNA(−) to the 100 µl of cDNA so that the final volume is no greater than 300 µl. Add 0.5 volume of 7.5 M ammonium acetate and 2× the aqueous volume of 95% ethanol. Place in a dry ice/ethanol bath for 15 minutes, and centrifuge for 20 minutes at 4°C, 14 Krpm in a microfuge.
2. Remove all the ethanol by aspiration. Allow the pellet to dry at room temperature. Resuspend the pellet in 120 µl of DEPC-treated H_2O.
3. Prepare this hybridization mixture in a 700 µl microfuge tube:

120.0 µl	cDNA/mRNA(−)
20.0 µl	4 M phosphate buffer
10.0 µl	2% SDS
10.0 µl	0.1 M EDTA
160.0 µl	Total

4. Layer 100 µl of mineral oil onto the reaction.
5. Place the tube in a boiling water bath (in a boiling rack) for 5 minutes, then transfer to a water bath at 68°C. Allow the mixture to hybridize to an equivalent R_0t of between 1500–3000 moles·seconds/liter. When calculating R_0t, be sure to include the correction factor appropriate for the concentration of phosphate buffer (PB) in your hybridization reaction. For a table of correction values, see Britten et al., 1974. The calculation for our hybridization reaction is shown here. [(250 µg/160 µl) × (1 mole nucleotide/346 g) × 5.82 (correction factor for utilizing 0.5 M PB vs 0.12 M PB) × (15.9 hr) × (3600 sec/hr)] = 1500 mole·seconds/liter. A minimum of 15.9 hours of hybridization is required under the conditions above to attain the desired equivalent R_0t. (See Britten et al., 1974.)
6. After hybridization, transfer the aqueous contents into a tube containing 0.5 ml of 60 mM PB, 0.1% SDS.

Hydroxylapatite Chromatography

1. Treat stock phosphate buffers with Chelex to remove metal ions.
2. Prepare hydroxylapatite column.
   a. Hydrate 20 g of HAP in 60 mM PB, 0.1% SDS. Heat for 15–30 minutes at 100°C. Allow the HAP to settle at room temperature, and remove buffer. Resuspend HAP in 60 mM PB, 0.1% SDS.
   b. Pour a slurry of HAP into a water-jacketed column to generate a packed volume of 1.0 ml. Maintain the water jacket at 60°C.
   c. Wash the column with 10 ml of 60 mM PB, 0.1% SDS. (Use air pressure to provide a gentle but steady flow.)
3. Load the sample onto the HAP bed a few drops at a time, avoiding the sides of the column, and gently force sample through by air pressure.
4. Wash the column with 2–4 ml of 60 mM PB, 0.1% SDS.
5. Elute the single-stranded cDNA by loading successive 1.0-ml aliquots of PB at a concentration determined by the manufacturer of the HAP (Bio-Rad) to be the optimum concentration for the elution of single-stranded nucleic acids (ssPB, usually in the range of 120–160 mM PB, pH 6.8), 0.1% SDS.
6. Elute the double-stranded nucleic acids (for analytical purposes) by one of three methods:
   a. Leave the column valve open, and increase the water-bath temperature to 97–98°C. One or two minutes after reaching the melting temperature, turn the water bath down to 60°C and elute the material with 6 successive 1-ml aliquots of ssPB.
   b. Add 10 successive 1-ml aliquots of double-stranded elution buffer (dsPB), again at the concentration determined by the manufacturer of the HAP (usually greater than 360 mM PB, pH 6.8).
   c. Dissolve the HAP column bed with 5.0 ml of 2M nitric acid.
7. Count 5 µl of each elution aliquot to determine the hybridization efficiency. (For similar RNAs, the double-stranded fraction should comprise 90–95% of the total counts/minute.)

Recovering cDNA

1. Combine the two or three single-stranded peak fractions into a polypropylene tube. Add an equal volume of sec-butyl alcohol (sec-butanol). Invert several times. Remove the upper (butanol) layer. Re-extract with successive aliquots of butanol, until the aqueous volume has been concentrated to 100 µl.

2. Remove the salts by either of two methods.
   a. Pass the aqueous sample through a spin column of G-50 Sephadex. Add 10 μg of carrier tRNA. Ethanol-precipitate with 7.5 M ammonium acetate. Resuspend the pellet in 20 μl of H_2O.
   b. Glass-bead purify (GENECLEAN, BIO-101): Bring the cDNA volume up to 200 μl with H_2O. Add 400 μl of sodium iodide solution and 10 μl of glass beads. Incubate for 5–10 minutes at room temperature or on ice, with frequent mixing to keep the beads in suspension. Pellet in a microfuge for 1 minute at 14 Krpm, then remove supernatant. Wash the beads 4 times with 200 μl of ethanol-wash solution. After the final centrifugation, remove all of the wash solution and let the glass beads air-dry for 10–20 minutes. Resuspend the beads in 25 μl of H_2O and incubate for 5 minutes at 45–55°C. Centrifuge for 2 minutes at 14 Krpm in a microfuge, and save the aqueous supernatant.

Rehybridization and HAP Chromatography

The initial subtractive hybridization reaction and subsequent HAP chromatography removed a majority of labeled cDNA(+) sequences present in both mRNA(−) and cDNA(+) populations. To remove the remainder of the sequences common to mRNA(−) and cDNA(+), you will rehybridize the single-stranded peak from the initial hybridization to more mRNA(−). This procedure is presented here.

1. Scale the second subtractive hybridization down to 40 μl of total volume using 60 μg of mRNA(−), and hybridize to an equivalent R_0t of 1500–3000 moles·sec/liter as in the prior section describing the initial hybridization.
2. Pass the hybridization mixture through 0.5 ml packed volume of HAP. Elute single-stranded fraction as described above. Count 5 μl of each fraction. (Single-stranded fraction should comprise 40–80% of the total counts/minute loaded.)
3. Combine the two or three single-stranded peak fractions and add them directly to the filter hybridization solution. (Add 1.3–3.3 $\times$ 10^5 cpm/cm^2 filters.) For filter-hybridization procedure, see Part III, in the section on hybridization conditions.

Hybridization of cDNA(+) to mRNA(+) (optional)

The two hybridization reactions above removed those sequences common to both mRNA(−) and cDNA(+). This optional hybridization of mRNA(+) to cDNA(+) removes any non-specific DNA sequences that remain in the subtracted cDNA(+) not common to the mRNA(+), resulting in the generation of a very clean hybridization probe.

1. Concentrate the subtracted cDNA(+) by sec-butanol (see the section on recovering DNA), and remove salts by passing through Sephadex G-50 spin column.
2. Add 10 μg of tRNA carrier, and ethanol-precipitate the mix with 7.5 M ammonium acetate. Resuspend the pellet in 20 μl of H_2O.
3. Hybridize 60 μg of the mRNA(+)—RNA from the same stock used to synthesize the cDNA(+)—to the subtracted cDNA to an equivalent R_0t of 1500–3000 mole · sec/liter, in a total volume of 40 μl.
4. Pass the hybridization mixture through a 0.5-ml packed volume of HAP.
5. Elute and discard the single-stranded nucleic-acid fraction.
6. Elute the mRNA/cDNA hybrid with dsPB. Count 5 μl of each fraction.
7. Combine the two or three peak fractions and extract with sec-butanol (see the section on recovering DNA). Remove salts by passage over a G-50

Sephadex spin column. Add 10 µg of carrier tRNA. Ethanol-precipitate with 7.5 M ammonium acetate. Resuspend pellet in 200 µl of TE.

8. Add 20 µl of 1 N NaOH. Incubate for 20 minutes at 68°C. Add 20 µl of 1 M Tris, pH 7.4.
9. Add subtracted cDNA probe to the filter-hybridization solution (1.3–3.3 × 10^5 cpm/cm^2 filters). For filter-hybridization procedure, see Part III in the section on hybridization conditions.

Subtracted cDNA Probe: Aqueous Hybridization and Biotin/Phenol Extraction

This protocol is presented as an alternative to HAP chromatography. With it, you use photobiotinylated mRNA(−) (PAB-RNA(−)) to hybridize to cDNA(+). The cDNA(+)/PAB-RNA(−) hybrids are separated from the cDNA by incubation with streptavidin followed by phenol extraction. The streptavidin binds the biotin on the PAB-RNA(−)/cDNA(+) hybrids, and these complexes partition into the organic phase. The aqueous phase contains the unhybridized cDNA, which is then isolated and upon which a second subtraction is performed. This subtracted cDNA(+) probe may be used at this time, or it can be hybridized with the same stock of mRNA(+) used to synthesize the cDNA(+), which results in the removal of any radiolabeled low-molecular-weight, non-sequence-specific DNA from the subtracted cDNA(+) probe. The resulting cDNA/PAB-RNA hybrids are isolated, after which the recovered cDNA is used to probe a cDNA library.

Additional Materials and Equipment (see the section on the generation of subtracted probes for primary list)

1. Photoactivatable biotin (PAB, Vector Laboratories, cat. # SP–1000).
2. Streptavidin (BRL, cat. # 5532LB).
3. Sunlamp (Time-A-Time, model RSK6A, General Electric).
4. HB buffer: Prepare a solution of 50 mM Hepes, pH 7.6, 2 mM EDTA, 500 mM NaCl.
5. Phenol: chloroform at 1:1, saturated with TE, pH 8.0 (P:C).

Method

Synthesize high-specific-activity first-strand cDNA from 20 µg of mRNA; (see the section on the generation of subtracted probes), hydrolyze RNA. Recover cDNA by ethanol precipitation and resuspend in H_2O.

mRNA(−) Biotinylation

1. Mix 80 µg of mRNA(−) in 20 µl of DEPC-treated H_2O with 130 µg of PAB in 130 µl of H_2O for a total volume of 150 µl.
2. Irradiate the solution in three opened microfuge tubes (50 µl/tube) on ice with a sunlamp, 10 cm above the tubes, for 15 minutes. (Wear protective goggles.) After irradiation, combine the biotinylated samples.
3. Remove unreacted PAB by adding 150 µl of 100 mM Tris-HCl, pH 9.0, 1 mM EDTA, and extracting twice with 300 µl of butanol. Remove the butanol by extraction with chloroform. Adjust the volume to 150 µl with DEPC-treated H_2O. Precipitate the RNA by adding 17 µl of 1.0 M sodium acetate and 445 µl of 95% ethanol. The RNA pellet should be orange-brown.
4. Resuspend the mRNA pellet in 20 µl of DEPC-treated H_2O, and repeat steps 1–3 twice to increase the density of the biotin residues.

Hybridization

1. To prepare for hybridization mix the high-specific-activity cDNA you have just synthesized with 60 µg of PAB-RNA, and ethanol-precipitate in the presence of ammonium acetate (see the section on the generation of subtracted probes).
2. Perform hybridization as described, hybridizing to an equivalent R_0t of 1500–3000 moles·sec/liter (see the section on the generation of subtracted probes; hybridization) in a volume of 40 µl.

Separation of cDNA(+) from cDNA(+)/PAB-RNA(−) Hybrids

1. Transfer the hybridization reaction (40 µl) to a microfuge tube containing 800 µl of HB buffer.
2. Add 40 µg of streptavidin to the reaction mix and incubate at room temperature for 5 minutes.
3. Extract the reaction with 800 µl of phenol:chloroform (P:C).
4. Save the aqueous phase (which contains the unhybridized cDNA(+)), and add 200 µl of HB and 2 µl of 20% SDS to the interface and organic phase. Vortex and spin.
5. Pool the aqueous phases (1000 µl total). Add 40 µg of streptavidin and repeat P:C extraction and back-extraction (add 200 µl of HB and 2 µl of 20% SDS to the interface and organic phase).
6. Pool the aqueous phases (now 1200 µl). Add 40 µg of streptavidin and repeat P:C extraction and back-extraction (add 200 µl of HB and 2 µl of 2% SDS to the interface and organic phase).
7. Pool the aqueous phases (1400 µl). Chloroform-extract and ethanol-precipitate.

Rehybridization and Biotin/Phenol Extraction

The initial subtractive hybridization reaction and streptavidin/biotin organic reaction removed a majority of the labeled cDNA(+) sequences present in both PAB-RNA(−) and cDNA(+) populations. To remove the remainder of the sequences common to PAB-RNA(−) and cDNA(+), you will rehybridize the single-stranded fraction to more PAB-RNA(−), then streptavidin-extract. This procedure is presented here.

1. Scale the second subtractive hybridization down to 8 µl total volume hybridizing the cDNA from the previous step to 12 µg of PAB-RNA(−). Hybridize to an equivalent R_0t of 1500–3000 moles·sec/liter (see the sections on the generation of subtracted probes; hybridization).
2. Perform streptavidin-P:C extraction.
3. Remove the salts by passing through Sephadex G-50 spin column.
4. Add 10 µg of tRNA carrier and ethanol-precipitate with 7.5 M ammonium acetate. Resuspend the pellet in 20 µl of H_2O.
5. Add probe to the filter-hybridization solution (1.3–3.3×10^5 cpm/cm² filters). For filter-hybridization procedure, see Part III, the section on hybridization conditions.

Hybridization of cDNA to mRNA (optional)

The two hybridization reactions above removed those sequences common to both PAB-RNA(−) and cDNA(+). This optional hybridization of PAB-RNA(+) and cDNA(+) removes any non-specific DNA sequences that remain in the subtracted cDNA(+) not common to the PAB-RNA(+) and the cDNA(+), resulting in the generation of a very clean hybridization probe.

1. Photobiotinylate 60 μg of the mRNA(+) stock used to synthesize the cDNA(+) probe.
2. Hybridize 60 μg of the PAB-RNA(+) with the twice-subtracted cDNA(+) to an equivalent R_0t of 1500–3000 moles·sec/liter, in a total volume of 40 μl (see the sections on the generation of subtracted probes; hybridization).
3. Transfer the hybridization reaction (40 μl) to a microfuge tube containing 800 μl of HB buffer.
4. Add 40 μg of streptavidin and incubate at room temperature for 5 minutes.
5. Extract the reaction with 800 μl of P:C, *but save the interface and P:C layers!*
6. Add 800 μl of HB buffer to the *interface* and P:C layers. Vortex. Discard aqueous layer. Repeat.
7. Transfer the *interface* to 800 μl of TE. Heat the mixture for 10 minutes at 95°C (this denatures the hybrids).
8. Add 40 μg of streptavidin and incubate at room temperature for 5 minutes.
9. Extract the reaction with 800 μl of P:C. Save the aqueous phase (which contains the released cDNA), and add to the filter-hybridization solution (1.3–3.3 × 10^5 cpm/cm² filters). For filter-hybridization procedure, see Part III, the section on hybridization conditions.

Subtracted cDNA(+) Probe: Phenol-Emulsion Reassociation-Technique (PERT) and HAP Chromatography

This procedure employs PERT during the hybridization of the cDNA(−) driver to the radiolabeled cDNA(+). With PERT, the hybridization reaction is assembled in an aqueous solution, after which phenol is added. The emulsion is continuously agitated at room temperature during the hybridization. For an unknown reason, the formation of this emulsion increases the rate of formation of DNA/DNA hybrids three orders of magnitude relative to aqueous hybridization. As a result, a hybridization reaction with an equivalent amount of cDNA(+) and driver cDNA(−) can be carried out in a volume ten times greater than an aqueous reaction, and the reaction will still occur 100 times faster. This is the real strength of this procedure. *This method does not work for the formation of RNA/DNA hybrids.* With this procedure, you do not rehybridize the cDNA(+) with the cDNA(−), nor do you rehybridize your subtracted cDNA(+) to mRNA(+), for obvious reasons. The single-stranded fraction remaining after PERT hybridization is still fractionated on HAP.

Additional Materials (see the section on the generation of subtracted probes for primary list)

1. Source of DNA for subtraction: Plasmid cDNA(−) library.
2. Hydroxylapatite (Bio-Rad HTP, cat. # 130-0520).
3. Phenol (Sigma, cat. # P1037).
4. NaSCN (Sigma, cat. # S7757).
5. Phosphate buffers (PBs), pH 6.8 (Chelex-treated).
6. Vortex shaker (Brinkmann 22363107, Baxter, cat. # C3514-31).
7. Water-jacketed column; 15 cm × 1 cm (I.D.) (Bio-Rad, cat. # 737-6200).
8. Circulating water bath (Haake or Lauda).
9. Sephadex G-50 spin columns (Worthington Biochemicals, cat. # LS4404).
10. Chelex 100 Resin (Bio-Rad, cat. # 143-2832).
11. Phenol: prepare a solution of phenol:chloroform:isoamyl alcohol(50:48:2).

Method

Synthesize high-specific-activity cDNA(+) from 20 µg of mRNA(+). Hydrolyze the RNA. Recover the cDNA(+) by ethanol precipitation and resuspend in 234 µl H$_2$O (see the section on the generation of subtracted probes).

Preparation of cDNA(−) Used for Subtraction

1. Digest 800 µg of the plasmid cDNA(−) library with enzyme(s) that will excise cDNA(−) inserts (in this example, the vector is about 3000 bp). Perform digestion at a DNA concentration of 200 µg/500 µl. Adjust volume of 800 µg digestion to 3.2 ml with H$_2$O.
2. To each of 10 tubes, add:

23.4 µl	cDNA(+) probe
320.0 µl	digested cDNA(−) library
343.4 µl	Total

3. Phenol-extract each tube once with 350 µl of phenol. Add 170 µl of 7.5 M ammonium acetate and 1000 µl of ethanol. Spin at 14 Krpm for 5 minutes.
4. Resuspend each pellet in 133 µl of TE.
5. Combine the pellets into four tubes of 333 µl. Add 170 µl of 7.5 M ammonium acetate, and ethanol-precipitate again.
6. Resuspend each of the four pellets in 200 µl of H$_2$O and combine.
7. Heat at 100°C for 5 minutes, then place on ice for 5 minutes (this denatures the DNA and prevents renaturation of the DNA prior to hybridization).
8. Dispense into 8 tubes:

100 µl	cDNA(+) probe/DNA(−) library
120 µl	5 M NaSCN
36 µl	90% phenol (phenol:H$_2$O,9:1(w/v))
144 µl	H$_2$O
400 µl	Total

9. Agitate on a vortex shaker for 48 hours at room temperature.

Preparation of Hydroxylapatite Column

1. Hydrate the HAP in 60 mM PB, 0.1% SDS. Heat for 15–30 minutes at 100°C. Allow the HAP to settle at room temperature and remove buffer. Resuspend in 60 mM PB, 0.1% SDS.
2. Pour a slurry of HAP into a water-jacketed column to generate a packed volume of 3.0 ml. Maintain the temperature of the water jacket at 60°C.

Preparation of Hybridized cDNA for HAP Chromatography

1. Chloroform-extract PERT reactions.
2. Add 1.0 ml of 95% ethanol to the aqueous layer, and precipitate as before.
3. Resuspend all the pellets in a total volume of 1.5 ml of 60 mM PB, 0.1% SDS.

HAP Chromatography

1. Equilibrate the HAP (at 60°C) with 10 ml of 60 mM PB, 0.1% SDS.
2. Load the 1.5 ml of DNA sample. Wash with 9.0 ml of 60 mM PB, 0.1% SDS.
3. Elute the ss cDNA with 10 successive 1-ml additions of ssPB, 0.1% SDS. Count 5 µl of each fraction.

4. To determine the ds cDNA fraction, dissolve HAP with 5.0 ml of 2 M nitric acid.
5. Pool the two or three single-stranded peak fractions. Add them directly to the filter-hybridization solution (1.3–3.3×10^5 cpm/cm^2 filter area). For filter-hybridization procedure, see Part III, the section on hybridization conditions.

GENERATION OF SUBTRACTED LIBRARIES

In this section, I describe two methods for the construction of subtracted cDNA libraries. These two methods are a combination of two of the methods for generating subtracted probes described in the previous section and the procedure for the construction of cDNA expression libraries in the section on library construction and amplification. The first method I describe employs aqueous hybridization followed by HAP chromatography to generate the subtracted cDNA (Davis et al., 1984; Davis, 1986). The next method employs aqueous hybridization followed by an organic extraction of the streptavidin/biotinylated driver/cDNA complexes (Sive and St. John, 1988; Sive et al., 1989).

The advantage of aqueous hybridization and HAP chromatography is that the method is tried and true. The disadvantage is that aqueous hybridization to high R_0t values takes several days, and HAP chromatography can be a nuisance. Aqueous hybridization with a photobiotinylated driver takes just as long, but the substitution of an organic extraction step for the HAP column is a major plus. In that both of these procedures work equally well, the selection of which to use is left to the reader. A diagram of the steps involved in the generation of subtracted libraries is shown in Figure 8-1.

Materials and Equipment

1. Nucleic acids
   a. 60 μg mRNA(+) for the synthesis of cDNA library. (See Part III, in the section on the analysis of RNA).
   b. 320 μg mRNA(−) for subtracting cDNA. (See Part III, in the section on the analysis of RNA).
   c. dGTP-tailed plasmid vector (see the section on library construction and amplification).
2. Enzymes
   a. DNA polymerase I (New England Biolabs, cat. # 209).
   b. MoMLV reverse transcriptase (BRL, cat. # 8025SA).
   c. RNasin (Promega, cat. # N2111).
   d. Terminal transferase (Ratliff Biochemicals).
   e. Mung-bean nuclease (New England Biolabs, cat. # 250).
   f. RNase H (BRL, cat. # 8041SA).
   g. AMV reverse transcriptase (Pharmacia, cat. # 27-0922-01).
   h. Klenow fragment of *E. coli* pol I (United States Biochemical, cat. # 70015).
   i. *E. coli* DNA Ligase (New England Biolabs, cat. # 205).
3. Chemicals
   a. 100 mM dATP, 100 mM dCTP, 100 mM dGTP, and 100 mM dTTP (Pharmacia, cat. # 27-2050-01, 27-2060-01, 27-2070-01, and 27-2080-01, respectively).
   b. p(dT)$_{12-18}$ (Pharmacia, cat. # 27-7858-01).
   c. Cacodylic acid, free acid (Sigma, cat. # C0125).
   d. tRNA (yeast, Sigma, cat. # R0128).
   e. β-NAD, grade 5 (Sigma, cat. # N1636).
4. Isotope [α−^{32}P] dCTP (400 Ci/mmole, NEN, cat. # NEG-013S).
5. Materials
   a. Micro tubes (Sarstedt).
   b. Glass beads (GENECLEAN, BIO-101).
   c. Seakem GTG agarose (FMC Bioproducts, cat. # 50072).
   d. SeaPrep agarose (FMC Bioproducts, cat. # 50302).
   e. Max Efficiency, DH5α-competent cells (BRL, cat. # 8258SA).
   f. S.O.C. medium (BRL, cat. # 5544SA).
   g. Water-jacketed column (Bio-Rad, cat. # 737-6200).
   h. Circulating water bath (Haake or Lauda).

i. Hydroxylapatite (HTP, Bio-Rad, cat. # 130-0520).
j. Chelex 100 resin (Bio-Rad, cat. # 143-2832).
k. Sephadex G-50 spin columns (Worthington Biochemicals, cat. # LS4404).
l. Siliconized micro tubes (see the section on library construction and amplification; cDNA synthesis).
m. LB-amp[50] plates (see the section on library construction and amplification).
n. Mineral oil (Sigma, cat. # M5904).

6. Solutions and buffers
   a. 10X annealing buffer (10X AB): Prepare a solution containing 0.1 M Tris-HCl, pH 7.6, 1.0 M NaCl, 10 mM EDTA.
   b. 10X terminal transferase buffer (10X TdT): Add 13.8 g of cacodylic acid and 3.0 g of Tris-base to 60 ml of H_2O. Adjust pH to 7.6 by the slow addition of solid KOH with constant stirring. Bring to final volume of 88 ml with H_2O. Chill to 0°C, then add 2.0 ml of 0.1 M DTT. Add 10 ml of 0.1 M $MnCl_2$ drop-wise with constant stirring. Store at 4°C in small tightly capped aliquots.
   c. 5X reverse transcriptase buffer (5X RTB): Prepare a solution containing 250 mM Tris-HCl, pH 8.3, 375 mM KCl, 15 mM $MgCl_2$, 10 mM DTT. Make this buffer fresh.
   d. 10X mung-bean nuclease buffer (10X MNB): Prepare a solution containing 300 mM sodium acetate, pH 5.0, 500 mM NaCl, 10 mM $ZnCl_2$, 50% glycerol.
   e. 5X second strand buffer (5X SSB): Prepare a solution containing 100 mM Tris-HCl, pH 7.5, 50 mM $MgCl_2$, 50 mM $(NH_4)_2SO_4$, 0.5 M KCl, 5 µg/ml BSA.
   f. DEPC-treated H_2O (see the section on library construction and amplification).
   g. 7.5 M ammonium acetate.
   h. Phenol: prepare a solution of phenol:chloroform:isoamyl alcohol (50:48:2).
   i. TE: prepare a solution containing 10 mM Tris-HCl, 1 mM EDTA, pH 8.0.
   j. 10 mM dNTPs: Combine 100 mM stocks of dATP, dCTP, dGTP, and dTTP and sterile H_2O to a stock 10 mM for each dNTP. Store in aliquots at −70°C.
   k. 5X AMV reverse transcriptase buffer (5XAMVB): 250 mM Tris-HCl, pH 8.3, 100 mM KCl, 50 mM $MgCl_2$, 25 mM DTT. Make this buffer fresh.
   l. Phospate buffer (PBs), pH 6.8 (see text for concentration to use). Treat stock buffers with Chelex to remove metal ions.

First-Strand Synthesis

1. Place 52 µg of mRNA(+) into a siliconized microfuge tube. Incubate at 68°C for 5 minutes, then place immediately on ice for at least 5 minutes.
2. Add in this order to the tube on ice:

267.0 µl	DEPC-treated H_2O containing 52 µg of mRNA(+)
14.0 µl	RNasin (40 U/µl)
104.0 µl	5X RTB
26.0 µl	10 mM dNTPs
52.0 µl	p(dT)$_{12-18}$ (100 pmole/µl)
5.0 µl	[α-^{32}P]dCTP (400–600 Ci/mmole, 10 µCi/µl)
52.0 µl	MoLV-RT (200 U/µl)
520.0 µl	Total

3. Upon completion of preparation of the first strand reaction in step 2, to a second tube on ice add:

2.0 µl	[α-^{32}P]dCTP
20.0 µl	first-strand reaction
22.0 µl	Total

4. Incubate both tubes at 37°C for 60 minutes. Place on ice.
5. To the second tube, add 5 µl of 0.5 M EDTA and 23 µl of H$_2$O. Store at −20°C for TCA and alkaline-gel analyses (see the section on library construction and amplification).
6. Base hydrolysis
   a. Add 50 µl of 1 N NaOH to the 500 µl reaction. Incubate for 20 minutes at 68°C.
   b. Add 50 µl of 1 M Tris, pH 7.4.
7. Removal of unincorporated dNTPs
   a. Add 300 µl of 7.5 M ammonium acetate.
   b. Dispense 450 µl into two tubes, then add 1.0 ml of 95% ethanol to each tube.
   c. Centrifuge at 4°C at 14 Krpm in a microfuge for 20 minutes.
   d. Resuspend the pellets in a total volume of 300 µl of TE. Add 150 µl of 7.5 M ammonium acetate and 1.0 ml of 95% ethanol.
   e. Centrifuge at 4°C at 14 Krpm in a microfuge for 20 minutes.
   f. Repeat steps (d) and (e) twice.
   g. Resuspend the pellet in 100 µl of DEPC-treated H$_2$O.

Hybridization

1. Add 250 µg of mRNA(−) to the 100 µl of cDNA so that the final volume is no greater than 300 µl. Add 0.5 volume of ammonium acetate and 3.0 volumes of 95% ethanol. Place in a dry-ice/ethanol bath for 15 minutes, and centrifuge for 20 minutes at 4°C at 14 Krpm.
2. Remove all of the ethanol by aspiration. Allow the pellet to dry at room temperature. Resuspend the pellet in 120 µl of DEPC-treated H$_2$O.
3. Prepare this hybridization mixture in a 700-µl autoclaved microfuge tube:

120.0 µl	cDNA/mRNA(−)
20.0 µl	4 M phosphate buffer
10.0 µl	2% SDS
10.0 µl	0.1 M EDTA
160.0 µl	Total

4. Add 100 µl of mineral oil.
5. Place the tube in a boiling water bath (in a boiling rack) for 5 minutes, then transfer to a water bath at 68°C. Allow the reaction to incubate to an equivalent R_0t of between 1500–3000 moles·seconds/liter. When calculating R_0t, be sure to include the correction factor appropriate for the concentration of phosphate buffer (PB) in your hybridization reaction. For a table of correction values, see Britten et al., 1974. The calculation for our hybridization reaction is shown here: [(250 µg/160 µl) × (1 mole nucleotide/346 g) × 5.82 (correction factor for utilizing 0.5 M PB vs 0.12 M PB) × (15.9 hr) × (3600 sec/hr)] = 1500 moles·seconds/liter. A minimum of 15.9 hours incubation is needed under the conditions above to attain the desired equivalent R_0t. (See Britten et al., 1974).
6. After hybridization, transfer aqueous contents into a tube containing 0.5 ml of 60 mM PB, 0.1% SDS.

Hydroxylapatite Chromatography

1. Preparation of hydroxylapatite column.
   a. Hydrate the HAP in 60 mM PB, 0.1% SDS. Heat for 15–30 minutes at 100°C. Allow the HAP to settle at room temperature and remove buffer. Resuspend in 60 mM PB, 0.1% SDS.
   b. Pour a slurry of HAP into a water-jacketed column to generate a packed volume of 1.0 ml. Maintain water jacket at 60°C.
   c. Wash the column with 10 ml of 60 mM PB, 0.1% SDS. (Use air pressure to provide a gentle but steady flow.)
2. Load the sample onto the HAP bed a few drops at a time, avoiding the sides of the column, and gently force the sample through by air pressure.
3. Wash the column with 2–4 ml of 60 mM PB, 0.1% SDS.
4. Elute the single-stranded cDNA by loading successive 1-ml aliquots of PB at a concentration determined by the manufacturer (Bio-Rad) to be the optimum concentration for the elution of single-stranded nucleic acids (usually in the range of 120–160 mM PB, pH 6.8).
5. Elute the double-stranded nucleic acids (for analytical purposes) by either of two methods:
   a. Leave the column valve open. Increase water-bath temperature to 97°–98°C. One or two minutes after reaching the melting temperature, turn water bath down to 60°C and elute the material with six successive 1-ml aliquots of single-stranded elution buffer.
   b. Add 10 successive 1-ml aliquots of double-stranded elution buffer, again at the PB concentration determined by the manufacturer (usually greater than 360 mM PB, pH 6.8).
6. Count 5 μl of each elution aliquot to determine the hybridization efficiency (for similar RNAs, the double-stranded fractions should comprise 90–95% of the total counts/minute).

Recovering cDNA

1. Combine the two or three single-stranded peak fractions into a polypropylene tube. Add an equal volume of sec-butyl alcohol (sec-butanol). Invert several times, and remove the upper (butanol) layer. Re-extract with successive aliquots of butanol until the aqueous volume has been concentrated to 100 μl.
2. Remove the salts by passage of the aqueous sample through a Sephadex G-50 spin column.
3. Add 10 μg of carrier tRNA. Ethanol-precipitate with 7.5M ammonium acetate. Resuspend the pellet in 20 μl of DEPC-treated H_2O.

Rehybridization and HAP Chromatography

1. Scale the second subtractive hybridization down to 40 μl total volume using 60 μg of mRNA(−), and hybridize to an equivalent R_0t of 1500–3000 moles·sec/liter (see the sections on the generation of subtracted probes; hybridization).
2. Pass the hybridization mixture through 0.5 ml packed volume of HAP (single-stranded fraction should be in the range of 40–80%).
3. Concentrate the subtracted cDNA by sec-butanol (see the section on recovering cDNA), and remove the salts by passing through Sephadex G-50 spin column.
4. Add 10 μg of tRNA carrier and ethanol-precipitate with ammonium acetate. Resuspend pellet in 20 μl of H_2O.

Second-Strand cDNA Synthesis (Two Methods)
Method I: Hairpin Priming of Second-Strand Synthesis

A characteristic of the molecular structure of single-stranded cDNAs is the presence of a few nucleotide residues at their extreme 3' ends that have folded back upon the single-stranded molecule to form a hairpin structure. The 3' end of this hairpin can serve as a primer for reverse transcriptase to synthesize a second strand, thus creating a double-stranded molecule whose strands are connected by a hairpin. To permit the insertion of this molecule into a plasmid vector, this hairpin must be digested with mung-bean nuclease. This digestion opens the hairpin structure and converts the double-stranded cDNA into a form that can function as a substrate for C-tailing at both ends in preparation for insertion into a G-tailed plasmid vector.

1. Prepare this reaction on ice:

20.0 μl	subtracted cDNA
10.0 μl	5X AMVB
2.5 μl	10 mM dNTPs
5.0 μl	[α-^{32}P] dCTP (10 μCi/μl)
10.0 μl	H$_2$O
2.5 μl	AMV-RT reverse transcriptase (200 U/μl)
50.0 μl	Total

2. Incubate for 1 hour at 37°C.
3. Add 10 units of Klenow. Incubate for 1 hour at 37°C.
4. Stop the reaction by adding 2 μl of 0.5 M EDTA.
5. Add 1 μl of the second-strand reaction to 49 μl of TE. TCA-precipitate 5 μl aliquots and count to determine the second-strand-synthesis yield.
6. Phenol-extract, ethanol-precipitate, and resuspend in 20 μl of H$_2$O.
7. Removing the hairpin with mung-bean nuclease:
    a. Prepare this reaction:

20.0 μl	ds cDNA
5.0 μl	10X MNB
24.5 μl	H$_2$O
0.5 μl	mung-bean nuclease (5 U/μl)
50.0 μl	Total

    b. Incubate reaction for 30 minutes at room temperature.
8. Purifying cDNA by glass beads (GENECLEAN, BIO-101):
    a. Add 100 μl of NaI solution, then add 5 μl of glass-bead slurry.
    b. Incubate at 0°C for 5–10 minutes, inverting the tube to keep glass beads in suspension.
    c. Pellet in microfuge for 1 minute at 14 Krpm, then remove supernatant.
    d. Wash the pellet with 200 μl of cold ethanol-wash solution. Pellet. Remove supernatant. Repeat three more times.
    e. Remove all residual ethanol. Air-dry for 10 minutes. Resuspend the pellet in 40 μl of H$_2$O (not TE!).
    f. Incubate for 5 minutes at 45–55°C.
    g. Centrifuge for 2 minutes at 14 Krpm in a microfuge. Save the supernatant, count 1 μl, and determine the concentration of the ds cDNA.

Method II: Gubler-Hoffmann

A highly efficient, alternative approach to the synthesis of a cDNA second strand is that developed by Gubler and Hoffman (1983). With this method, the mRNA component of a cDNA/mRNA hybrid molecule is nicked with RNase H, after which the nicks are employed by DNA polymerase I to prime the synthesis of the cDNA second strand. Once synthesis of the second strand is completed, the double-stranded molecule can serve as an appropriate template for C-tailing in preparation for insertion into a G-tailed plasmid vector.

I begin this procedure with the single-stranded fraction of subtracted cDNA(+) you have eluted from the HAP column. Because the Gubler/Hoffman method does not work on a single-stranded cDNA, you must first hybridize your subtracted cDNA(+) with mRNA(+) to create mRNA/cDNA hybrids. These hybrid molecules are appropriate substrates for the Gubler/Hoffman reaction. The procedures for hybrid formation and second-strand synthesis are provided here.

1. Hybridization of cDNA(+) to mRNA(+)
   a. Hybridize 60 µg of mRNA(+) (RNA from the same stock used to synthesize the cDNA) to the subtracted cDNA to an equivalent R_0t of 1500–3000 moles·sec/liter (see the sections on the generation of subtracted probe hybridization), in a total volume of 40 µl.
   b. Pass the hybridization mixture through a 0.5-ml packed volume of HAP.
   c. Elute and discard the single-stranded nucleic-acid fraction.
   d. Elute the mRNA(+)/cDNA(+) hybrid with the concentration of PB recommended by the manufacturer to elute double-stranded nucleic acids (usually in the range of 0.4–0.5 M PB).
   e. Combine the two or three peak fractions and concentrate with sec-butanol (see the section on recovering cDNA). Remove salts by passage over a Sephadex G-50 spin column.
   f. Add 10 µg of carrier tRNA. Ethanol precipitate with 7.5 M ammonium acetate. Resuspend the pellet in 20 µl of H_2O.
2. Second-strand synthesis
   a. Prepare this reaction:

20.0 µl	mRNA(+)/cDNA(+) hybrid
8.0 µl	5X SSB
1.0 µl	RNase H (2 U/µl)
1.0 µl	DNA Pol I (10 U/µl)
1.3 µl	10 mM dNTPs
1.0 µl	[α-^{32}P]dCTP (10 µCi/µl)
1.0 µl	6.0mM β-NAD
1.0 µl	E.coli ligase (1U/µl)
5.7 µl	H_2O
40.0 µl	Total

   b. Incubate at 16°C for 2 hours.
   c. Remove 1 µl of the reaction, then add to 14 µl of TE. Spot 5 µl on GF filters and determine the second-strand yield by TCA-precipitation analysis and counting.
   d. Add 10 µl of TE to the reaction, then glass-bead purify the double-stranded cDNA (see the section on recovering cDNA).
   e. Determine the amount of ds cDNA (see the section on library construction and amplification; cDNA synthesis).

dCTP-Tailing of ds cDNA with Terminal Transferase

1. Obtain 0.1 µg of ds cDNA for the production of the library and 0.1 µg of ds cDNA to optimize the terminal-transferase-tailing reaction.

2. Optimizing the tailing of cDNA
   a. Prepare this reaction:

6.5 µl	10X TdT
1.0 µl	1 mM dCTP
X µl	ds cDNA (0.1 µg)
Y µl	H$_2$O
64.5 µl	Total

   b. Prewarm the tube at 37°C for 15 minutes.
   c. Prepare tubes labeled 45″, 60″, 75″, 90″, 105″, and 120″, and add 78 µl of 1 mM EDTA stop solution to each tube.
   d. Add 60 units of terminal transferase ($\simeq$ 0.5µl) to the reaction cocktail. Incubate at 37°C. Remove 10 µl at each time point and pipette directly into the stop solution.
   e. For each time point, mix:

44.0 µl	C-tailed cDNA time point
1.0 µl	G-tailed vector
5.0 µl	10X AB
50.0 µl	Total

   f. Anneal and transform DH5α (see the section on library construction and amplification). The optimal time point for dCTP-tailing of cDNA is that time point that results in the highest number of bacterial transformants.
   g. Prepare the same cocktail as in step (a). Prewarm the reaction as before, then add 60 units of terminal transferase. Incubate at 37°C for the determined optimal time period, and immediately terminate the reaction by adding 500 µl of 1 mM EDTA.
   h. Perform a large-scale transformation and solid-state amplification (see the section on library construction and amplification).

Size Fractionation of cDNA Library

1. Amplify the cDNA library in 1.0 liter of 2X LB (100 µg/ml amp) + 0.3% SeaPrep agarose (see the section on library construction and amplification).
2. Pellet the bacteria and extract the plasmid DNA.
3. Fractionate the cDNA-library plasmid DNA on a 0.8% SeaKem agarose gel, and cut the gel at the appropriate molecular-weight cutoff (e.g., a molecular weight of approximately 250–500 b.p.) greater than the vector monomer to eliminate plasmid clones containing very small, potentially truncated, cDNA inserts.
4. Glass-bead purify the fractionated plasmid cDNA. Transform competent bacteria. Suspend in 1.0 liter of 2X LB (100µg/ml amp) + 0.3% SeaPrep agarose (see the section on library construction and amplification).

Constructing a Subtracted cDNA Library: Biotin/Phenol Extraction

This protocol utilizes photobiotinylated mRNA(−) (PAB-RNA) to hybridize to cDNA(+). The cDNA/PAB-RNA hybrids are separated from the cDNA by incubation with streptavidin, then phenol extracted. The aqueous phase contains the unhybridized cDNA(+), which is then isolated, after which a second subtraction is performed. The recovered cDNA is made double-stranded, tailed, and cloned as described in the section on library construction and amplification.

Additional Materials and Equipment (see the section on the generation of subtracted libraries for primary list)

1. Photoactivatible biotin (PAB, Vector Laboratories, cat. # SP-1000)
2. Streptavidin (BRL, cat. # 5532LB; Vector Laboratories, cat. # SA-5000).
3. Sunlamp (Time-A-Time, model RSK6A, General Electric).
4. HB buffer: 50 mM Hepes, pH 7.6, 2 mM EDTA, 500 mM NaCl.
5. Phenol:chloroform at 1:1, saturated with TE, pH 8.0.

mRNA(−) Biotinylation

1. Mix 80 μg of mRNA(−) in 20 μl of H_2O with 130 μg of PAB in 130 μl for a total volume of 150 μl.
2. Irradiate the solution in three open microfuge tubes (50 μl/tube) on ice with a sunlamp (10 cm above the tubes) for 15 minutes. (Wear protective goggles.) After irradiation, combine the biotinylated samples.
3. Remove unreacted PAB by adding 150 μl of 100 mM Tris, pH 9.0, 1 mM EDTA, and extracting twice with 300 μl of sec-butanol. Remove the butanol by extracting twice with chloroform. Adjust the volume to 150 μl with DEPC-treated H_2O. Precipitate the RNA by adding 17 μl of 1.0 M sodium acetate and 445 μl of 95% ethanol. The RNA pellet should be orange-brown.
4. Resuspend the mRNA(−) pellet in 20 μl of DEPC-treated H_2O, and repeat steps 1–3 twice to increase the density of the biotin residues.

Hybridization

1. Synthesize first-strand cDNA(+) and prepare for hybridization (see the section on subtractive hybridization; cDNA synthesis). Mix with 60 μg PAB-RNA(−). Ethanol-precipitate with 7.5 M ammonium acetate.
2. Perform hybridization (see the section on subtractive hybridization; hybridization).

Separation of cDNA(+) from cDNA(+)/PAB-RNA(−) Hybrids

1. Transfer the hybridization reaction (40 μl) to a microfuge tube containing 800 μl of HB buffer.
2. Add 40 μg of streptavidin to the reaction mix and incubate at room temperature for 5 minutes.
3. Extract the reaction with 800 μl of phenol:chloroform (P:C).
4. Save the aqueous phase (which contains the unhybridized cDNA), and add 200 μl of HB and 2 μl of 20% SDS to the interface and organic phase. Vortex, and spin.
5. Pool the aqueous phases (1000 μl total). Add 40 μg of streptavidin. Repeat P:C extraction and back-extraction, adding 200 μl of HB and 2 μl of 20% SDS to the interface and organic phase.
6. Pool the aqueous phases (now 1200 μl) and add 40 μg of streptavidin. Repeat P:C extraction and back-extraction (again adding 200 μl of HB and 2 μl of 20% SDS to the interface and organic phase).
7. Pool the aqueous phases (1400 μl), chloroform-extract, and ethanol-precipitate.

Rehybridization and Biotin/Phenol Extraction

1. Scale down a second subtractive hybridization reaction to 8 μl total volume using 12 μg of PAB-RNA(−). Hybridize to an equivalent R_0t of 1500–3000 moles·sec/liter (see the section on subtractive hybridization; hybridization).

2. Perform streptavidin-P:C extraction (see the section on subtractive hybridization; separation of cDNA(+) from cDNA(+)/PAB-RNA(−) hybrids, above).
3. Remove the salts by passing through a Sephadex G-50 spin column.
4. Add 10 μg of tRNA carrier and ethanol-precipitate with ammonium acetate. Resuspend pellet in 20 μl of H_2O.
5. To complete the synthesis of a subtracted cDNA library, synthesize the second strand cDNA; see method I: hairpin priming of second-strand synthesis.

HYBRIDIZATION ANALYSIS OF PLASMID cDNA LIBRARIES

Once a subtracted cDNA probe or a subtracted cDNA library has been generated, the probe can be employed to screen a cDNA library for differentially expressed clones. Likewise, a subtracted library will have to be plated and analyzed further. Procedures for the plating of plasmid libraries, the generation of filter replicas of those colonies, and the preparation for hybridization with probes or with subtracted probes are described here.

Materials

1. Nitrocellulose filters (BA85, 0.45-μm pore size, 82 mm in diameter, Schleicher and Schuell, cat. # 20440).
2. Whatman filter paper, 90-mm circles (cat. # 1001 090).
3. Whatman 3MM chromatography paper.
4. Porcelain Buchner funnel (diameter of plate is 100 mm; Baxter, cat. # F7300-6).
5. Filter-adapter set (Neoprene rubber; Baxter, cat. # F7350-10).
6. Vacuum flask (2 liters; Baxter, cat. # F4380-2L).
7. 2X LB broth.
8. LB-amp[50] plates (see the section on library construction and amplification).
9. Solution I: 10% SDS.
10. Solution II: 0.5 N NaOH, 1.5 M NaCl.
11. Solution III: 1.0 M Tris, pH 7.4, 1.5 M NaCl.
12. Solution IV: (2X SSC): 300 mM NaCl, 30 mM trisodium citrate, pH 7.0.
13. 95% ethanol.
14. Filter forceps (Millipore).
15. 2 glass plates, about 8" × 8".
16. Hypodermic needle, 18 gauge.
17. Cafeteria-style serving trays.

Method

1. Plating the cDNA library
    a. Label 50 nitrocellulose filters with a permanent marker. (Always wear gloves when handling filters.)
    b. Thaw a glycerol stock of a plasmid cDNA library that has been previously titered for the number of colony-forming units (cfu)/ml.
    c. Dilute the plasmid cDNA library in 2X LB medium to produce 500 ml of a plating suspension at a final concentration of 400 cfu/ml. Keep the suspension on ice to prevent the cells from dividing.
    d. Assemble the plating apparatus by inserting a Buchner funnel into the mouth of a 2-liter vacuum flask (using the funnel adapters) and attaching vacuum tubing to the side-arm of the flask. Attach the other end of the tubing to the vacuum source.
    e. Place two Whatman filter circles on the bottom of the Buchner funnel. Pass 200 ml of 95% ethanol to sterilize the assembly, then pass 400 ml of LB medium to wash away the ethanol.
    f. With the vacuum turned off, place a nitrocellulose filter on top of the Whatman filter cushion with a sterile filter forceps. Be sure to wear gloves to avoid filter contamination.
    g. Pipet 5–8 ml of 2X LB media (without cells) on top of the nitrocellulose filter so that the filter is wet. Occasionally, the outer 1-mm perimeter of the filter will not wet. This is of no concern.

h. Turn on the vacuum for a few seconds to pull the medium through the filter. This will wash away any surfactants left in the filter. Turn off the vacuum.
i. Pipet 5 ml of the plating suspension directly onto the nitrocellulose filter. If necessary, tilt the funnel assembly slightly to spread the suspension evenly over the filter.
j. Turn on the vacuum for a few seconds to pull the medium through the filter.
k. Turn off the vacuum. Carefully lift the nitrocellulose filter with sterile forceps and place it on the agar surface of a LB-amp^{50} plate with the cell side up.
l. Repeat steps (f)–(k) with the remaining filters.
m. Incubate the plates upside down at 37°C overnight, until the colonies (2000/filter) have each reached a diameter of 1.0 mm. Place the plates at room temperature and proceed to processing the filters.

2. Replica plating of cDNA library colonies
   a. Label a duplicate set of nitrocellulose filters and place each one on the surface of a fresh LB-amp^{50} plate. The filters should be allowed to wet completely on the agar surface before proceeding.
   b. Cut Whatman 3MM paper to produce four 6" squares.
   c. Preparing a master filter duplicate of a colony-laden filter:
      i. Place one glass plate on the laboratory bench, and place two Whatman filter squares on top of the glass plate.
      ii. Using forceps, place a colony-laden filter on top of two Whatman 6" squares, colony side up. Discard the LB-amp^{50} plate.
      iii. Place the corresponding duplicate, wetted filter directly on top of the colony-laden filter, using two pairs of filter forceps. Take care not to slide the filters against each other once they are in contact.
      iv. Place two Whatman 6" squares on top of the nitrocellulose filters, then place the second glass plate on top of the whole assembly to create a "sandwich."
      v. With even pressure, carefully press the glass plate down on the sandwich. The glass plate can be rotated 90° in each direction.
      vi. Remove the glass plate and the top two Whatman squares from the sandwich. Pierce the pressed nitrocellulose filters with a hypodermic needle to produce orientation marks. *This is extremely important!* Without these marks, you will not be able to align your master filters containing bacterial cells with the positive signals from the corresponding hybridized filters.
      vii. Using forceps, peel apart the nitrocellulose filters. Place the original filter, colony side up, on Whatman 3MM paper to dry. Place the second filter on the same LB-amp^{50} agar plate it was wetted on—this is the master plate. Incubate the master plates upside down at 37°C for 16 hours. Post incubation, store at 4°C.

3. Processing colony-laden filters
   a. Cut two pieces of Whatman 3MM paper so that the sheets will fit into plastic serving trays, creating two layers of paper. Prepare four trays.
   b. Label the trays I–IV and soak the 3MM sheets with Solutions I–IV, respectively. Roll disposable 10-ml pipettes over the surfaces of the wetted papers to eliminate any air bubbles, then pour off the excess liquid from each tray.
   c. Transfer the colony-laden filters to tray I (colony side up), and allow them to sit for 5 minutes. Use forceps for transfer.

d. Transfer the filters to tray II, and allow them to sit for 10–15 minutes.
e. Transfer the filters to tray III, and allow them to sit for 2 minutes.
f. Transfer the filters to tray IV, and allow them to sit for 1 minute.
g. Transfer the filters to Whatman 3MM paper to air-dry.
h. Interleave the dried nitrocellulose filters between successive sheets of Whatman 3MM paper so that no filter touches another filter. Place the stack of filters/3MM papers into a vacuum drying oven. Bake in a vacuum at 80°C for 2 hours.
i. After baking, the filters can be prehybridized or stored at room temperature.
j. Refer to Part III in the section on hybridization conditions for specific hybridization protocols.

9

RETROVIRUS-MEDIATED GENE TRANSFER

Background information on the engineering of retroviral vectors and retrovirus packaging lines can be found in Part I in the chapter on retroviral vectors.

There are dramatic differences between viral- and DNA-mediated gene transfer, both in the manner in which the genetic material is introduced into the cell and in the level of expression one obtains from each integrated copy of the transferred gene. To appreciate the advantages of retrovirus-mediated gene transfer, we will re-examine the phenomenon of DNA-mediated gene transfer, then compare.

In DNA-mediated gene transfer, the object of the procedure is to chemically create a situation where purified DNA can be efficiently transferred from the outside of a cell to the nucleus, and in a manner such that, upon arrival, the DNA has remained intact and can serve as an efficient transcriptional template. In the case of stable expression, the DNA must exist in a form that will allow reasonably efficient integration of the foreign DNA into the chromosome of the host cell and subsequently serve as an efficient transcriptional template.

Recombinant-retrovirus-mediated gene transfer offers two key advantages when compared to DNA-mediated gene transfer. Although infectious viral genomes introduced stably into cells in culture can manifest position effects, several lines of evidence support the contention that exogenous genes introduced via *infection* tend to express themselves at consistently higher levels than their transfected counterparts. Furthermore, infectious recombinant-retrovirus stocks can be generated that are of sufficiently high titer to permit the stable transformation of an entire population of cells. Here is how it is done.

GENERATION OF HIGH-TITER HELPER-VIRUS-FREE RECOMBINANT-RETROVIRUS STOCKS

Perhaps the most surprising observation one can make about the generation of high-titer retroviral stocks is the tremendous variation in viral titer that is manifest in concurrently derived clonal isolates derived from a single transfection or transinfection experiment. Variations of 3–4 orders of magnitude are not uncommon, even when the copy number of the integrated genomes is identical. Even a single integrated copy can manifest this range of viral production. The simplest interpretation of this phenomenon is that it, like so many other gene-transfer methods, is also subject to chromosomal position effects. This means that after transfection or transinfection, the experimenter should isolate 10–20 virus-producing clones to facilitate the isolation of one or two producer clones that generate consistently high viral titers.

Transfection and transinfection are two different methods for generating cell lines that stably produce infectious recombinant retroviruses. With transfection, one merely transfers the recombinant retroviral genome into one of a variety of retroviral packaging cell lines (see Part I, in the section on retroviral vectors). Post transfection with either the calcium phosphate method or the electroporation method, transfected cells are subjected to drug selection to facilitate the isolation of viral producers, and the drug-resistant clones are picked, expanded, and evaluated for virus titer as described below.

The process of transinfection involves the transient transfer and subsequent expression of a recombinant retroviral genome into either an ecotropic or amphotropic retroviral packaging cell line. Twenty-four hours post transfection by the calcium phosphate method, the culture supernatant is transferred to either an untransfected ecotropic or amphotropic retroviral packaging cell line. If the cell line that was originally transfected was ecotropic, then one can transinfect an amphotropic packaging line, and vice versa. This is possible because ecotropic and amphotropic viruses belong to different interference groups (see Part I in the section on retroviral vectors). As is the case with the standard transfection method, one should isolate 10–20 virus producers to facilitate the isolation of one or two high-titer producers.

Transfected and transinfected packaging cells can produce titers as high as 1×10^7 colony forming units (cfu)/ml. Media containing viral particles can be removed from producer cultures and frozen at $-70°C$. Retroviral particles stored in this manner retain their bioactivity for years.

Materials

Transfection Solutions

Prepare these solutions fresh.

Chemicals

2.0 M $CaCl_2 \cdot 2\ H_2O$ (Sigma, cat. # C7902)
0.5 M HEPES, free acid (Sigma, cat. # H9136)
2.0 M NaCl (Sigma, cat. # S5886)
0.15 M Na_2HPO_4 (Sigma, cat. # S5136)

Polybrene stock (100X): Prepare a stock solution of 400 μg/ml of polybrene (Sigma, cat. # P4515) in PBS. Filter-sterilize and store at 4°C.

2X HBS Solution

5.0 ml of 0.5 M HEPES.
6.25 ml of 2.0 M NaCl.
0.5 ml of 0.15 M Na_2HPO_4.
Add H_2O to ~45 ml. Adjust pH to 7.05–7.10 with 1 N NaOH. Bring to 50 ml final. Filter-sterilize.

DNA Samples

Sterilize plasmid DNA samples by bringing aqueous solution to 0.1 M NaCl and precipitating with two volumes of ethanol. Resuspend the DNA pellet in sterile TE (10 mM Tris, 1 mM EDTA, pH 7.5) at a concentration of 500–1000 μg/ml.

Transinfection

Day 1:

Plate retroviral packaging cells at 5×10^5 cells/60-mm dish.

Day 2:

Feed recipient cultures with 4.0 ml of fresh medium. Transfect with the calcium-phosphate-precipitation procedure:

For each plasmid DNA sample, prepare a DNA/CaCl$_2$ solution by mixing 25 μl of 2.0 M CaCl$_2$ and 10 μg of plasmid DNA. Bring to a total volume of 200 μl with dH$_2$O.

For each sample, add 200 μl of 2X HBS to a clear plastic tube. Add 200 μl of DNA/CaCl$_2$ solution drop by drop to the 2X HBS, with constant agitation. After incubating at room temperature undisturbed for 30 minutes, add 400 μl of the fine precipitate to the recipient packaging line cells and swirl the medium to mix.

Day 3:

Aspirate medium containing DNA. Feed cells with fresh medium. Plate corresponding packaging cells for infection on Day 4 at 5×10^5 cells/60-mm dish.

Day 4:

After no more than 16 hours after feeding on Day 3, remove the viral supernatant from the transfected culture and centrifuge for 5 minutes at 3000 × g at room temperature. Feed the recipient cells with 4 ml of fresh medium supplemented with polybrene to a final concentration of 4μg/ml. Infect the recipient cells with samples of the virus-containing medium. We generally infect multiple plates with ever-increasing amounts (1μl, 10μl, 100 μl) of virus-containing media. After virus addition, swirl the polybrene-supplemented medium in the culture dishes to mix the virus. Incubate the infected cultures overnight at 37°C.

Day 5:

Trypsinize the *transinfected* cells 1:20 into selection medium. Change the selection medium every 3–4 days until colonies have formed.

TITRATION AND ANALYSIS OF RECOMBINANT-RETROVIRUS STOCKS

Once you have generated retrovirus-producer colonies by transfection or transinfection, you will want to titer the production of virus by each clone. Titration of retroviruses is a simple procedure, and the method is provided here. Perhaps the most important consideration in virus titration is that the cells you infect be rapidly growing. The number and density of cells we plate, indicated below, is ideal for virus titration. There is an important trick to maintain viral titer in your best virus-producing clone: maintain the producer cells in selection media to insure that both the transfected/infected recombinant retroviral genome and the transfected, enfeebled packaging genome are retained. We use PE501 and PA317 routinely, and we carry them in HAT medium because the enfeebled packaging genome was co-transfected into them with the HSVtk gene and HAT selection kills those PE501/PA317 cells that have lost the co-transfected tk gene. We observe that the loss of the tk gene in these cell lines usually accompanies the loss of the transfected, enfeebled packaging genome. Thus, we indirectly select for the retention of the transfected, enfeebled packaging genome and ensure that few, if any, of the cells in our producer cultures are, in fact, non-producers. Prior to a virus harvest, we transfer our producers to HT medium to wean the cells off HAT. We pass the cells three times in HT and twice in DMEM prior to harvest. Selection for the packaging-competent, dominant-selectable-marker-encoding, recombinant retrovirus is removed at least one passage prior to virus harvest. A similar approach makes good sense for maintaining all of the packaging lines described in this manual.

Materials

Polybrene stock (100X): 400 μg/ml of polybrene (Sigma, cat. # P4515) in PBS. Filter-sterilize and store at 4°C.

Method

Day 1:

Plate viral producers at 2×10^5 cells/60-mm dish.

Day 4:

Change media of producers. Plate recipient cells at 5×10^5 cells/60-mm dish.

Day 5:

After a 16-hr. incubation, harvest the viral supernatant and centrifuge at $3000 \times g$ for 5 minutes to remove cells and debris. Feed recipient cells with 4 ml of fresh medium supplemented with polybrene to a final concentration of 4 μg/ml. To titer your virus stock, assume you have a virus titer of 1×10^8 colony-forming units (cfu)/ml. Prepare a dilution series (10^{-1} to 10^{-9}) of virus and infect with 100 μl of diluted virus. Incubate the infected cells overnight at 37°C.

Day 6:

Trypsinize cells and seed 1:20 into appropriate selection medium. Feed cultures with selection medium every three days. When colonies appear, stain as described in the section on the propagation of cell lines. Virus titer is 20 $\times$ dilution $\times$ number of colonies.

10

PCR-BASED EXPRESSION

This technique was discussed in general in Part I of this manual. The specific details of the procedure are described here. The example I describe was chosen because it involves many of the types of problems one can run into when one chooses to pursue this approach. The solutions to these problems are also presented. In these experiments, we assembled a functional cDNA clone encoding a human leukocyte interferon a/d hybrid protein. We chose to synthesize the gene this way, rather than assemble it after PCR of genomic DNA or RNA, because it is an artificial hybrid cDNA composed of sequences of two naturally occurring interferon genes, and this approach was simpler than cloning out the two cDNAs encoding the sequences and subsequently assembling them. The completed product was a 591-bp fragment that could be further utilized by insertion into an expression plasmid or used in PCR-based expression.

For information regarding the relationship and numbering of the synthetic oligonucleotides employed in this experiment, see Figure 10-1.

PCR-BASED GENE ASSEMBLY

Materials

1. Sixteen oligonucleotides 51 nucleotides in length, with complementary 15-nucleotide overlapping sequences.
2. 10X PCR buffer: 500 mM KCl, 200 mM Tris-HCl, pH 8.4, 25 mM $MgCl_2$, 1 mg/ml nuclease-free BSA (Pharmacia, cat. # 27-8914-01).
3. 10 mM dNTPs: Combine 100 mM stocks of dATP, dCTP, dGTP, and dTTP (Pharmacia, cat. # 27-2050-01, 27-2060-01, 27-2070-01, and 27-2080-01, respectively) and dilute with 10 mM Tris-HCl, pH 7.0.
4. Taq polymerase (Perkin-Elmer/Cetus).
5. Thermal cycler (Perkin-Elmer/Cetus).
6. Polyacrylamide-gel-electrophoresis reagents.

Gene Synthesis, Part 1

The first step in the gene-construction process is the generation, in seven separate reactions, of the 159-bp PCR products A, B, C, D, E, F, and G. Each PCR product is constructed from four 51-nucleotide-long synthetic oligonucleotides that overlap each other by 15 nucleotides. The PCR-reaction components and

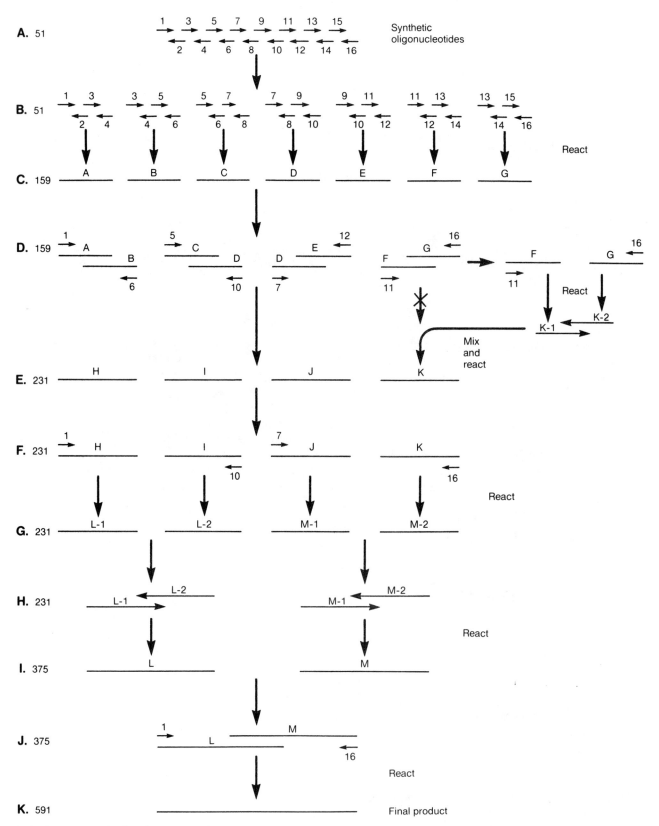

FIGURE 10-1 *PCR-based gene assembly.* The strategy for PCR-based gene assembly described in the text is depicted here. **A.** The numbering and orientation of the 16 overlapping 51-mers used to synthesize the gene. **B.** The numbering and orientation of the subsets of the 51-mers used to generate the seven 159-mers shown in **C. D.** The numbering and orientation of the components used to generate the next generation of 231-mers, shown in **E. F.** The numbering and orientation of the components used to generate greater quantities of the 231-mers, shown in **G,** which are ultimately combined and reacted, as shown in **H,** to form the 375-mers shown in **I. J.** The orientation and labeling of the components used to generate the 591-base pair final product shown in **K.**

TABLE 10-1 Oligonucleotides to Add to PCR to Generate A, B, C, D, E, F, and G.

PCR Product	Outer 5' Oligo (400 pmoles)	Inner 5' Oligo (5 pmoles)	Inner 3' Oligo (5 pmoles)	Outer 3' Oligo (400 pmoles)
A	1	2	3	4
B	3	4	5	6
C	5	6	7	8
D	7	8	9	10
E	9	10	11	12
F	11	12	13	14
G	13	14	15	16

conditions are given here. The quantities of each synthetic oligonucleotide to add to each of the seven PCR reactions are indicated in Table 10-1.

1. Prepare this PCR to generate A, B, C, D, E, F, and G (see Figure 10-1):

5 µl	10X PCR buffer
0.5 µl	10 mM dNTPs
4 µl	outer 5' oligonucleotide
4 µl	inner 5' oligonucleotide
4 µl	inner 3' oligonucleotide
4 µl	outer 3' oligonucleotide
27.5 µl	dH$_2$O
1 µl	Taq polymerase (5 U/µl)
50 µl	Total

   Layer 50 µl of mineral oil (Sigma, cat. # M5904).
2. PCR amplification:
   a. Perform 10 cycles of:
      - 95°C, 1 minute.
      - 45°C, 1 minute.
      - 72°C, 1 minute.
   b. At the end of these 10 cycles add five additional units of Taq polymerase and continue for 10 more cycles.

   Upon completion of PCR, isolate the products as described here.

Gel Isolation of PCR Products

1. Electrophorese each reaction on an 8% polyacrylamide gel. For each reaction, load these samples on the gel in the following order:
   a. DNA molecular-weight marker.
   b. 5X PCR reaction.
   c. 5X PCR reaction.
2. Slice the gel between lanes (b) and (c). Set lane (c) aside. Stain lanes (a) and (b) with ethidium bromide (5 µg/ml in dH$_2$O) to visualize molecular-weight markers and the PCR products in lane (b). Excise the 159-bp (or other appropriate size) fragment in lane (b).
3. Align lanes (a) and (b) with the unstained lane (c). Excise the region of lane (c) that corresponds to the 159-bp (or other appropriate size) fragment in lane (b).
4. Crush the excised gel region with a Pasteur pipette that is heat-sealed at the tip.

TABLE 10-2 Oligonucleotides to Add to PCR to Generate H, I, J, and K.

PCR Product	Outer 5' Oligo (100 pmoles)	Fragment 1 (8 µl)	Fragment 2 (8 µl)	Outer 3' Oligo (100 pmoles)
H	1	A	B	6
I	5	C	D	10
J	7	D	E	12
K	11	F	G	16

5. Add 100 µl of dH$_2$O to the slurry and incubate at room temperature overnight.
6. Centrifuge slurry at 14 Krpm in a microfuge for 5 minutes.
7. Save the supernatant, which contains the PCR product, for the next stage in the assembly process.

Gene Synthesis, Part 2

The second step in the gene-construction process is the generation, in four separate reactions, of the 231-bp PCR products H, I, J, and K. Each PCR product is constructed from two of the fragments you just eluted and the two oligonucleotides that constitute the extreme ends of the desired 231-bp product. The PCR-reaction components and conditions are given here. The quantities of each synthetic oligonucleotide and each eluted fragment to add to each of the four PCR reactions are indicated in Table 10-2.

1. Prepare this PCR to generate H, I, J, and K (see Figure 10-1):

5	µl	10X PCR buffer
0.5	µl	10 mM dNTPs
4	µl	outer 5' oligonucleotide
8	µl	eluted fragment 1, termed "fragment 1"
8	µl	eluted fragment 2, termed "fragment 2"
4	µl	outer 3' oligonucleotide
19.5	µl	dH$_2$O
1	µl	Taq polymerase (5 U/µl)
50	µl	Total

    Layer 50 µl of mineral oil (Sigma, cat. # M5904).
2. PCR amplification:
    a. Perform 10 cycles of:
       - 95°C, 1 minute.
       - 45°C, 1 minute.
       - 72°C, 1 minute.
    b. At the end of these 10 cycles add five additional units of Taq polymerase and continue for 10 more cycles.

Upon completion of PCR, isolate the 231-bp products as described in gene synthesis, part 1.

Note: In gene synthesis, part 2, we noticed that the synthesis of product K did not work well. When this type of problem arises, we usually break the PCR reaction into two parts, a left-side PCR reaction composed of the outer 5' oligonucleotide (11) and eluted fragment 1 (F), reaction K-1, and a right-side PCR reaction composed of the outer 3' oligonucleotide (16) and eluted fragment 2 (G), reaction K-2.

1. Prepare these PCRs to generate K-1 and K-2:

    Fragment K Half-Reactions

	K-1		K-2
5 μl	10X PCR buffer	5 μl	10X PCR buffer
8 μl	fragment F	8 μl	fragment G
4 μl	outer 5' oligonucleotide 11 (100 pmoles/μl)	4 μl	outer 3' oligonucleotide 16 (100 pmoles/μl)
0.5 μl	10 mM dNTPs	0.5 μl	10 mM dNTPs
31.5 μl	dH$_2$O	31.5 μl	dH$_2$O
1 μl	Taq polymerase (5 U/μl)	1 μl	Taq polymerase (5 U/μl)
50 μl	Total	50 μl	Total

    Layer 50 μl of mineral oil (Sigma, cat. # 5904).
2. PCR amplification:
    a. Perform 10 cycles of:
        - 95°C, 1 minute.
        - 45°C, 1 minute.
        - 72°C, 1 minute.
    b. **Do not gel-purify these fragments.**

1. Prepare this PCR to generate fragment K:

    24.5 μl K-1 reaction
    24.5 μl K-2 reaction
    1 μl Taq polymerase (5 U/μl)
    50 μl Total

    Layer 50 μl of mineral oil (Sigma, cat. # M5904).
2. PCR amplification:
    a. Perform 10 cycles of:
        - 95°C, 1 minute.
        - 45°C, 1 minute.
        - 72°C, 1 minute.
    b. Upon completion of PCR, isolate the 231-bp product as described in Gene Synthesis, part 1.

Gene Synthesis, Part 3

The third step in the gene-construction process is the assembly of fragments H and I to form M and, in a separate reaction, the assembly of fragments J and K to form L. First, we synthesize L-1, L-2, M-1, and M-2. Then we assemble L-1 and L-2 into L and M-1 and M-2 into M.

1. Prepare this PCR to generate L-1, L-2, M-1, and M-2:

    Fragment L Half-Reactions

	L-1		L-2
5 μl	10X PCR buffer	5 μl	10X PCR buffer
8 μl	fragment H	8 μl	fragment I
4 μl	outer 5' oligonucleotide 1 (100 pmoles/μl)	4 μl	outer 3' oligonucleotide 10 (100 pmoles/μl)
0.5 μl	10 mM dNTPs	0.5 μl	10 mM dNTPs
31.5 μl	dH$_2$O	31.5 μl	dH$_2$O
1 μl	Taq polymerase (5 U/μl)	1 μl	Taq polymerase (5 U/μl)
50 μl	Total	50 μl	Total

Fragment M Half-Reactions

	M-1		M-2
5 µl	10X PCR buffer	5 µl	10X PCR buffer
8 µl	fragment J	8 µl	fragment K
4 µl	outer 5' oligonucleotide 7 (100 pmoles/µl)	4 µl	outer 3' oligonucleotide 16 (100 pmoles/µl)
0.5 µl	10 mM dNTPs	0.5 µl	10 mM dNTPs
31.5 µl	dH$_2$O	31.5 µl	dH$_2$O
1 µl	Taq polymerase (5 U/µl)	1 µl	Taq polymerase (5 U/µl)
50 µl	Total	50 µl	Total

Layer 50 µl of mineral oil (Sigma, cat. # 5904).

2. PCR amplification:
   a. Perform 10 cycles of:
      - 95°C, 1 minute.
      - 45°C, 1 minute.
      - 72°C, 1 minute.
   b. **Do not gel-purify these fragments.**

1. Prepare this PCR to generate fragments L and M:

24.5 µl	L-1 reaction	24.5 µl	M-1 reaction
24.5 µl	L-2 reaction	24.5 µl	M-2 reaction
1 µl	Taq polymerase (5 U/µl)	1 µl	Taq polymerase (5 U/µl)
50 µl	Total	50 µl	Total

   Layer 50 µl of mineral oil (Sigma, cat. # M5904).

2. PCR amplification:
   a. Perform 10 cycles of:
      - 95°C, 1 minute.
      - 45°C, 1 minute.
      - 72°C, 1 minute.
   b. Upon completion of PCR, isolate the 375-bp products on a 6% polyacrylamide gel as described in Gene Synthesis, part 1.

Gene Synthesis, Part 4

The fourth step in the gene-construction process is the assembly of fragments L and M into the complete hybrid alpha interferon gene.

1. Prepare this PCR to generate the complete synthetic gene:

10 µl	10X PCR buffer
4 µl	outer 5' oligonucleotide 1
24 µl	fragment L
24 µl	fragment M
4 µl	outer 3' oligonucleotide 16
1 µl	10 mM dNTPs
31 µl	dH$_2$O
2 µl	Taq polymerase (5 U/µl)
100 µl	Total

   Layer 50 µl of mineral oil (Sigma, cat. # 5904).

2. PCR amplification:
   a. Perform 10 cycles of:
      - 95°C, 1 minute.
      - 45°C, 1 minute.
      - 72°C, 1 minute.
   b. Upon completion of PCR, phenol-extract and ethanol-precipitate reaction. Electrophorese products on a 1-percent agarose gel. Purify the 591-bp synthetic gene by glass-bead purification (see the section on subtractive hybridization; recovering cDNA). The gene is now ready for further biochemical manipulations.

PCR-BASED GENE EXPRESSION

This technique is similar to Synthesis 4 above. With PCR-based gene expression, we use PCR to attach a eukaryotic expression element to a structural gene, one preferably, but not necessarily, also encoding a polyadenylation site. We generate microgram quantities of material to employ in transient or stable-transfection gene-transfer experiments. Thus, we eliminate the need to subclone the structural gene or expression element, or both, into a plasmid and subsequently identify and purify the recombinant plasmid of choice. Although this procedure has some disadvantages, it is very quick, and the time savings are substantial when compared to other technologies. We have employed this technique in our laboratory with a good deal of success.

This technique involves two PCR-synthesis steps. The first synthesis creates overlapping, complementary sequences on the two DNA fragments to be joined. The second synthesis serves to join the two fragments, generating

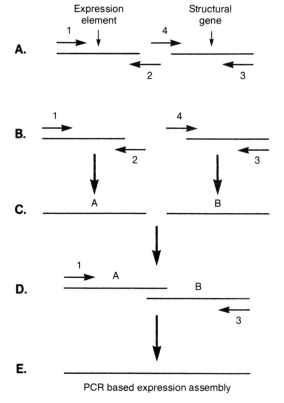

FIGURE 10-2 *PCR-based gene expression.* The strategy for PCR-based gene expression described in the text is depicted here. First, the expression element and the structural gene to be expressed are prepared for joining. The DNA encoding the expression element is mixed with the 5' amplimer, 1, and the 3' amplimer, 2, and a PCR is performed to produce product A. Simultaneously the DNA encoding the structural gene is mixed with the 5' amplimer, 4 (which shares overlapping sequences with amplimer 2), and the 3' amplimer, 3, and a PCR is performed to produce product B as shown in **A**, **B**, and **C**. Products A and B, which now share overlapping sequence information at their 3' and 5' ends, respectively, are mixed with each other and with amplimers 1 and 3, and a PCR is performed to produce the completed PCR-based expression assembly as shown in **D** and **E**. This DNA product can now be introduced into cells in culture to express the structural gene of interest.

the completed product ready for DNA transfection. For further information regarding the numbering and relationship of the synthetic nucleotides and DNA fragments employed in this procedure, see Figure 10-2.

In practice, one examines the 3' sequence of the expression element to be joined and the 5' sequence of the structural gene to be added. An oligonucleotide amplimer (oligonucleotide 1) encoding the 5' end of the expression element is synthesized, as is a corresponding 3' oligonucleotide amplimer, (oligonucleotide 2) encoding both the 3' end of the expression element fragment and the 5' end of the fragment encoding the structural gene. Similarly, an oligonucleotide amplimer (oligonucleotide 3) encoding the 3' end of the structural gene fragment, and a corresponding 5' oligonucleotide amplimer (oligonucleotide 4) encoding both the 5' end of the fragment containing the structural gene and the 3' end of the expression element and complementary to oligonucleotide 2, are synthesized.

Oligonucleotides 1 and 2 are mixed with the fragment encoding the expression element, and PCR is carried out under standard conditions. The reactions are subjected to electrophoresis, and the appropriate-sized fragment is eluted as described for PCR-based gene synthesis. This fragment, fragment A, is set aside. Another PCR reaction containing oligonucleotides 3 and 4 and the fragment encoding the structural gene is performed, and, upon completion, the resultant PCR fragment is purified as described in the procedure for gene synthesis, part 1. This fragment is called fragment B.

To complete the assembly process, fragments A and B are mixed with oligonucleotides 1 and 3, and another PCR is carried out. The reactions are fractionated on an agarose gel, and the DNA fragment of the correct molecular size is glass-bead purified (see the sections on subtractive hybridization; recovering cDNA). This purified DNA fragment can then be used to transfect cells in culture by either calcium phosphate, DEAE dextran, or electroporation DNA-transfer methodologies.

PART II REFERENCES

AARONSON, S., TODARO, G., and FREEMAN, A. 1970. Human sarcoma cells in culture identified by colony-forming activity on monolayers of normal cells. *Exp. Cell Res.* **6**:1.

ARUFFO, A. and SEED, B. 1987. Molecular cloning of a CD28 cDNA by a high-efficiency COS cell expression system. *Proc. Natl. Acad. Sci. U.S.A.* **84**:8573.

ASHMAN, C. R., and DAVIDSON, R. L. 1985. High spontaneous-mutation frequency of BPV shuttle vector. *Somat. Cell. Mol. Genet.* **11**:499.

BEACH, L. and PALMITER, R. 1981. Amplification of the metallothionein-1 gene in cadmium-resistant mouse cells. *Proc. Natl. Acad. Sci. U.S.A.* **78**:2110.

BEBBINGTON, C. and HENTSCHEL, C. 1987. The use of vectors based on gene amplification for the expression of cloned genes in mammalian cells. In *DNA Cloning, Volume III, a practical approach*. Ed. D. M. Glover. Oxford: IRL Press.

BEHRINGER, R. R., MATHEWS, L. S., PALMITER, R. D., and BRINSTER, R. L. 1988. Dwarf mice produced by genetic ablation of growth-hormone-expressing cells. *Genes and Dev.* **2**:453.

BONTHRON, D., HANDIN, R., KAUFMAN, R., WASLEY, L., ORR, E., MITSOK, L., EWENSTEIN, B., LOSCALZO, J., GINSBERG, D., and ORKIN, S. 1986. Structure of pre-pro-von Willebrand factor and its expression in heterologous cells. *Nature* **324**:270.

BORRELLI, E., HEYMAN, R., HSI, M., and EVANS, R. M. 1988. Targeting of an inducible toxic phenotype in animal cells. *Proc. Natl. Acad. Sci. U.S.A.* **85**:7572.

BREITMAN, M. L., CLAPOFF, S., ROSSANT, J., TSUI, L. C., GOLDE, L. M., MAXWELL, I. H., and BERNSTEIN, A. 1987. Genetic ablation: Targeted expression of a toxin gene causes micropthalmia in transgenic mice. *Science* **238**:1563.

BRITTEN, R., GRAHAM, D., and NEUFIELD, B. 1974. Analysis of repeating DNA sequences by reassociation. *Methods Enzymol* **29**:363.

CALOS, M. P., LEBKOWSKI, J. S., and BOTCHAN, M. R. 1983. High mutation frequency in DNA transfected into mammalian cells. *Proc. Natl. Acad. Sci. U.S.A.* **80**:3015.

CARTIER, M., CHANG, M., and STANNERS, C. 1987. Use of the *Escherichia coli* gene for asparagine synthetase as a selective marker in a shuttle vector capable of dominant transfection and amplification in animal cells. *Mol. Cell. Biol.* **7**:1623.

CHIANG, T. and MCCONLOGUE, L. 1988. Amplification and expression of heterologous ornithine decarboxylase in Chinese hamster cells. *Mol. Cell. Biol.* **8**:764.

CHEN, C. and OKAYAMA, H. 1987. High-efficiency transformation of mammalian cells by plasmid DNA. *Mol. Cell. Biol.* **7**:2745.

CHOI, K., CHEN, C., KRIEGLER, M., and RONINSON, I. 1988. An altered pattern of cross resistance in multi-drug-resistant cells results from spontaneous mutations in the mdr-1 (p-glycoprotein) gene. *Cell* **53**:519.

COLBERE-GARAPIN, F., CHOUSTERMAN, S., HORODNICEANU, F., KOURILSKY, P., and GARAPIN, A. C. 1979. Cloning of the active thymidine kinase gene of herpes simplex virus type 1 in *Escherichia coli* K-12. *Proc. Natl. Acad. Sci. U.S.A.* **76**:3755.

D'ANDREA, A. D., LODISH, H. F., and WONG, G. G. 1989. Expression cloning of the murine erythropoetin receptor. *Cell* **57**:277.

DAVIS, M., NIELSEN, E., STEINMETZ, M., PAUL, W., and HOOD, L. 1984. Cell-type-specific cDNA probes and the murine 1 region: the localization and orientation of A-alpha[d]. *Proc. Natl. Acad. Sci. U.S.A.* **81**:2194.

DAVIS, M. 1986. Subtractive cDNA hybridization and the T-cell receptor genes. In *Handbook of Experimental Immunology*. Ed. D. Weir. Oxford: Blackwell Scientific.

DE SAINT VINCENT, B., DELBRUCK, S., ECKHART, W., MEINKOTH, J., VITTO, L., and WAHL, G. 1981. The cloning and reintroduction into animal cells of a functional CAD gene, a dominant amplifiable genetic marker. *Cell* **27**:267.

EVANS, G. A. 1989. Dissecting mouse development with toxigenics. *Genes Dev.* 3: 259.

FURMAN, P. A., MCGUJIRT, P. V., KELLER, P. M., FYFE, J. A., and ELION, G. B. 1980. Inhibition by acyclovir of cell growth and DNA synthesis of cells biochemically transformed with herpes virus genetic information. *Virology* **102**: 420.

GRAHAM, F. L. and VAN DER EB, A. J. 1973. A new technique for the assay of infectivity of human adenovirus IV DNA. *Virology* **52**:456.

GRITZ, L. and DAVIES, J. 1983. Plasmid-encoded hygromycin-B resistance: The sequence of hygromycin-beta-phosphotransferase gene and its expression in *E. coli* and *S. cerevisiae*. *Gene* **25**:179.

GUBLER, U. and HOFFMAN, B. 1983. A simple and very effective method for generating cDNA libraries. *Gene* **25**:263.

HARTMAN, S. and MULLIGAN, R. 1988. Two dominant-acting selectable markers for gene-transfer studies in mammalian cells. *Proc. Natl. Acad. Sci. U.S.A.* **85**:8047.

HEDRICK, S., COHEN, D., NIELSON, E., and DAVIS, M. 1984. Isolation of cDNA clones encoding T-cell-specific membrane-associated proteins. *Nature* **308**:149.

HEYMAN, R. A., BORRELLI, E., LESLEY, J., ANDERSON, D., RICHMOND, D. D., BAIRD, S. M., HYMAN, R., and EVANS, R. M. 1989. Thymidine kinase obliteration: Creation of transgenic mice with controlled immunodeficiencies. *Proc. Natl. Acad. Sci. U.S.A.*

HIRT, B.J. 1967. Selective extraction of polyoma DNA from infected mouse cell cultures. *J. Mol. Biol.* **26**:365.

JOLLY, D., OKAYAMA, H., BERG, P., ESTY, A., FILPULA, D., BOHLEN, P., JOHNSON, G., SHIVELY, J., HUNKAPILLAR, T., and FRIEDMANN, T. 1983. Isolation and characterization of a full-length expressible cDNA for human hypoxanthine phosphoribosyltransferase. *Proc. Natl. Acad. Sci. U.S.A.* **80**:477.

KANE, S., TROEN, B., GAL, S., UEDA, K., PASTAN, I., and GOTTESMAN, M. 1988. Use of cloned multi-drug resistance gene for coamplification and overproduction of major excreted protein, a transformation-regulated secreted acid protease. *Mol. Cell. Biol.* **8**:3316.

KAUFMAN, R., MURTHA, P., INGOLIA, D., YEUNG, C., and KELLEMS, R. 1986. Selection and amplification of heterologous genes encoding adenosine deaminase in mammalian cells. *Proc. Natl. Acad. Sci.* **83**:3136.

KAUFMAN, R. and SHARP, P. 1982. Amplification and expression of sequences cotransfected with a modular dihydrofolate reductase complementary DNA gene. *J. Mol. Biol.* **159**:601.

KAUFMAN, R. 1990. Selection and coamplification of heterologous genes in mammalian cells. In *Methods in Enzymology*, Vol. 185. Ed. D.V. Goeddel. San Diego: Academic Press, Inc.

KOHNE, D., LEVISON, S., and BYERS, M. 1977. Room-temperature method for increasing the rate of DNA reassociation by many thousandfold: the phenol emulsion reassociation technique. *Biochem.* **16**:5329.

KRIEGLER, M., PEREZ, C., and BOTCHAN, M. 1983. Promoter substitution and enhancer augmentation increase the penetrance of the SV40 A gene to levels comparable to that of the Harvey murine sarcoma virus ras gene in morphologic transformation. In "Gene Expression," UCLA Symposium on Molecular and Cellular Biology, New Series, Vol. 8. Eds. D. Hamer and M. Rosenberg. New York: Alan R. Liss.

LANDEL, C. P., ZHAO, J., BOK, D., and EVANS, G. A. 1988. Lens-specific expression of recombinant ricin induces developmental defects in the eyes of transgenic mice. *Genes and Dev.* **2**:1168.

LOWY, I., PELLICER, A., JACKSON, J., SIM, G. K., SILVERSTEIN, S., and AXEL, R. 1980. Isolation of transforming DNA: Cloning the hamster aprt gene. *Cell* **22**:817.

MACPHEARSON, I. and MONTAGNIER, L. 1964. Agar suspension culture for the selective assay of cells transformed by polyoma virus. *Virology* **23**:291.

MAITLAND, N. J. and McDOUGALL, J. K. 1977. Biochemical transformation of mouse cells by fragments of herpes simplex virus DNA. *Cell* **11**:233.

MANSOUR, S. L., THOMAS, K. R., and CAPECCHI, M. R. 1988. Disruption of the proto-oncogene int-2 in mouse-embryo-derived stem cells: a general strategy for targeting mutations to non-selectable genes. *Nature* **336**:348.

MARTIN, G. and EVANS, M. 1975. Differentiation of clonal lines of teratocarcinoma cells: formation of embryoid bodies in vitro. *Proc. Natl. Acad. Sci. U.S.A.* **78**:1441.

MULLIGAN, R., and BERG, P. 1981. Selection for animal cells that express the *E. coli* gene coding for xanthine-guanine phosphoribosyltransferase. *Proc. Natl. Acad. Sci. U.S.A.* **78**:2072.

MURRAY, A., DROBETSKY, E., and ARRAND, J. 1984. Cloning the complete human adenine phosphoribosyltransferase gene. *Gene* **31**:233.

PALMER, T., HOCK, R., OSBOURNE, W., and MILLER, A. 1987. Efficient retrovirus-mediated transfer and expression of a human adenosine deaminase gene in diploid skin fibroblasts from an adenosine-deaminase-deficient human. *Proc. Natl. Acad. Sci. U.S.A.* **84**:1055.

PALMITER, R. D., BEHRINGER, R. R., QUAIFE, C. J., MAXWELL, F., MAXWELL, I. H., and BRINSTER, R. L. 1987. Cell lineage ablation in transgenic mice by cell-specific expression of a toxin gene. *Cell* **50**:435.

PATTERSON, D. and CARNWRIGHT, D. 1977. Biochemical genetic analysis of pyrimidine biosynthesis in mammalian cells. Isolation of a mutant defective in the early steps of de novo pyrimidine biosynthesis. *Somat. Cell. Genet.* **3**:483.

RAZZAQUE, A., MIZUSAWA, H., and SEIDMAN, M. M. 1982. Rearrangement and mutagenesis of a shuttle-vector plasmid after passage in mammalian cells. *Proc. Natl. Acad. Sci. U.S.A.* **80**:3010.

RISSER, R. and POLLACK, R. 1974. A non-selective analysis of SV40 transformation of mouse 3T3 cells. *Virology* **59**:477.

RUIZ, J. and WAHL, G. 1986. *Escherichia coli* aspartate transcarbamylase: a novel marker for studies of gene amplification and expression in mammalian cells. *Mol. Cell. Biol.* **6**:3050.

SCHIMKE, R. 1984. Gene amplification in cultured animal cells. *Cell* **37**:705.

SCHIMKE, R. 1988. Gene amplification in animal cells. *J. Biol. Chem.* **263**:5989.

SIMONSEN, C. and LEVINSON, A. 1983. Isolation and expression of an altered mouse dihydrofolate reductase cDNA. *Proc. Natl. Acad. Sci. U.S.A.* **80**:2495.

SIVE, H. and ST. JOHN, T. 1988. A simple subtractive hybridization technique employing photoactivatible biotin and phenol extraction. *Nuc. Acids Res.* **16**:10937.

SIVE, H., HATTORI, K., and WEINTRAUB, H. 1989. Progressive determination during formation of the anteroposterior axis in *Xenopus laevis*. *Cell* **58**:171.

SOUTHERN, P. and BERG, P. 1982. Transformation of mammalian cells to antibiotic resistance with a bacterial gene under the control of the SV40 early region promoter. *J. Mol. Appl. Gen.* **1**:327.

STAMBROOK, P., DUSH, M., TRILL, J., and TISHFIELD, J. A. 1984. Cloning of a functional human adenosine deaminase phosphoribosyltransferase (APRT) gene: Identification of a restriction-fragment-length polymorphism and preliminary analysis of DNAs from APRT-deficient families and cell mutants. *Som. Cell. Mol. Genet.* **4**:359.

SUBRAMANI, S., MULLIGAN, R., and BERG, P. 1981. Expression of the mouse dihydrofolate reductase complementary deoxyribonucleic acid in simian virus 40 vectors. *Mol. Cell. Biol.* **1**:854.

SUGDEN, B., MARSH, K., and YATES, J. 1985. A vector that replicates as a plasmid and can be efficiently selected in B-lymphoblasts transformed by Epstein-Barr virus. *Mol. Cell. Biol.* **5**:410.

TIMBLIN, C., BATTEY, J., and KUEHL, W. 1990. Applications for PCR technology to subtractive cDNA cloning: identification of genes expressed specifically in murine plasmacytoma cells. *Nuc. Acids Res.* **18**:1587.

TRAVIS, G., and SUTCLIFFE, J. 1988. Phenol-emulsion-enhanced DNA-driven subtractive cDNA cloning: Isolation of low-abundance monkey-cortex-specific mRNAs. *Proc. Natl. Acad. Sci.* **85**:1696.

WARE, L. and AXELRAD, A. 1972. Inherited resistance to N-tropic and B-tropic murine leukemia virus in vitro: Evidence that congenic mouse strains sim and sim.r differ at the FV-1 locus. *Virology* **50**:339.

WIGLER, M., SILVERSTEIN, S., LEE, L. S., PELLICER, A., CHENG, V. C., and AXEL, R. 1977. Transfer of purified herpes virus thymidine kinase gene to cultured mouse cells. *Cell* **11**:223.

WILSON, R. and BEBBINGTON, C. (Inventors), The University Court of the University of Glasgow (Applicants), Patent Title: Recombinant DNA sequences, vectors containing them, and method for the use thereof. International Publication Number: WO 87/04462. International Application Number: PCT/GC87/00039. International Publication Date: July 30, 1987.

VOGELSTEIN, B. and GILLESPIE, D. 1979. Preparative and analytical purification of DNA from agarose. *Proc. Natl. Acad. Sci. U.S.A.* **76**:615.

WONG, G., WITEK, J. S., TEMPLE, P. A., WILKENS, K. M., LEARY, A. C., LUXENBERG, D. P., JONES, S. S. BROWN, E. L., KAY, R. M., ORR, E. C., SHOEMAKER, C., GOLDE, D. W., KAUFMAN, R. J., HEWICK, R. M., WANG, E. A., and CLARK, S. C. 1985. Human GM-CSF: Molecular cloning of the complementary DNA and purification of the natural and recombinant proteins. *Science* **228**:810.

YEUNG, C., INGOLIA, E., BOBONIS, C., DUNBAR, B., RISER, M., SICILIANO, M., and KELLEMS, R. 1983. Selective overproduction of adenosine deaminase in cultured mouse cells. *J. Biol. Chem.* **258**:8338.

III

ASSAYS FOR GENE TRANSFER AND EXPRESSION

The protocols described in Part II of this manual enabled you to transfer genes of interest into cells in culture by a variety of methods. Part III of this manual contains experimental protocols that will enable you to monitor the transfer and expression of the transferred genes as well as the post-translational modification of the proteins expressed from these genes. What distinguishes the protocols presented here are their *extreme sensitivity* and their ease of *reproducibility*. Part III begins with assays for gene transfer including the detection of transferred DNA by both Southern analysis and PCR-based techniques. This is followed by assays for gene expression including the analysis of RNA by both Northern analysis and PCR-based techniques. Also included in this section are several protocols for the generation of DNA and RNA hybridization probes for both isotopic and non-isotopic detection of the transferred genes or their mRNA transcripts. This part closes with protocols for the analysis of the post-translational modification of expressed proteins including procedures for analyzing the state of glycosylation or acylation of a given protein.

11

ASSAYS FOR GENE TRANSFER

There are, of course, several methods that may be employed to monitor and quantitate successful gene transfer. One can assay for the presence of transferred DNA by a Southern blot of genomic DNA or, if the transferred DNA has some unique sequence feature, by the polymerase chain reaction (PCR) carried out on genomic DNA. One may also monitor successful gene transfer by assaying for mRNA production by the exogenous DNA through Northern analysis of cellular RNA or by RNA-driven PCR. Similarly, protein production can be monitored through the application of either immune precipitation or Western techniques, as well as immunofluorescence analysis or bioassay where appropriate. In this section on assays for gene transfer, I discuss and describe Southern- and PCR-based DNA-detection techniques. I will leave RNA- and protein-detection methods for the next section.

For maximum sensitivity and speed, PCR is the method of choice for assaying for gene transfer. With PCR, one can, in a matter of hours, unequivocally determine if gene transfer has occurred. Although PCR can also be used to quantitate gene transfer, Southern-blot analysis is less cumbersome to use quantitatively. The major drawback of Southern blotting is that one must first begin with purified DNA. On the other hand, PCR analysis of genomic DNA can be carried out on crude cell homogenates.

Generally speaking, one can derive more general information from Southern-blot analysis. A Southern can provide a wealth of map information, including information on junction fragments—that is, regions of the host chromosomal DNA flanking the integrated DNA. Such an analysis can reveal the arrangement of the integrated DNA in the host chromosome as well as provide accurate information about copy number of the integrated DNA.

By employing the techniques described here, you can achieve maximal sensitivity in such experiments with both isotopic and non-isotopic detection techniques. One major advantage of the non-isotopic, chemiluminescent technologies I describe are the remarkably short exposure times—on the order of seconds—required to detect a signal detectable by isotopic means after exposures of hours or days.

Both conventional Southern-type analysis and PCR-based gene detection analysis are performed on genomic DNA isolated from the cells of interest. We use the procedure described here to derive genomic DNA from both substrate-attached and suspension cell cultures. We have found that this procedure serves to generate very high molecular weight DNA that can, if necessary, be subsequently sheared by passage through a hypodermic needle. This DNA is ideal for the analyses described in this section.

PREPARATION OF GENOMIC DNA

Materials

1. Cell scrapers (Razor Policemen): Wrap a single-edge razor (VWR, cat. # TR 70106-50) lengthwise with tape (Baxter, cat. # L1600-75) so that the cutting edge is flush with the tape. Autoclave the scrapers in a glass petri dish or in aluminum foil.
2. Proteinase K (Boehringer Mannheim, cat. # 745723): Dissolve proteinase K in water at a final concentration of 20 mg/ml. Store at $-20°C$.
3. Large-bore glass pipettes (Baxter, 5-ml pipette, cat. # P4646-5W, 10-ml pipette, cat. # P4646-10W).
4. Lysis buffer: 10 mM Tris-HCl, pH 8.0, 5 mM EDTA, 0.5% (w/v) SDS, and 200 μg/ml proteinase K. Add the proteinase K to the lysis buffer just before use.
5. Phenol: Add 8-hydroxyquinoline (Sigma, cat. # H6878) to liquefied phenol (Sigma, cat. # P1037) to a concentration of 0.1% (w/v). Extract the phenol twice with an equal volume of 1.0 M Tris (pH 8.0). After the final extraction, mix the phenol with chloroform (Aldrich, cat. # 27,063-6) and isoamyl alcohol (Sigma, cat. # I1885) at a ratio of 50:48:2, respectively. Equilibrate the final phenol mixture with an equal volume of 100 mM Tris-HCl, pH 8.0, and 0.2 % (v/v) 2-mercaptoethanol (Sigma, cat. # M3148), and store at 4°C in a brown jar.
6. Dulbecco's phosphate buffered saline (PBS; Sigma, cat. # D5780).
7. TE: 10 mM Tris-HCl, 1 mM EDTA, pH 8.0.

Method

1. Rinse the confluent monolayers of cells twice with PBS.
2. Add lysis buffer to the monolayers (0.75–1.0 ml/100-mm dish) and incubate the dishes overnight at 37°C in a tissue-culture incubator.
3. Gently scrape the cell lysates into polypropylene test tubes with the autoclaved scrapers.
4. Gently extract the cell lysates once with phenol by slowly inverting the test tube.
5. Transfer the aqueous phase to dialysis tubing by using a wide-bore pipette.
6. Dialyze the DNA six times against 4 liters of TE at 4°C.
7. Dilute an aliquot of the DNA stock 1:10 in TE and shear by passing through a 22-gauge needle 12 times.
8. Determine the concentration of the sheared DNA sample spectrophotometrically (260 nm). An absorbance of 1.0 is equivalent to a concentration of 25 μg/ml DNA in a sample of total nucleic acids (DNA and RNA) extracted from mammalian cells.

Note: To extract genomic DNA from cells grown in suspension, collect the cells by centrifugation, wash the cells twice with cold PBS, resuspend the cells at 4×10^7 cells/ml in TE, and add 10 volumes of lysis buffer. Incubate the suspension overnight at 37°C, and proceed with steps 4–8.

SOUTHERN-TYPE TECHNIQUES

We use two very similar methods of Southern transfer in our lab. One is a modification of Maniatis et al. (1982) incorporating nitrocellulose filters. The other method is a modification of the Gene Screen Plus™ method using Gene Screen Plus™ transfer membranes. The basic differences between the two methods are reagents and treatment of the filters/membranes. The principle of DNA transfer is the same.

Method I

Materials

1. 1.5 M NaCl, 1.5 N NaOH.
2. 1 M Tris-HCl, pH 8.0, 1.5 M NaCl.
3. 20X SSC: 3 M sodium chloride, 0.3 M trisodium citrate, pH 7.0.
4. Glass Pyrex baking dishes.
5. Whatman 3MM filter paper.
6. Schleicher & Schuell BA85 nitrocellulose filter.

Method I:

1. Prepare an agarose gel in a horizontal gel apparatus. The percentage of agarose the gel should contain depends on the size of the DNA fragment you intend to resolve in the gel.
2. Obtain 0.2–0.5 μg of restriction endonuclease digested recombinant plasmid DNA, which is more than sufficient to allow inserted DNA sequences to be easily detected by Southern hybridization. For restriction endonuclease digested mammalian genomic DNA, run 10–30 μg per lane in order to detect sequences that occur as a single-copy in a haploid genome.
3. After electrophoresis, stain the gel with 0.5 μg/ml ethidium bromide for 30 minutes, and photograph the gel on an ultraviolet light box with a fluorescent ruler (Sigma, cat. # R8008) alongside the gel. This will help orient the DNA banding and migration pattern on the filter with respect to the gel.
4. Transfer the gel to a glass baking dish. Denature the DNA by soaking the gel in several volumes of 1.5 M NaCl/1.5 N NaOH for 1 hour at room temperature, with constant motion.
5. Pour off the denaturing solution and neutralize the gel by soaking it in several volumes of 1 M Tris-HCl (pH 8.0)/1.5 M NaCl for 1 hour at room temperature, with constant motion.
6. To set up the blot, we use the acrylic gel tray used to form the gel. Other supports can be used, such as a stack of glass plates. Place the gel support upside down in a glass Pyrex baking dish. Add 20X SSC to within ~1 cm of the top of the tray (Figure 11-1).
7. Wet a piece of Whatman 3MM paper approximately twice as long and of the same width as the gel. Wrap this over the support. This will function as a wick. Squeeze out any air bubbles that may have formed underneath the paper.
8. Cut a piece of nitrocellulose filter approximately 2 mm wider and 2 mm longer than the gel. Place a pencil mark in the lower right-hand corner of the filter to help orient the filter after hybridization. Prewet the nitrocellulose filter in H_2O, then add 2X SSC until the filter is completely wet. Note: *Always* handle nitrocellulose with gloves. Oil from hands prevents wetting of the filter, thus interfering with the binding of nucleic acids to the filter in those areas.

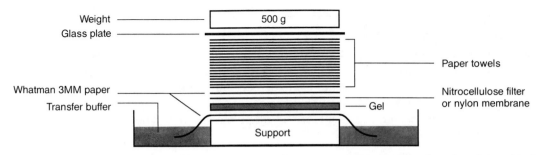

FIGURE 11-1 *Southern/Northern apparatus.* The arrangement of components for the transfer of nucleic acids from agarose gels to nitrocellulose filters and nylon membranes is depicted as described in the text. [Reprinted from Sambrook et al. (1989), *Molecular Cloning: A Laboratory Manual,* Cold Spring Harbor Press, Fig. 9.35.]

9. Invert the gel onto the Whatman wick so as to place the open wells in contact with the Whatman 3MM paper wick. Gently squeeze out any air bubbles.
10. Overlay the gel with the prewet nitrocellulose filter so as to position the pencil mark in the lower left-hand corner of the inverted gel. Make sure the entire gel is covered with the filter. Remove any air bubbles that may have become trapped between the filter and the gel.
11. Overlay the nitrocellulose filter with 2–3 pieces of Whatman 3MM paper (prewet with 2X SSC) cut to the same size as the gel. Remove any air bubbles.
12. Cut a stack of paper towels 6–8 cm high and ~2–3 mm smaller in both dimensions than the previous Whatman 3MM paper. Place on top of the Whatman 3MM paper.
13. Place a glass plate on top of the stack of paper towels, followed by a 500-g weight. The object is to set up a flow of liquid from the reservoir (in the glass baking dish) through the gel and the nitrocellulose filter, so that DNA fragments are eluted from the gel and deposited onto the nitrocellulose filter.
14. Cover the blotting gel with plastic wrap to prevent evaporation of the 20X SSC transfer solution.
15. The efficiency of DNA transfer is largely a function of the size of the DNA. As a matter of convenience, we allow the transfer to proceed for ~16 hours (overnight). Note: Replace the paper towels as they become wet.
16. Remove the paper towels and Whatman 3MM paper above the gel. Turn the dehydrated gel and filter over and lay them, gel side up, on a dry sheet of Whatman 3MM paper. Mark the positions of the gel slots on the filter with soft pencil.
17. Peel off the filter, rinse the filter with 2X SSC in a squirt bottle, and allow it to air-dry.
18. Bake the filter between two pieces of Whatman 3MM paper for 1.5–2 hours at 80°C in a vacuum oven.

Method II

Gene Screen Plus™ (NEN, cat. # NEF-984) is a synthetic transfer membrane that exhibits greater strength than nitrocellulose, does not require baking for fixing nucleic acids onto the membrane, and exhibits a higher binding capacity than nitrocellulose. When used in combination with non-isotopically labeled probes, nylon membranes display lower background. These features represent significant improvements over conventional nitrocellulose membranes.

Reagents

1. 0.4 N NaOH, 0.6 M NaCl.
2. 1.5 M NaCl, 0.5 M Tris-HCl, pH 7.5.
3. 10X SSC: 1.5 M NaCl, 0.15 M sodium citrate, pH 7.0.
4. 0.4 N NaOH.
5. 0.2 M Tris-HCl, pH 7.5, 2X SSC.

Gene Screen Plus™ Southern Transfer

(Note: Gene Screen Plus™ membranes have a natural curl to them when they are dry. The convex side is side A, and the concave side is side B.)

1. Electrophorese digested DNA into the gel. After electrophoresis, stain with 0.5 µg/ml ethidium bromide for 30 minutes and photograph gel with a fluorescent ruler beside the gel to facilitate analysis of the DNA-migration pattern.
2. Incubate the gel in 0.4 N NaOH, 0.6 M NaCl for 30 minutes, with constant rocking motion.
3. Pour off previous solution and incubate gel in a solution containing 1.5 M NaCl, 0.5 M Tris-HCl, pH 7.5 for 30 minutes at room temperature, with constant rocking motion.
4. Cut a Gene Screen Plus™ membrane 2 mm longer and 2 mm wider than the agarose gel. Label side B with a soft pencil mark in the lower right-hand corner. Prewet the membrane in deionized H$_2$O, then soak it for 15 minutes in 10X SSC. Note: Gloves must be worn whenever handling the membrane.
5. Assemble the gel support and set up the blot according to the same protocol used for nitrocellulose (Figure 11-1). Side B of the membrane should be in contact with the gel. Use 10X SSC as the transfer solution.
6. Allow the transfer to occur for 16 hours or overnight.
7. Remove the paper towels and Whatman 3MM paper. Mark the wells with a soft pencil as needed to orient the membrane.
8. Peel off the membrane and immerse in 0.4 N NaOH for 30–60 seconds.
9. Remove the membrane and immerse in a solution containing 0.2 M Tris-HCl, pH 7.5, 2X SSC for 2–3 minutes.
10. Place the membrane onto a piece of Whatman 3MM paper, with the transferred-DNA-side up. Allow membrane to dry at room temperature. *Do not* bake the membrane.
11. Proceed with prehybridization.

PCR-BASED TECHNIQUES

The detection of the presence or absence of nucleic-acid sequences has been enormously simplified by the introduction of the polymerase chain reaction (Mullis and Faloona, 1987; Saiki et al., 1988).

Here, a brief description is given of the use of PCR in the detection of DNA sequences, in either cell lysates or high-molecular-weight DNA isolated from tissue or cells containing transferred genes. In addition, some technical improvements are described.

The sensitivity of nucleic-acid detection by PCR depends upon a number of factors, including cycling conditions, integrity of the target sequences being used as templates, sequences of the primers being employed to direct amplification, and buffer conditions.

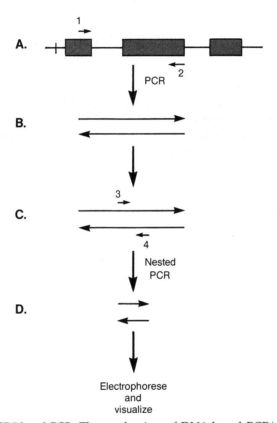

FIGURE 11-2 *PCR/Nested PCR.* The mechanism of DNA-based PCR/nested PCR described in the text is depicted here. First, the target DNA is mixed with amplimers 1 and 2, which flank the DNA of interest, and a PCR is performed (**A**) to produce the double-stranded product depicted in (**B**). This product can be analyzed after electrophoresis by either ethidium bromide staining or Southern hybridization analysis, or it can be used in a second type of PCR/nested PCR. The double-stranded product of (**B**) is mixed with internal amplimers 3 and 4, which are homologous to DNA sequences internal to the 5' and 3' ends of the original PCR product (**C**), and a PCR is performed, resulting in the nested PCR product (**D**). This product can be directly visualized after electrophoresis and ethidium bromide staining. Nested PCR serves as a double check for the specificity of the initial PCR reaction products and, upon completion, eliminates the need for Southern hybridization analysis of the PCR products.

In general, oligonucleotides 18–25 bases in length are most suitable for PCR, and, although a good rule of thumb is to design the oligonucleotides with 50 percent GC content, the actual amplification efficiency must be determined empirically. Unique primer sequences will anneal only to the target sequence and no mispriming will result. Unfortunately, you will not know if your PCR primer sequence represents a sequence unique to the target or if it is present elsewhere in the genomic DNA under study. In our laboratory, different primer pairs are first optimized to yield single-fragment products as defined by ethidium staining of agarose gels after electrophoresis. Usually 30–40 pmol of each oligonucleotide is used per reaction. The annealing temperature is optimized first, then the length of the annealing step is limited to further reduce non-specific PCR bands. In addition, the number of actual target molecules should be titrated, an excess usually resulting in the appearance of non-specific PCR products. Such optimization should lead to an increased stringency in the PCR, resulting in either (a) no requirement for subsequent hybridization because ethidium-bromide-stained agarose gels resolve the reaction products into single-product bands, or (b) a non-stringent and non-radioactive hybridization with an oligonucleotide complementary to a sequence between the amplimers.

Nested-primer amplification (Mullis et al., 1986; see Figure 11-2) will further increase the level of stringency and simplify detection. The nested-primer method involves the synthesis of two opposing primers that lie within the region synthesized by the first set of primers. The use of a 1/100th aliquot of the first reaction as a template for nested-primer amplification boosts the ability to detect sequence-specific targets because the inner primers introduce an increased level of sensitivity, both through their unique sequences and by the fact that the initial amplification results in the synthesis of an amplified target for the second, inner reaction.

Thus, two possible protocols may be followed in optimizing target-sequence detection: (1) a single reaction run for 35–50 cycles at a high stringency, with an aliquot analyzed thereafter on an ethidium-bromide-stained agarose gel with eventual hybridization, or (2) a nested-primer amplification yielding only a single detectable band on an agarose gel after ethidium-bromide staining without any further hybridization. Protocols for both approaches are provided here.

Cell-Lysate Work-Up

Materials

1. Proteinase K (Boehringer Mannheim, cat. # 745 723): Dissolve proteinase K at a concentration of 20 mg/ml in sterile water. Store in aliquots at $-20°C$.
2. Proteinase K digestion buffer: Prepare a solution containing 50 mM Tris-HCl, pH 8.5, 1 mM EDTA, 0.5% Tween 20 (Sigma, cat. # P1379).

Method

1. Pellet approximately 10^4 mammalian cells by centrifugation, and wash twice in PBS. Add to the washed pellet 100 μl of proteinase K digestion buffer, supplemented with proteinase K at a final concentration of 200 μg/ml.
2. Incubate for at least 2 hours at 55°C or overnight at 37°C.

3. Centrifuge the cell lysate to remove cell debris.
4. Incubate the lysate supernatant at 95°C for 8–10 minutes to denature the proteolytic enzymes and nucleases.
5. An aliquot of this supernatant may be used directly for PCR. The supernatant can be stored for future use at −20°C.

PCR Conditions

Generally, more than one reaction is run simultaneously. Thus, for the sake of convenience and to reduce the risk of cross-contamination, a PCR "cocktail" is prepared including dNTPs, the appropriate buffer, and the oligonucleotides required for amplification. For example, for 10 PCR reactions, aliquot 100 μl of a 10X PCR buffer (10X = 500 mM KCl, 200 mM Tris (pH 8.4) and 25 mM $MgCl_2$ 1 mg/ml BSA (nuclease free)), 10 μl of a 10 mM dNTP stock solution containing a mixture of all four deoxynucleotide triphosphates, and 300–500 pmol of each oligonucleotide, and adjust the cocktail to 1 ml volume (i.e., 100 μl per PCR reaction). On addition of the target nucleic acid, the tube is incubated at 95°C for 5 minutes to ensure denaturation of the template. The thermal cycler is allowed to cool to 80°C (for this, an additional file must be added to the amplification profile), at which time one unit of Taq polymerase is added. Subsequently, the tube is cycled at three temperatures. The temperatures are chosen so as to optimize the *in vitro* synthesis of template. In this respect, the annealing temperature is most critical, since specific priming reduces non-specific products and, hence, "wasted" synthesis of products by Taq polymerase.

A typical temperature-cycling profile will involve 30 seconds at 95°C (Note: the cycler must remain over 92°C actual temperature for 30 seconds), 30 seconds at an annealing temperature, usually between 50–72°C; and 30 seconds and longer at 72°C the actual time depending upon the size of the template being synthesized. As mentioned above, this temperature is arrived at empirically. Sequential PCRs may be run at ever-increasing annealing temperatures, then these products run side-by-side on agarose gels to decide upon an optimal annealing temperature.

Since PCR-product sizes range usually from 100–1,000 base pairs, 2–4% agarose gels are run to size the products. We use 3% NuSieve (FMC, cat. # 50082) and 1% SeaKem (FMC, cat. # 50074) agarose. However, normal agarose between 2–4% will size PCR products adequately.

Materials

1. 10X PCR Buffer: Prepare a solution containing 500 mM KCl, 200 mM Tris-HCl, pH 8.4, 25 mM $MgCl_2$, 1 mg/ml nuclease-free BSA (Pharmacia, cat. # 27-8914-01).
2. 10 mM dNTP stock: Prepare a solution containing 10 mM dATP, 10 mM dCTP, 10 mM dGTP, and 10 mM dTTP. Prepare from 100 mM stocks (Pharmacia, cat. # 27-2050-01, 27-2060-01, 27-2070-01, and 27-2080-01, respectively), diluting with 10 mM Tris-HCl, pH 7.5.
3. PCR primers: 30–50 pmoles/μl, containing both the 5'- and 3'-amplimers.
4. Taq polymerase (5 U/μl, Perkin-Elmer Cetus).
5. Mineral oil (Sigma, cat. # M5904).
6. Microfuge tubes: Use tubes certified for use in the thermal cycler.
7. 3% NuSieve/1% SeaKem agarose (FMC Corporation, cat. # 50092).
8. 6X loading buffer (6X LB): 0.25% bromophenol blue, 0.25% xylene cyanol, 30% glycerol in water.

9. TEA electrophoresis buffer: 40 mM Tris-HCl, pH 7.5, 1 mM EDTA, 5 mM sodium acetate.

Method

1. Prepare this PCR:

   - 1 µl   cell lysate
   - 10 µl 10X PCR buffer
   - 10 µl 10 mM dNTPs
   - 1 µl   upstream primer (30–50 pmoles/µl)
   - 1 µl   downstream primer (30–50 pmoles/µl)
   - 77 µl H$_2$O
   - 100 µl Total

   Layer 100 µl of mineral oil.

2. PCR amplification
   a. Incubate the reaction at 95°C for 5 minutes.
   b. Let the reaction cool to 80°C, then add one unit of Taq polymerase.
   c. Perform 25–35 cycles:
      - 95°C, 30 seconds (denaturing step).
      - 50–72°C, 30 seconds (annealing step).
      - 72°C, 30–120 seconds (extension step).
3. Extract the mineral oil by adding 300 µl of TE-saturated chloroform.
4. Extract any residual chloroform by ether extraction.
5. Remove any residual ether by heating the tube, open, at 68°C for 5 minutes.
6. Add 2 µl of 6X LB to 10 µl of PCR reaction and load into a composite 3% NuSieve/1% SeaKem agarose gel in TEA buffer.
7. Electrophorese, stain with 10 µg/ml ethidium bromide for 10 minutes, destain with H$_2$O for 20 minutes, and visualize on a UV light box. Blot the DNA, if desired.

Nested-PCR Conditions

For nested-PCR amplification, 1 µl from the *outer* PCR reaction is removed for the subsequent *inner*, or *nested*, PCR amplification. The sample from the *outer* reaction replaces the cell-lysate component, and the nested PCR primers replace the original PCR primers. Otherwise the PCR is identical (see Figure 11-2).

Blotting Procedure

Follow the procedures preceding this section for Southern blotting (see Southern-type techniques).

Probe Synthesis

Follow the protocol for labeling oligonucleotides with modified dUTP via terminal deoxynucleotidyl transferase (see the sections on synthesis of nucleic-acid probes for hybridization analysis; synthesis of non-isotopically labeled nucleic-acid probes).

Hybridization

Follow hybridization protocols for non-isotopic nucleic-acid probes (see the sections on hybridization conditions for nucleic-acid probes; hybridization conditions for non-isotopically labeled nucleic-acid probes).

Detection

Follow the protocol for Enhanced Chemiluminescence (ECL) (See the section on non-isotopic detection utilizing biotin-labeled probes: enhanced chemiluminescence (ECL) detection).

SYNTHESIS OF NUCLEIC-ACID PROBES FOR HYBRIDIZATION ANALYSIS

Synthesis of Radiolabeled Nucleic-Acid Probes

Synthesis of Radiolabeled DNA Probe: Nick Translation

Of the many methods for radiolabeling nucleic-acid probes, nick translation of DNA is the old standby, a classic. With the procedure presented here (Sambrook et al., 1989), high-specific-activity probes can be generated from plasmid, phage, or purified fragment DNA with very little work.

Materials

1. 10X nick-translation buffer (10X NTB): Prepare a solution containing 500 mM Tris-HCl, pH 7.5, 100 mM MgSO$_4$, 1 mM dithiothreitol, 500 μg/ml BSA (DNase and RNase-free, Sigma, cat. # B8894, 20 mg/ml). Aliquot and store at $-20°$C.
2. 100 mM dATP, 100 mM dGTP, and 100 mM dTTP (Pharmacia, cat. # 27-2050-01, 27-2070-01, and 27-2080-01, respectively).
3. Nick-translation-grade Polymerase I (BRL, cat. # 8162 SA): 0.4 U/μl DNA polymerase I and 40 pg/μl DNase I.
4. [α-^{32}P]dCTP (> 3000 Ci/mmole; 10 μCi/μl in aqueous solution; NEN, cat. # NEG-0137 or Amersham, cat. # PB.10475).
5. Sephadex G-50 spin column (Worthington Biochemicals, cat. # LS4404) or a push-column (Stratagene, cat. # 400701).

Method

1. Prepare a stock of unlabeled 20 mM dNTPs (minus dCTP) by adding 5μl of 100 mM dATP, 5 μl of 100 mM dGTP, 5 μl of 100 mM dTTP, and 10 μl of H$_2$O. This mixture is referred to as "3dNTPs."
2. Prepare this reaction:

5	μl	10X NTB
2	μl	3 dNTPs
10	μl	[α-^{32}P]dCTP (> 3000 Ci/mmole)
X	μl	dsDNA, 1 μg
Y	μl	H$_2$O, to bring to 45 μl
5	μl	nick-translation-grade-Pol I
50	μl	Total

3. Incubate at 16°C for 60 minutes.
4. Terminate the reaction with 5 μl of 0.5 M EDTA.
5. Separate the radiolabeled DNA from the unincorporated dNTPs by
   a. Centrifugation through a Sephadex G-50 spin column, or
   b. Passing through a Stratagene push column.
6. To use this radiolabeled DNA as a probe, it must be denatured (100° C for 5 minutes and then placed on ice immediately for 5 minutes) before adding to filter hybridization solution.

Note: Pol I and DNase I are sensitive to contaminants. Use "clean" DNA. Supercoiled DNA is the best substrate for nick translation. DNA fragments shorter than 2.0 kb should be labeled by the random-primer method.

Synthesis of Radiolabeled DNA Probe: Primer Extension via Random Priming

The radiolabeling of DNA by primer extension with random primers (modified from Keller and Manak, 1989) can be used to generate nucleic-acid probes with specific activities of 10^9 cpm/µg. The specific activities achieved with this method represent a substantial improvement over nick translation. The use of randomly primed high-specific-activity probes in hybridization analysis can dramatically reduce autoradiographic exposure times.

Materials

1. 10X random priming buffer (10X RPB): Prepare a solution containing 500 mM Tris-HCl, pH 6.6, 100 mM $MgCl_2$, 10 mM DTT, 500 µg/ml BSA (DNase and RNase-Free, Sigma, cat. # B8894).
2. p(dN)$_6$ (Pharmacia, cat. # 27-2166-01).
3. 100 mM dATP, 100 mM dGTP, and 100 mM dTTP (Pharmacia, cat. # 27-2050-01, 27-2070-1, and 27-2080-01, respectively).
4. [α-^{32}P]dCTP (>3000 Ci/mmole; 10 µCi/µl in aqueous solution; New England Nuclear or Amersham).
5. Sephadex G-50 spin column (Worthington Biochemicals, cat. # LS4404) or a push column (Stratagene, cat. # 400701).
6. Klenow fragment of *E. coli* Pol I (United States Biochemical, cat. # 70015).

Method

1. Prepare a "3 dNTPs" stock as in step (1) of previous instructions with 20 mM each of dATP, dGTP, and dTTP in H_2O.
2. Heat-denature 200 ng of double-stranded DNA (linear or closed circular) in 10 µl of H_2O, by boiling 5 minutes and placing immediately on ice for at least 5 minutes.
3. Prepare this reaction:

10 µl	Heat-denatured DNA, 200 ng
5 µl	10X RPB
5 µl	p(dN)$_6$, 2 mg/ml
1 µl	3 dNTPs, 20 mM
28 µl	[α^{32}P]dCTP (> 3000 Ci/mmole)
1 µl	Klenow (5 U/µl)
50 µl	Total

4. Incubate at 25°C for 3 hours.
5. Terminate the reaction with 5 µl of 0.5 M EDTA.
6. Separate the radiolabeled DNA from the unincorporated dNTPs by
   a. Centrifugation through a Sephadex G-50 spin column, or
   b. Passing through a Stratagene push column.
7. To use this radiolabeled DNA as a probe, it must be denatured (100°C for 5 minutes, and placed on ice immediately for 5 minutes) before adding to the filter-hybridization solution.

   Note: This high-specific-activity probe should be used immediately because of the rapid damage incurred by radioautolysis.

Synthesis of Radiolabeled Oligonucleotide Probe with Polynucleotide Kinase

The availability of synthetic oligonucleotides has changed the complexion of radiolabeled-probe synthesis. By this procedure (Carter et al., 1985), you can kinase an oligonucleotide to very high specific activity with a minimum of work. We use such probes routinely for screening cDNA libraries for plasmid clones of known sequences. You should keep in mind that the efficiency of kination can vary from oligonucleotide to oligonucleotide.

Materials

1. 10X kinase buffer (10X KB): Prepare a solution containing 500 mM Tris-HCl, pH 8.0, 100 mM MgCl$_2$, 50 mM DTT, 1 mM EDTA.
2. Oligonucleotide, unphosphorylated.
3. [γ-^{32}P]ATP (>3000 Ci/mmole, in aqueous solution; NEN, cat. # NEG-002H or Amersham, cat. # PB.10218).
4. T4 polynucleotide kinase (New England Biolabs, cat. # 201).
5. Stratagene push column (Stratagene, cat. # 400701).

Method

1. Prepare this reaction:

1.0 µl	oligonucleotide (10 pmoles/µl)
2.0 µl	10X KB
16.0 µl	[γ-^{32}P]ATP
1.0 µl	polynucleotide kinase (1–5 U/µl)
20.0 µl	Total

2. Incubate at 37°C for 45 minutes, then at 68°C for 10 minutes.
3. Purify the probe by passing through a Stratagene push column.

Synthesis of Radiolabeled RNA Probe: In Vitro Transcription

Although this procedure is a bit more complex than those cited previously, radiolabeled probes generated by *in vitro* transcription have several important qualities. Such probes are strand-specific, and they can be of very high specific, activity. If you desire a strand-specific probe, this is the method of choice. This method requires that the DNA insert for which you intend to generate a probe be inserted into a vector carrying an SP6 or a T7 promoter.

Materials

1. 5X transcription buffer: Prepare a solution containing 200 mM Tris-HCl, pH 7.5, 30 mM MgCl$_2$, 10 mM spermidine, 50 mM NaCl.
2. 10 mM ATP in H$_2$O, pH 7.0; 10 mM CTP in H$_2$O, pH 7.0; 10 mM GTP in H$_2$O, pH 7.0; 10 mM UTP in H$_2$O, pH 7.0 (Promega provides a kit, cat. # P1121).
3. 0.1 M DTT.
4. RNasin (Promega, 40 U/µl, cat. # N2111).
5. RQ1, RNase-free DNase (Promega, 1 U/µl, cat. # M6101).
6. H$_2$O, DEPC-treated (see the section on library construction and amplification).
7. T7 or SP6 RNA polymerase (Promega, 15–20 U/µl, cat. # P2073 and # P1083, respectively).
8. Linearized plasmid: Linearize the DNA 3' to the cloned insert with a restriction endonuclease that produces blunt ends or 5' overhangs. Phenol-extract and ethanol-precipitate the DNA. Resuspend the DNA pellet at a final concentration of 1 µg/ml in DEPC-treated H$_2$O.
9. [α-^{32}P] CTP (400–3000 Ci/mmole; NEN, Amersham).
10. Dilute 10 mM CTP to 0.1 mM with DEPC-treated H$_2$O.

Method

1. Prepare this reaction at room temperature:

4.0 μl	5X transcription buffer
2.0 μl	0.1 M DTT
0.5 μl	RNasin (40 U/μl)
1.0 μl	10 mM ATP
1.0 μl	10 mM GTP
1.0 μl	10 mM UTP
2.4 μl	0.1 mM CTP
5.0 μl	[α-^{32}P] CTP (400–3,000 Ci/mmole; 50 μCi)
1.0 μl	H$_2$O, DEPC-treated
1.1 μl	Linearized plasmid DNA (1μg/μl)
1.0 μl	SP6 or T7 RNA pol (15–20 U/μl)
20.0 μl	Total

2. Incubate for 60 minutes at 37°C.
3. Add 1 unit of RQ1 RNase-free DNase. Incubate for 15 minutes at 37°C.
4. Add 80 μl DEPC-treated H$_2$O.
5. Extract the transcription reaction with 100 μl of phenol: chloroform: isoamyl alcohol at 50:48:2.
6. Extract the reaction with 100 μl of chloroform.
7. Add 50 μl of 7.5 M ammonium acetate and 300 μl of ice-cold ethanol. Mix.
8. Incubate at −20°C for 30 minutes. Centrifuge at 4°C for 20 minutes in a microfuge.
9. Resuspend the pellet in 100 μl of DEPC-treated H$_2$O.
10. Repeat steps 7–9.
11. More than 99% of unincorporated [α-^{32}P]CTP should be removed by these multiple ethanol precipitations.
12. Add the probe to filter-hybridization solution immediately, using tRNA (wheat germ, Sigma, cat. # R7876) as carrier (200 μg/ml).

Non-isotopically Labeled Probes

Labeling of DNA with Biotinylated dUTP: Nick Translation

Of the many methods for labeling nucleic-acid probes, nick translation of DNA is the old standby, a classic. With the procedure presented here (a combination of Sambrook et al., 1989, and Keller and Manak, 1989), biotinylated probes for non-isotopic detection can be generated from plasmid, phage, or purified fragment DNA.

Materials

1. 10X nick translation buffer (10X NTB): Prepare a solution containing 500mM Tris-HCl, pH 7.5, 100mM MgSO$_4$, 1mM DTT, 500 μg/ml BSA (Sigma, cat. # B8894, 20 mg/ml). Aliquot. Store at −20°C.
2. 100 mM dATP, 100 mM dCTP, and 100 mM dGTP (Pharmacia, cat. # 27-2050-01, 27-2060-01, and 27-2070-01).
3. Nick-translation-grade Polymerase I (BRL, cat. # 8162 SA): 0.4 U/μl DNA polymerase I and 40 pg/μl DNase I.
4. 0.3 mM biotinylated dUTP: prepare a solution that is 0.2 mM dTTP (Pharmacia, cat. # 27-2080-01) and 0.1 mM Biotin-11-dUTP (Sigma, cat. # B7645; Enzo Diagnostics, cat. # NU-806) in 10 mM Tris-HCl, pH 7.5.
5. Sephadex G-50 spin column (Worthington Biochemicals, cat. # LS4404) or a push column (Stratagene, cat. # 400701).

Method

1. Prepare a stock of unlabeled 20 mM dNTP (minus dTTP) by adding 5 µl of 100 mM dATP, 5 µl of 100 mM dCTP, 5 µl of 100 mM dGTP, and 10 µl of H_2O. This mixture is referred to as "3 dNTPs."
2. Prepare this reaction:

5 µl	10X NTB
2 µl	3 dNTPs
34 µl	0.3 mM biotinylated dUTP
1 µl	dsDNA, 1 µg/µl
3 µl	H_2O
5 µl	nick-translation grade pol I
50 µl	Total

3. Incubate at 16°C for 60 minutes.
4. Terminate the reaction with 5 µl of 0.5 M EDTA.
5. Separate the DNA from the unincorporated dNTP by either:
   a. Centrifugation through Sephadex G-50 spin column, or
   b. Passing through Stratagene push column.
6. To use this labeled probe, it must be denatured (100°C for 5 minutes, and placed on ice for 5 minutes) before adding to filter hybridization mix.

Note: Pol I and DNase I are sensitive to contaminants. Use "clean" DNA. Supercoiled DNA is the best substrate for nick translation. DNA fragments shorter than 2.0 kb should be labeled by the random-priming method.

Labeling of DNA with Biotinylated dUTP: Primer Extension via Random Priming

The labeling of DNA by primer extension with random primers (modified from Keller and Manak, 1989) can be used to generate nucleic-acid probes containing biotin residues for use in non-isotopic detection of nucleic-acid sequences. The amount of biotinylated dUTP incorporated into the probe with this method represents a substantial improvement over nick translation. The use of randomly primed highly substituted probes in hybridization analysis can dramatically reduce film-exposure times.

Materials

1. 10X random priming buffer (10X RPB): Prepare a solution containing 500 mM Tris-HCl, pH 6.6, 100 mM $MgCl_2$, 10 mM DTT, 500 µg/ml BSA (DNase and RNase-free, Sigma, cat. # B8894).
2. p(dN)$_6$ (Pharmacia, cat. # 27-2166-01).
3. 100 mM dATP, 100 mM dCTP, and 100 mM dGTP (Pharmacia, cat. # 27-2050-01, 27-2060-01, and 27-2070-01, respectively).
4. 0.3 mM biotinylated dUTP: prepare a solution that is 0.2 mM dTTP (Pharmacia, cat. # 27-2080-01) and 0.1 mM Biotin-11-dUTP (Sigma, cat. # B7645; Enzo Diagnostics, cat. # NU-806) in 10 mM Tris-HCl, pH 7.5.
5. Sephadex G-50 spin column (Worthington Biochemicals, cat. # LS4404) or push column (Stratagene, cat. # 400701).
6. Klenow fragment of *E. coli* Pol I (United States Biochemical, cat. # 70015).

Method

1. Prepare a "3 dNTPs" stock with 10 mM each of dATP, dGTP, and dCTP in 10 mM Tris-HCl, pH 7.5.

2. Heat-denature 200 ng of double-stranded DNA (linear or closed circular) in 10 µl of H_2O by boiling 5 minutes and placing immediately on ice for at least 5 minutes.
3. Prepare this reaction:

10 µl	Heat denatured DNA, 200 ng
5 µl	10X RPB
5 µl	$p(dN)_6$, 2 mg/ml
2 µl	3 dNTPs, 10 mM
7 µl	0.3 mM modified dUTP
20 µl	H_2O
1 µl	Klenow (5 U/µl)
50 µl	Total

4. Incubate at 25°C for 3 hours.
5. Terminate the reaction with 5 µl of 0.5 M EDTA.
6. Separate the labeled DNA from the unincorporated dNTPs by
   a. Centrifugation through a Sephadex G-50 spin column, or
   b. Passage through a Stratagene push column.
7. Probe can be stored at −20°C (for at least 12 months).
8. This probe must be denatured (100°C for 5 minutes, on ice immediately for 5 minutes) before adding to hybridization solution.

Labeling of DNA with Biotinylated dUTP: PCR

With this method, you can prepare very large quantities of a sequence-specific probe containing biotinylated dUTP using PCR. This procedure (Lo et al., 1990) has all of the benefits associated with PCR and works quite well.

Materials

1. 10X PCR buffer: Prepare a solution containing 500 mM KCl, 200 mM Tris-HCl, pH 8.4, 25 mM $MgCl_2$, 1 mg/ml BSA (DNase and RNase-free, Pharmacia, cat. # 27-8914-01).
2. 10X dNTP stock: 10 mM dATP, 10 mM dCTP, 10 mM dGTP, and 7.2 mM dTTP. (Prepare from 100 mM dNTP stock solutions from Pharmacia, cat. # 27-2050-01, 27-2060-01, 27-2070-01, and 27-2080-01, respectively, diluting with 10 mM Tris-HCl, pH 7.5.)
3. 0.3 mM biotinylated dUTP: prepare a solution of 0.3 mM Biotin-11-dUTP (Sigma, cat. # B7645; Enzo Diagnostics, cat. # NU-806) in 10 mM Tris-HCl, pH 7.5.
4. PCR primers (50 pmoles/µl).
5. Taq polymerase (5 U/µl, Perkin-Elmer Cetus).
6. Mineral oil (Sigma, cat. # M5904).
7. Microfuge tubes: Use tubes certified for use in thermal cycler.
8. Glass beads (GENECLEAN, BIO-101).
9. 3% NuSieve/1% SeaKem agarose (FMC Corporation, cat. # 50092).
10. 6X loading buffer (6X LB): 0.25% bromophenol blue, 0.25% xylene cyanol, 30% glycerol in H_2O.
11. TEA electrophoresis buffer: 40 mM Tris-HCl, pH 7.5, 1 mM EDTA, and 5 mM sodium acetate.

Method

1. Prepare this PCR:

1 µl	target DNA (plasmid DNA, 1-5 ng/µl)	
10 µl	10X PCR buffer	
10 µl	10X dNTP stock	
17 µl	0.3 mM biotinylated dUTP	
1 µl	upstream primer (50 pmoles/µl)	
1 µl	downstream primer (50 pmoles/µl)	
1 µl	Taq polymerase (5 U/µl)	
59 µl	H$_2$O	
100 µl	total	

    Layer 100 µl of mineral oil.

2. PCR Amplification
    a. Denature the target DNA by incubating at 95°C for 8 minutes.
    b. Perform 25 cycles of
        - 95°C, 2 minutes
        - cooling 1 minute to 55°C
        - 55°C, 2 minutes
        - heating 1 minute to 72°C
        - 72°C, 3 minutes
        - heating 1 minute to 95°C
    c. Perform a 26th cycle, but stop the cycle after reaching the 72°C extension step. Incubate for 7 minutes.
3. Purification of PCR Fragment
    a. Extract the mineral oil by adding 300 µl of TE-saturated chloroform.
    b. To the aqueous layer, add 50 µl of 7.5 M ammonium acetate and 300 µl of ethanol.
    c. Incubate in an ethanol/dry-ice bath for 20 minutes, or at −70°C for 1 hr.
    d. Spin in microfuge at 4°C and 14 Krpm for 30 minutes.
    e. Resuspend the pellet in 10 µl of TE and 2 µl of 6X LB. Load a 3% NuSieve/1% SeaKem agarose gel in TEA buffer. Electrophorese the PCR products in TEA buffer.
    f. Stain the gel with ethidium bromide.
    g. Glass-bead purify the PCR fragment (see Part II, library construction and amplification, fractionation and purification of cDNA). Elute in 50 µl of H$_2$O.
    h. Store the DNA probe at −20°C for up to one year.
4. Boil probe for 10 minutes, then quick-chill on ice for 5 minutes, before adding to filter-hybridization solution.

Note: PCR conditions should be optimized for the production of a single PCR product. The above amplification procedure is a suggested starting point.

Labeling of Oligonucleotides with Biotinylated dUTP: Tailing

The tailing of oligonucleotides with biotinylated dUTP is the non-isotopic equivalent of the kination of oligonucleotides with ^{32}P-ATP. With this technique (modified from Keller and Manak, 1989), you add modified (biotinylated) dUTP to the 3' end of the oligonucleotide with terminal deoxynucleotide transferase. This procedure works very well and is very simple and quick. If you intend to probe filters carrying nucleic-acid sequences with oligonucleotides non-isotopically, this is the method of choice.

Materials

1. 5X terminal deoxynucleotide transferase buffer (5X TdT): 500 mM potassium cacodylate, pH 7.0, 1 mM $CoCl_2$, and 1.0 mM DTT.
2. Terminal deoxynucleotide transferase (Ratliff Biochemicals).
3. Modified dUTP: Dilute to 0.3 mM with 10 mM Tris-HCl, pH 7.5.
   a. Biotin-11-dUTP (Sigma, cat. # B7645; Enzo Diagnostics, cat. # NU-806).
4. Stratagene push column.

Method

1. Prepare this reaction:

1 µl	oligonucleotide (100 pmoles/µl)
3 µl	0.3 mM biotinylated dUTP
2 µl	5X TdT
3 µl	H_2O
1 µl	terminal transferase (3–5 units)
10 µl	Total

2. Incubate at 37°C for 20 minutes.
3. Add 1 µl of 0.5 M EDTA and 39 µl of H_2O to reaction.
4. Separate unincorporated dNTPs from labeled oligonucleotides by passing the reaction through a Stratagene push column.

Synthesis of Non-Isotopically Labeled RNA Probes with Biotinylated UTP

Although this procedure is a bit more complex than those cited previously, biotinylated probes generated by *in vitro* transcription have several important qualities. Such probes are strand-specific and they can be very highly substituted. If you desire a strand-specific probe for non-isotopic detection, this is the method of choice. This method requires that the DNA insert for which you intend to generate a probe be inserted into a vector carrying an SP6 or a T7 promoter.

Materials

1. 10X transcription buffer: Prepare a solution containing 240 mM Tris-HCl, pH 7.5, 30 mM $MgCl_2$, 10 mM spermidine.
2. 0.1 M DTT.
3. 10 mM ATP in H_2O, pH 7.0; 10 mM GTP in H_2O, pH 7.0; 10 mM CTP in H_2O, pH 7.0; 10 mM UTP in H_2O, pH 7.0 (Promega provides these in a kit, cat. # P1221).
4. 5 mM modified UTP: prepare a solution of 5 mM Bio-11-UTP (Enzo Diagnostics, cat. # NU-815; Sigma, cat. # B5770) in 10 mM Tris-HCl, pH 7.5.
5. RNasin (Promega, cat. # N2111).
6. T7 or SP6 RNA polymerase (Promega, cat. # P2703 and # P1083, respectively).
7. Linearized plasmid: Linearize the DNA 3' to the cloned insert with a restriction endonuclease that produces blunt ends or 5' overhangs. Phenol-extract and ethanol-precipitate the linearized DNA. Resuspend the DNA pellet at a final concentration of 1 µg/ml in DEPC-treated H_2O.
8. RQ1 RNase-free DNase (Promega, cat. # M6101).
9. 3 M sodium acetate.
10. DEPC-treated H_2O (see Part II, the section on library construction and amplification).

Method

1. Prepare this reaction at room temperature. Add reagents in this order:

5.0 µl	10X transcription buffer
5.0 µl	0.1 M DTT
20.3 µl	H$_2$O, DEPC-treated
1.3 µl	RNasin (40 U/µl)
2.5 µl	10 mM ATP
2.5 µl	10 mM CTP
2.5 µl	10 mM GTP
1.7 µl	10 mM UTP
1.7 µl	5 mM modified UTP
2.5 µl	linearized plasmid DNA (1µg/ml)
5.0 µl	T7 or SP6 RNA polymerase (10 U/µl)
50.0 µl	Total

2. Incubate at 40°C for 60 minutes.
3. Add 2.5 units of RQ1 RNase-free DNase to the transcription reaction.
4. Incubate at 37°C for 15 minutes.
5. Add 5 µl of 3 M sodium acetate and 138 µl of ethanol. (No phenol extraction is necessary).
6. Incubate for 30 minutes at −20°C. Centrifuge at 4°C in microfuge for 20 minutes. Decant the supernatant. Allow the pellet to air-dry.
7. Resuspend air-dried pellet in 100 µl of DEPC-treated H$_2$O.
8. Store the probe at –70°C.

Photobiotin Labeling of DNA and RNA

This method of nucleic-acid modification by photobiotinylation (McInness et al., 1987) for non-isotopic hybridization analysis is very quick and highly useful. With this technique, you can generate large quantities of modified nucleic acid. You can even directly modify purified mRNA for use as a probe, if you wish.

Materials

1. Photobiotin acetate (Clontech, cat. # 5000-1; Vector Laboratories, cat. # SP1000; BRL, cat. # 8186SA): Dissolve in H$_2$O at a concentration of 1 mg/ml. The stock solution can be stored at −20°C *protected from light* for up to 6 months.
2. Sunlamp: General Electric model RSM/H equipped with a 275W bulb.
3. RNA or DNA sample at a concentration of 1 µg/µl in H$_2$O or 0.1 mM EDTA. (Use DEPC-treated H$_2$O for RNA sample.)
4. Sec-butanol.
5. 100 mM Tris-HCl, pH 9.0, 1 mM EDTA.
6. 3 M sodium acetate, pH 5.2.

Method

1. Mix 25 µl of photobiotin stock with 25 µl of nucleic-acid stock in a sterile microfuge tube.
2. Place tube, with lid open, in an ice bath, 10 cm below the sunlamp. Irradiate for 15–20 minutes.
3. Add 50 µl of 100 mM Tris-HCl, pH 9.0, 1 mM EDTA.
4. Extract twice with 100 µl of sec-butanol. (The final volume of the aqueous phase will be ~35 µl.)
5. Add 5 µl of 3 M sodium acetate and 100 µl of ethanol.
6. Incubate for 30 minutes at −70°C or overnight at −20°C.

7. Centrifuge for 25 minutes at 4°C and 14 Krpm in a microfuge.
8. Wash the pellet with cold 70% ethanol and air-dry. (The pellet should be orange-brown in color.)
9. Dissolve the pellet in 0.1 mM EDTA, pH 8.0, and store at −20°C (for up to 8 months).
10. Measure absorbance at 260 nm to determine probe concentration.

Note: 1. The nucleic-acid sample must be free of any contaminating organic material such as DNA, RNA, proteins, *or buffers such as Tris.* 2. Do not irradiate samples in volumes greater than 50 µl, otherwise the light intensity will be attenuated, which will lower the biotinylation efficiency.

NON-ISOTOPIC DETECTION

Non-Isotopic Detection Utilizing Biotin-Labeled Probes: Colorimetric Detection

In this section, I discuss non-isotopic detection of biotinylated probes post hybridization (For hybridization conditions, see the section on hybridization conditions for nucleic-acid probes). Two techniques are presented here. The first, colorimetric detection (Helmuth, 1990), is far less sensitive than the second. I present it here because it is commonly used and does not require X-ray film or a developer to obtain a hard copy of the data. For many purposes, colorimetric detection is sufficient.

Materials

1. Southern or Northern filters hybridized with biotinylated probes and washed. (Keep filters moist if reprobing is desired.)
2. Streptavidin horseradish peroxidase conjugate (1 mg/ml, Vector Laboratories, cat. # SA-5004; 1 mg/ml, Boehringer Mannheim, cat. # 1089–153).
3. Buffer 1: PBS, 100 mM NaCl, 5% Triton X-100 (Boehringer Mannheim, cat. # 789-704).
4. Buffer 2: PBS, 100 mM NaCl, 5% Triton X-100, 1 M urea (Sigma, cat. # U5378), 1% dextran sulfate (Sigma, M.W. 500,000, cat. # D8906.)
5. Buffer 3: 100 mM sodium citrate (pH 5.0).
6. 3,3',5,5'-tetramethylbenzidine (TMB, Boehringer Mannheim, cat. # 784-974): 2 mg/ml in 100% ethanol. (Store in an opaque container at 4°C. This stock will remain stable for 2 months.)
7. Hydrogen peroxide (Sigma, cat. # H1009).
8. Buffer 4: 19 parts buffer 3 and one part 2 mg/ml TMB.
9. Buffer 5: 2000 parts buffer 4 and one part 3% hydrogen peroxide.

Method

1. Incubate the filter in Buffer 1 (10 ml/100 cm^2), supplemented with streptavidin horseradish peroxidase conjugate at a final concentration of 5 ng/ml, for 10 minutes at room temperature, with moderate shaking.
2. Incubate the filter in Buffer 2 (10 ml/100 cm^2) for 5 minutes at room temperature, with moderate shaking.
3. Wash the filter in Buffer 3 for 5 minutes at room temperature, with moderate shaking. Prepare Buffer 4 during the Buffer 3 wash step. Keep the TMB solution away from light to avoid photo-oxidation.
4. Incubate the filter in Buffer 4 (20 ml/100 cm^2) for 10 minutes at room temperature, with moderate shaking. Cover the dish with foil. Prepare Buffer 5 during the Buffer 4 wash step. Keep Buffer 5 covered.
5. Develop the filter in Buffer 5 at room temperature, with moderate shaking. Keep this buffer covered. Development time is generally 1–15 minutes.
6. Stop the color development by washing the filter with 3 changes of Buffer 3 during a 60-minute incubation at room temperature.
7. Photograph the filter for a permanent record. The filter can be stored permanently in Buffer 3 in a heat-sealed bag covered with foil, with a minimal reduction of the signal.

Note: 1. Keep TMB and TMB solutions away from light to prevent photo-oxidation. 2. Faint signal: Try increasing probe concentration, increasing hybridization time, and decreasing stringency of hybridization/wash conditions. 3. High background: Try decreasing probe concentration, decreasing hybridization time, increasing stringency of hybridization/wash solutions, and decreasing development time.

Non-isotopic Detection Utilizing Biotin-Labeled Probes: Enhanced Chemiluminescence (ECL) Detection

This method of non-isotopic detection (a modification of Helmuth, 1990) with enhanced chemiluminescence (ECL) is remarkably sensitive and is the method of choice in our laboratory. With this procedure, you can vary exposure times (unlike the colorimetric method) to obtain the optimal signal. *Exposure times of a few seconds are not uncommon.*

Materials

1. Southern or Northern filters hybridized with biotinylated probes and washed. (Keep filters moist if reprobing is desired.)
2. Streptavidin horseradish peroxidase conjugate (1 mg/ml, Vector Laboratories, cat. # SA-5004; 1 mg/ml, Boehringer Mannheim, cat. # 1089-153).
3. Buffer 1: PBS, 100 mM NaCl, 5% Triton X-100 (Boehringer Mannheim, cat. # 789-704).
4. Buffer 2: PBS, 100 mM NaCl, 5% Triton X-100, 1 M urea (Sigma, cat. # U5378), 1% dextran sulfate (Sigma, M.W. 500,000, cat. # D8906).
5. ECL Gene Detection System, Reagents 1 and 2 (Amersham, cat. # RPN. 2105).
6. Hyperfilm-ECL (Amersham, cat. # RPN. 2103).

Method

1. Incubate the filter in Buffer 1 (10 ml/100 cm^2), supplemented with the streptavidin horseradish peroxidase conjugate at a final concentration of 5 ng/ml, for 30 minutes at room temperature, with moderate shaking.
2. Wash the filter in Buffer 1 for 5 minutes at room temperature, with moderate shaking. REPEAT.
3. Wash the filter twice in Buffer 2 for 5 minutes each at room temperature, with moderate shaking.
4. Incubate the filter in a 10-ml solution consisting of 5 ml of ECL Reagent 1 and 5 ml of ECL Reagent 2 for 1 minute at room temperature, with moderate shaking.
5. Cover the moist filter carefully with Saran wrap and expose for 5 minutes to Hyperfilm-ECL (Kodak XAR-5 film may also be used, but Hyperfilm-ECL is much more transparent, post processing, than is Kodak XAR-5).
6. Depending on the signal intensity desired, the exposure can be repeated for longer or shorter time periods.

Note: ECL signal has a half-life between 20–30 minutes. For single-copy gene detection, the first 60 minutes of exposure will provide the strongest signal.

HYBRIDIZATION CONDITIONS FOR NUCLEIC-ACID PROBES

Hybridization Conditions for Radiolabeled Nucleic-Acid Probes

The hybridization conditions described next are appropriate for both nitrocellulose filters and nylon membranes, as well as for the analysis of processed colony-laden filters of cDNA plasmid libraries by hybridization of probes. The hybridization conditions for radiolabeled and non-isotopically labeled probes are slightly different, and therefore nearly identical protocols are provided for both. With non-isotopic probes, the transfer of nucleic acid to nylon membranes results in lower background.

Materials

1. 50X Denhardt's: Prepare a solution containing 5 g of Ficoll (Sigma, cat. # F2637), 5 g of polyvinylpyrrolidone (Sigma, cat. # P5288), and 5 g of bovine serum albumin (Sigma, cat. # A7888) in 500 ml H_2O. Filter through a 0.45-μ filter and store at $-20°C$.
2. Denatured, sheared salmon-sperm DNA:
    a. Dissolve salmon-sperm DNA (Type III, sodium salt, Sigma, cat. # D1626) in H_2O at 10 mg/ml.
    b. Stir the solution on a magnetic stirrer from 4–6 hours at room temperature.
    c. Adjust the solution to 0.1 M NaCl. Extract once with TE-saturated phenol and once with TE-saturated phenol:chloroform.
    d. Recover the aqueous phase. Shear the DNA by passing 12–15 times through an 18-gauge hypodermic needle.
    e. Add 2 volumes of 95% ethanol and centrifuge at 4°C for 30 minutes at 12,000g.
    f. Redissolve the pellet at approximately 10 mg/ml in H_2O, but determine the concentration exactly by spectrophotometry.
    g. Boil for 10 minutes and aliquot. Store at $-20°C$.
    h. Heat DNA at 100 °C for 5 minutes, then quick-chill in ice water before adding to hybridization solutions.
3. 20X SSC: Dissolve 175.3 g of NaCl and 88.2 g of sodium citrate in 800 ml of H_2O. Adjust pH to 7.0 with 1 N NaOH. Adjust volume to 1.0 liter. Sterilize by autoclaving.
4. Formamide (Fluka, cat. # 47670; Aldrich, cat. # 22, 119-8): If any yellow color is present, mix 50 ml of formamide with 5 g of mixed bed, ion-exchange resin (Bio Rad, AG 501-X8(D), cat. # 142-6425). Stir for 30–45 minutes at room temperature on a magnetic stirrer. Filter twice through Whatman No. 1 filter paper. Aliquot and store at $-20°C$.
5. Seal-A-Meal bags and heat-sealer (Sears).
6. Prehybridization/hybridization solutions: Southern blotting
    a. 68°C hybridizations: 6X SSC, 5X Denhardt's; 0.5% SDS (nylon filters 1.0% SDS); 100 μg/ml denatured, sheared salmon-sperm DNA.
    b. 42°C hybridizations: 6X SSC; 5X Denhardt's, 0.5% SDS (nylon filters 1.0% SDS); 100 μg/ml denatured, sheared salmon-sperm DNA; and 50% formamide.
7. Prehybridization/hybridization solutions: Northern blotting
    a. 68°C hybridizations: 6X SSC, 5X Denhardt's, 0.5% SDS (nylon filters 1.0% SDS), 20 mM $NaPO_4$, pH 7.0, 5 mM EDTA, 200 μg/ml tRNA (wheat germ, Sigma, cat. # R7876).
    b. 42°C hybridizations: Same as above but with 50% formamide.

8. 2X SSC.
9. Low-stringency wash solution: 2X SSC, 0.1% SDS (1% SDS for nylon filters).
10. High-stringency wash solution: 0.1X SSC, 0.1% SDS (0.5% SDS for nylon filters).

Method

1. Electrophorese the nucleic acids and transfer them to either nitrocellulose filters or nylon membranes following the protocols described in the section on Southern-type techniques.
2. Place the filter or membrane into a heat-sealable bag and add the appropriate pre-hybridization solution at 200 µl/cm² of filter. Squeeze out as many bubbles as possible, then seal the bag.
3. Incubate 1–2 hours submerged in a water bath:
   a. 68°C for aqueous hybridization.
   b. 42°C for 50% formamide hybridization.
4. Denature double-stranded probes by heating for 5 minutes at 100°C. Quick-chill by placing in ice water.
5. Cut off the corner of a bag, add probe, then reseal. For nylon membranes, change the prehybridization solution completely, add probe, then reseal.
6. For safety's sake, either place the bag in a plastic box or seal the bag in a second bag to contain any radioactive hybridization solution that might leak. Submerge the bag and its container in a water bath set to the appropriate temperature, and incubate for 12–16 hours for aqueous solutions and 24 hours for formamide solutions. Note: Oligonucleotide probes should be hybridized in aqueous solutions at 10°C below their T_m's.
7. Remove the bag, cut one corner, and pour the hybridization solution into a radioactive waste container. Cut three sides of the bag and immediately slip the filter or membrane into a tray of 2X SSC at room temperature. (Do not let the filter dry during the washing steps.)
8. After 5 minutes, wash the filter or membrane in low-stringency wash solution for 30 minutes at 68°C (for oligonucleotide probes, wash at 5°C below T_m), with gentle agitation. REPEAT.
9. Wash the filter or membrane with high-stringency wash solution for 30 minutes at 68°C (for oligonucleotide probes, wash at 5°C below T_m), with gentle agitation. REPEAT.
10. Use a hand-held mini-monitor to determine roughly the amount of background radioactivity on the filter or membrane. In single-copy gene analysis of mammalian genomic DNA, you may not be able to detect any radioactivity with a mini-monitor from a bona fide signal.
11. Remove excess liquid by blotting the filter or membrane on Whatman 3MM paper.
12. Place the damp filter or membrane into a heat-sealable bag and seal. Mount the bag on pre-cut Whatman 3MM filter paper, and mark with phosphorescent ink (Ult-emit, New England Nuclear, cat. # NEF-980).
13. Expose the filter or membrane to X-ray film (Kodak XAR-5 or Amersham Hyperfilm-ECL, cat. # RPN 2103) with an intensifying screen at −70°C to obtain an autoradiograph.

Hybridization Conditions for Non-Isotopically Labeled Nucleic-Acid Probes

The hybridization conditions for non-isotopically labeled probes are described here. These conditions are appropriate for the analysis of processed, colony-laden filters of cDNA plasmid libraries by hybridization of probes. The hybridization conditions for radiolabeled and non-isotopically labeled probes are slightly different, and therefore nearly identical protocols are provided for both. Solution formulations are not repeated in this section because they are identical with those in the section on hybridization conditions for radiolabeled nucleic acid probe; with non-isotopic probes, the transfer of nucleic acid to nylon membranes results in lower background.

Materials

1. Nylon membranes: Nytran (Schleicher & Schuell), Magnagraph (Micron Separations, Inc.), Biodyne A (Pall Biosupport Corp.).
2. 50X Denhardt's.
3. Denatured Salmon sperm DNA.
4. 20X SSC.
5. Formamide.
6. Seal-A-Meal bags and heat-sealer (Sears).
7. Prehybridization/hybridization solutions: Southern Blotting.
8. Prehybridization/hybridization solutions, Northern Blotting.
9. 2X SSC.
10. Low-stringency wash solution. 2.0X SSC, 1% SDS.
11. High-stringency wash solution. 0.1X SSC, 0.5% SDS.

Method

1. Electrophorese the nucleic acids and transfer them to nylon membranes following protocols described in the section on Southern-type techniques.
2. Place a filter into a heat-sealable bag and add the appropriate prehybridization solution at 200 ml/cm^2 of filter. Squeeze out as many bubbles as possible, then seal the bag.
3. Prehybridize by incubating the bag for 1–2 hours submerged in a water bath:
   a. 68°C for aqueous solutions.
   b. 42°C for 50% formamide solutions.
4. Denature double-stranded probes by heating for 5 minutes at 100°C, then quick-chill by placing in ice water.
5. Cut corner of bag, pour out prehybridization solution, and replace it with fresh hybridization solution, at 5 ml/100-cm^2 filter, containing probe at a concentration of 50–200 ng/ml. Submerge the bag in a water bath set to the appropriate temperature and incubate for 12–16 hours for aqueous solutions and 24 hours for formamide solutions. Note: Oligonucleotide probes should be hybridized in aqueous solutions at 10°C below their T_m's.
6. Remove the bag, cut off one corner, and pour out the hybridization solution. (Hybridization solutions containing DNA probes may be stored at −20°C and reused several times. Just before reuse, denature the probe by heating the hybridization solution to 95°C.)
7. After a 5-minute wash with 2X SSC, wash filter in 2X SSC, 0.1% SDS for 30 minutes at 68°C (for oligonucleotide probes, wash at 5°C below T_m), with gentle agitation. REPEAT.
8. Wash the filter in 0.1X SSC, 0.1% SDS for 30 minutes, at 68°C (for oligonucleotide probes, wash at 5°C below T_m), with gentle agitation. REPEAT.
9. The filter can be used directly after the washing steps for immunological detection, or stored after air-drying.

10. Note: 1. Low sensitivity: a. Check the labeling efficiency of the probe. b. Decrease the incubation temperature to 60°C for aqueous hybridization and 37°C for formamide hybridization. c. Increase probe concentration to 200 ng/ml. 2. High background: Repurify the probe by phenol/chloroform extraction and/or ethanol precipitation.

12
ASSAYS FOR GENE EXPRESSION

The expression of a given gene is initially manifest as an RNA transcript, which is subsequently translated into a protein molecule. That protein molecule may possess characteristics such as antigenicity or bioactivity that enable the experimenter to monitor its production. Described next are the most sensitive, reliable assays currently available for monitoring the presence of specific RNAs, including Northern- and PCR-based methods, as well as methods for monitoring the presence of a given protein, including immunoprecipitation, Western, and immunofluorescence analysis. Also included are the quickest, most sensitive isotopic and non-isotopic detection methods for both specific RNAs and specific proteins. The protein-detection method facilitates the detection of nanogram quantities of proteins in Western blotting experiments with exposure times of but a few seconds. Lastly, procedures for the analysis of post-translational modification of proteins are described.

ANALYSIS OF RNA (NORTHERN- AND PCR-BASED TECHNIQUES)

In addition to the analysis of gene expression after gene transfer, several of the protocols in this manual require purified mRNA of very high quality. Execution of the procedures presented here will result in the generation of very high quality mRNA, mRNA that will give a clear signal on a Northern blot, can serve as an excellent target molecule for PCR analysis, or can be used for the generation of hybridization probes and subtracted probes, cDNA libraries and subtracted libraries. I emphasize the use of RNase-free reagents because this is the secret of success.

RNase-Free Techniques

To avoid exogenous sources of RNase contamination, we routinely incorporate certain precautionary measures when working with RNA. These include the following:

1. All reagents used in the isolation or analysis of RNA are maintained separately from general-use stocks. They are handled with gloves and measured out with heat-baked spatulas onto disposable plastic weigh boats.
2. All reagents requiring water are made with DEPC-treated H_2O (see Materials) and stored in heat-baked glassware or sterile disposable plasticware.

3. Glassware is baked at 180°C dry heat for 8–10 hours.
4. Disposable, sterile plasticware such as pipette tips, tubes, and pipettes are used exclusively.
5. A designated electrophoresis apparatus is used for only RNA analysis and is pretreated with 0.1% DEPC for 1 hour, then rinsed with copious amounts of sterile DEPC-treated H$_2$O prior to use.
6. Gloves are always worn whenever working with RNA, RNA reagents, or equipment, and they are changed frequently, as hands are a major source of RNase contamination.

Chemical Precautions

Many of the chemicals used in RNA processing are known to be hazardous. A high level of caution should be exercised with the following chemicals, as they are, at the very least, strong irritants.

1. Formaldehyde.
2. Formamide.
3. Guanidine isothiocyanate.
4. Diethylpyrocarbonate (DEPC).
5. 2-mercaptoethanol.

To maintain laboratory safety, a fume hood is essential for most procedures involving the manipulation of RNA.

Total mRNA Isolation

The first step in the isolation of mRNA is the isolation of total RNA (mRNA, ribosomal RNA, tRNA, and other types of RNA). The procedure I describe here works well because it incorporates a powerful, fast-acting protein (RNase) denaturant, guanidine isothiocyanate (GITC). While there may be quicker RNA isolation procedures, execution of this procedure ensures maximum recovery of intact RNA molecules, so I think the extra work is worth it.

Materials

1. DEPC-Treated H$_2$O: Used in preparation of all reagents requiring H$_2$O. Add diethylpyrocarbonate (Sigma, cat. # D5758) to 0.1% (v/v) to sterile H$_2$O. Mix well, and allow to stand overnight. Autoclave to dissipate the DEPC.

   Note: DEPC reacts rapidly with amines and cannot be used to treat solutions containing buffers such as Tris. DEPC-treated H$_2$O must be autoclaved prior to reagent preparation.

2. GITC solution: 5 M guanidine isothiocyanate (International Biotechnologies, Inc., cat. # IB05100), 25 mM sodium citrate (Sigma, cat. # C8532), pH 7.0, 0.5% N-lauroylsarcosine (Sigma, cat. # L5125). Store at room temperature.
3. GITC-BME 5%: Add 5 ml of 2-mercaptoethanol (Sigma, cat. # M3148) to 95 ml of GITC solution. Prepare this solution just prior to use.
4. 20 mM Tris-HCl, 2 mM EDTA, pH 7.5.
5. 5.7 M CsCl (International Biotechnologies, Inc., cat. # 37062) in 20 mM Tris-HCl, pH 7.5, 2 mM EDTA.
6. 40% w/v CsCl in 20 mM Tris-HCl, pH 7.5, 2 mM EDTA.
7. Formaldehyde (Baker, cat. # 2106-11).

8. Formamide (Fluka, cat. # 47670; Aldrich, cat. # 22, 119-8): If any yellow color is present, deionize by stirring 500 ml of formamide with 50 g of AG501-X8(D) resin (Bio-Rad, cat. # 142-6425) on a magnetic stirrer for 1 hour. Filter through Whatman #1 filter paper. Repeat. Store at $-20°C$.
9. Cell homogenizer (Brinkman).
10. High-speed ultracentrifuge.
11. Rotors and polyallomer tubes (Beckman): SW 60 Ti, Tube, cat. # 328874. SW 41 Ti, Tube, cat. # 331372. SW 28, Tube, cat. # 326823.

Method

1. Clean the cell-homogenizer probe with 0.2 N NaOH at 65°C for 35 min. Rinse with copious amounts of sterile H_2O. Rinse until the pH of the rinse solution is neutral. No residual NaOH should remain.
2. Soak the probe in GITC-BME solution for 30 minutes prior to use. Run probe at high speed to break up any debris that may have clung to the probe.
3. Pellet the cells to be homogenized in a sterile 50-ml polypropylene conical tube. Decant off all the supernatant. Loosen cell pellets by tapping. (Note: Store the pellets on ice until ready to process.)
4. *Pour* the GITC-BME solution into the conical tube. Add 15 ml of GITC-BME solution per 2.5 ml of cell-pellet volume. Begin homogenizing with 1–2- second pulses. Continue with three 10–20-second pulses. Homogenate will appear clear.
5. Rinse the probe with several changes of H_2O. Store the probe immersed in H_2O.
6. Spin the homogenate in a table-top centrifuge at ~2.5 Krpm to dissipate the foam that has formed.

At this point you will centrifuge the homogenate through a cesium chloride (CsCl) step gradient. For quantitative information regarding gradient construction, see Table 12-1.

7. To construct the gradient, overlay the 5.7 M CsCl with 40% w/v CsCl, then slowly overlay the cushion with the cell homogenate. Overlay the cell homogenate with additional GITC-BME as needed to bring the total volume within 2–3 mm of the top of the tube.
8. Spin for about 16 hours at the appropriate speed. *Aspirate* off 75% of the gradient. *Pour* off the last 25% of the gradient. Invert the tubes and allow the tubes to drip-dry. Cut off the lower "cup" from the end of the centrifuge tube with a sterile scalpel or razor blade and wipe the area around the perimeter of the glassy-appearing RNA pellet with a sterile cotton swab. Resuspend the RNA pellet in DEPC-treated H_2O and measure O.D. at 260 and 280 nm. A 260/280 ratio of 1.8 or greater indicates that the RNA preparation is protein-free. 1 $O.D._{260}$ = 40 μg/ml RNA. Maintain the resuspended RNA on ice.

TABLE 12-1

Rotor Type	Volume of 5.7 M CsCl Cushion	Volume of 40% w/v CsCl Added	Volume of Homogenate Overlaid	Centrifugation Speed
SW60	0.75 ml	0.5 ml	3 ml	35 Krpm
SW41	2 ml	1.5 ml	7.5 ml	35 Krpm
SW28	10 ml	6 ml	15 ml	26 Krpm

9. Process RNA for poly (A)$^+$ selection immediately or freeze under ethanol (0.3 M sodium acetate and 2.5X volume 100% ethanol) and store at $-70°C$.

Poly (A)$^+$ RNA Selection

Having completed the isolation of total RNA, you will want to isolate the messenger RNA, which is primarily poly (A)$^+$ RNA, from the remainder of the molecules. The use of oligo (dT) cellulose for affinity chromatography of poly (A)$^+$ RNA is the method of choice and is described here:

Materials

1. *2X loading buffer:* 40 mM Tris-HCl, pH 7.6, 1 M NaCl, 2 mM EDTA, 0.2% SDS.
2. 0.1 N NaOH.
3. *Loading buffer (low-salt):* 40 mM Tris-HCl, pH 7.6, 0.1 M NaCl, 2 mM EDTA, 0.2% SDS.
4. *Elution buffer:* 10 mM Tris-HCl, pH 7.5, 1 mM EDTA, 0.05% SDS.
5. 3 M sodium acetate, pH 5.2.
6. 100% ethanol.
7. 70% ethanol.
8. Oligo(dT) cellulose, Type 2 (Collaborative Research, cat. # 20002).
9. Quik-Sep columns (Isolab, cat. # QS-Q).

Method

1. Equilibrate the oligo(dT) cellulose in sterile loading buffer.

    Note: Binding affinity is batch-specific; however, generally speaking, 1 g will bind $\approx$2–2.5 mg of Poly(A)$^+$ RNA.

2. Pour 1.0 ml packed volume of oligo(dT) cellulose in the Quik-Sep column.
3. Wash the column with three column volumes each of
    a. sterile H$_2$O, DEPC-treated.
    b. 0.1 N NaOH.
    c. sterile H$_2$O, DEPC-treated.
    Check that the pH of the column effluent is less than 8.
4. Wash the column with five volumes of sterile loading buffer.
5. Dissolve the total RNA in sterile water. Heat to 65°C for 5 minutes. Add an equal amount of 2X loading buffer and cool the sample to room temperature.
6. Apply the sample to the column and collect the flow-through.
7. Heat the flow-through to 65°C, cool, and reapply to the column.
8. Wash the column with 5–10 column volumes of loading buffer, followed by four column volumes of low-salt loading buffer.

    Note: The first RNA to elute off the column will be the poly (A)$^-$ fraction. The poly (A)$^+$ fraction will elute with the no-salt elution buffer.

9. Elute the poly (A)$^+$ RNA with 2–3 column volumes of sterile elution buffer.

    Note: The eluted poly (A)$^+$ RNA can be selected again on oligo(dT)-cellulose by adjusting the NaCl concentration of the eluted RNA to 0.5 M and repeating steps 5–7.

    Note: We generally elute into Falcon 2059 polypropylene tubes (17 × 100 mm).

10. Add sodium acetate (3 M, pH 5.2) to a final concentration of 0.3 M. Precipitate the RNA with 2.5 volumes of 100% ethanol at $-20°C$ overnight. Spin the precipitated RNA at 10 Krpm (12,000 × g) for about 30 minutes.

11. Aspirate off the supernatant. Air-dry the pellet. Resuspend the RNA pellet in an appropriate volume of DEPC-treated H_2O. Determine concentration of RNA by spectrophotometry. RNA is most stable when stored under ethanol at $-70°C$.

Northern Blotting

Northern analysis of RNA is the RNA equivalent of Southern analysis of DNA. In fact, after electrophoresis of the RNA of interest in an agarose gel as described here, the transfer of the nucleic acid and the subsequent hybridization with either radiolabeled or non-isotopically labeled probe is identical to that described for Southern analysis. *Remember that formaldehyde Northern gels are very brittle, much more brittle than their Southern counterparts.*

Reagents

1. Formaldehyde, 36–37% filtered.
2. Formamide, deionized.
3. Agarose (SeaKem, GTG, FMC, cat. # 50074).
4. 20X gel-running buffer: 400 mM $NaPO_4$ (use Na_2HPO_4, Sigma, cat. # S3264, and NaH_2PO_4, Sigma, cat. # S3139), 20 mM EDTA, pH 7.2.
5. 1X gel-running buffer: 20 mM $NaPO_4$, 1 mM EDTA, pH 7.2, 10% v/v formaldehyde.
6. 20X SSC.
7. DEPC-treated H_2O.
8. 6X loading buffer: 0.25% bromophenol blue, 0.25% xylene cyanol, 15% Ficoll (type 400) in gd H_2O.
9. Glass pyrex baking dishes, baked.
10. Fume hood.
11. Electrophoresis box and power source.
12. Circulating pump for gel box.
13. Nitrocellulose (Schleicher and Schuell, various sizes).
14. Baking oven.

Method

(Note: Due to the noxious nature of formaldehyde, it is imperative that the formaldehyde gel be poured, run, and blotted in a fume hood.)

1. To prepare 120 ml of 1.5% agarose include:

1.8 g	agarose
6 ml	20X running buffer
94 ml	H_2O
100 ml	Total

   Mix and melt. Cool to 60°C. Add 20 ml formaldehyde, and swirl to mix. Avoid creating bubbles. Pour the gel into a horizontal gel apparatus.

2. To denature the RNA, add the reagents to the RNA* so that the final concentration is 50% formamide, 6% formaldehyde, 1X running buffer.
   (*We generally run 1–5 μg poly $(A)^+$ RNA per lane.)

   Heat the RNA samples at 60°C for 10 minutes, then quick-chill on ice. Add one-fifth volume of 6X loading buffer. Load the sample into the submerged gel.

3. Recirculate the gel-running buffer to avoid the generation of a pH-gradient.
4. Run the gel at 3–4 V/cm. This generally takes about 8 hours for a 12-cm-length gel. (Or run overnight at 1–2 V/cm).

Transfer of RNA from Agarose Gels to Nitrocellulose

The transfer of RNA to nitrocellulose is identical to the transfer of DNA to a nitrocellulose filter (see the section on assays for gene transfer; Southern type techniques). For a diagram of the transfer apparatus, see Figure 11-1.

1. After electrophoresis, rinse the gel briefly in 0.1X SSC.
   Note: *Formaldehyde gels are brittle; handle with care.*
2. Cut a piece of Whatman 3MM paper approximately twice as long and of the same width as the width of the gel. This paper will function as a wick. Lay this paper over the platform that will support the gel during blotting. (We generally use the acrylic gel form used in forming the gel turned upside down, as the support is about the same size as the gel.)
3. Place the support and Whatman 3MM into a glass pyrex baking dish. Wet the wick with 20 ml of 20X SSC and squeeze out any air bubbles between the support/wick interface. Add 20X SSC to within ~1 cm of the top of the support.
4. Gently invert the gel so the open wells are in contact with the 3MM Whatman paper. Squeeze out any air bubbles between the gel/wick interface.
5. Cut a piece of nitrocellulose approximately 2 mm wider and 2 mm longer than the gel. Label the lower right-hand side of the nitrocellulose with a pencil mark. This will help orient the filter later. Pre-wet the nitrocellulose with 0.1X SSC for 5–10 minutes prior to use.
6. Overlay the gel with the nitrocellulose filter such that the pencil mark is in contact with the lower *left*-hand corner of the gel. Make sure the entire gel is evenly covered by the filter. Remove all air bubbles that may have formed between the gel and the filter.
7. Wet 2–3 pieces of Whatman 3MM paper with 2X SSC (cut to the *same* size as the gel) and overlay on top of the nitrocellulose filter. Remove all air bubbles that may have formed between the nitrocellulose and the Whatman paper.
8. Cut a stack of paper towels 5–8 cm high and ~2–3 mm smaller than the previous Whatman 3MM paper. Place the stack on top of the Whatman 3MM paper.
9. Place a glass plate on top of the stack of papers, followed by a 500-g weight. The object is to set up a flow of liquid from the reservoir (in the glass baking dish) through the gel and the nitrocellulose filter, so that RNA molecules are eluted from the gel and deposited onto the nitrocellulose filter.
10. Cover the blotting gel assembly with plastic wrap to prevent evaporation of the 20X SSC transfer solution during transfer of the RNA to the filter.
11. Allow the transfer to proceed for about 20 hours. Replace the paper towels as they become wet.
12. Remove the paper towels and the 3MM paper above the gel. Turn the dehydrated gel and filter over and lay them, gel side up, on a dry sheet of 3MM paper. Mark the positions of the gel slots on the filter with a soft pencil.
13. Peel off the filter and allow it to air-dry.
14. Bake the filter between pieces of Whatman 3MM paper for 1.5–2 hours at 80°C in a vacuum oven.
15. Proceed with hybridization; see Part III in the section on assays for gene transfer.

ANALYSIS OF RNA (NORTHERN- AND PCR-BASED TECHNIQUES)

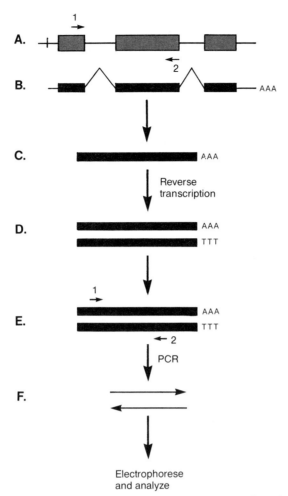

FIGURE 12-1 *RNA PCR*. The mechanism of RNA-based PCR described in the text is depicted here. The gene of interest (**A**) is transcribed in the nucleus of the cell to produce an unspliced, pre-mRNA sequence (**B**). This pre-mRNA sequence is spliced to generate a mature mRNA molecule (**C**), which serves as the target molecule for RNA PCR. The mRNA is mixed with either oligo (dT) or with random hexamer primers which, after annealing with the mRNA, prime the synthesis of cDNA by reverse transcriptase (**D**). In this example, oligo (dT) has been used to prime cDNA synthesis. The RNA/DNA hybrid product in (**D**) is mixed with the amplimers 1 and 2 and a PCR is performed (**E**). The resultant products (**F**) are analyzed by electrophoresis and either ethidium bromide staining or Southern hybridization analysis.

PCR Analysis of RNA

An alternative, less quantitative approach to Northern analysis of RNA is the use of the PCR. With this technique (Kawasaki, 1990) you first reverse-transcribe the mRNA, priming synthesis with a random hexamer or oligo (dT), and then in the same reaction buffer carry out the PCR with primers homologous to the target cDNA to amplify the reverse transcript (see Figure 12-1). Upon completion of these reactions, you analyze the products as described in the section on assays for gene transfer: PCR-based techniques.

Materials
1. 10X PCR buffer: 500 mM KCl, 200 mM Tris-HCl, pH 8.4, 25 mM $MgCl_2$, 1 mg/ml nuclease-free BSA (Pharmacia, cat. # 27-8914-01).

2. 10 mM dNTPs: 10 mM dATP, dCTP, dGTP, and dTTP, made by diluting 100 mM dNTPs stocks (Pharmacia, cat. # 27-2050-01, 27-2060-01, 27-2070-01, and 27-2080-01, respectively) with 10 mM Tris-HCl, pH 7.5.
3. RNasin (Promega, 20–40 U/μl, cat. # N2111).
4. p(dN)$_6$: (Pharmacia, cat. # 27-2166-01). Prepare as a 100 pmole/μl solution in TE (10 mM Tris-HCl, pH 8.0, 1 mM EDTA).
5. Moloney Murine leukemia virus reverse transcriptase (MoMLV-RT)(BRL, cat. # 8025 SA), 200 U/μl.
6. PCR primers: 18–22 bases in length, 10–100 pmole/μl in TE.
7. Taq polymerase (Perkin-Elmer Cetus), 5 U/μl.
8. Mineral oil (Sigma, cat. # M5904).
9. 3% NuSieve/1% SeaKem agarose (FMC Corporation, cat. # 50092).
10. Microfuge tubes: Use tubes specified for thermal cycler.

Method

1. Prepare this reverse transcription reaction:

2.0 μl	10 mM dNTPs
2.0 μl	10X PCR buffer
1.0 μl	p(dN)$_6$ (100 pmole/μl)
0.5 μl	RNasin (40 U/μl)
8.5 μl	DEPC-treated H$_2$O
5.0 μl	RNA sample (1 mg total or 20-30 μg poly (A)$^+$ RNA; roughly equivalent to 50,000–100,000 typical mammalian cells)
1.0 μl	MoMLV RT (200 units)
20.0 μl	Total

2. Incubate for 10 minutes at 23°C; then for 30–60 minutes at 42°C; then for 10 minutes at 95°C; then quick-chill on ice.
3. Prepare this PCR:

20 μl	reverse transcriptase reaction
8 μl	10X PCR buffer
1 μl	upstream primer (50 pmole/μl)
1 μl	downstream primer (50 pmole/μl)
8 μl	10 mM dNTPs
61 μl	H$_2$O
1 μl	Taq pol (5 U/μl)
100 μl	Total

   Layer 100 μl of mineral oil.
4. PCR amplification
   a. Perform 20–50 cycles of:
   - 95°C, 30 seconds.
   - cool for 1 minute to 55°C.
   - 55°C, 30 seconds.
   - heat at least 30 seconds to 72°C.
   - 72°C, 30 seconds.
   - heat at least 60 seconds to 95°C.
5. Analysis of products
   a. Extract the mineral oil from the sample by adding 300 μl of TE-saturated chloroform.
   b. Save the aqueous layer. Load 5–10 μl in a 3% NuSieve/1% SeaKem agarose composite gel on 8–10% polyacrylamide gels.

ANALYSIS OF PROTEIN

In most cases the phenotype of a transferred gene is manifested by the protein encoded by that gene. Here we present methods for analyzing the production of the protein of interest by three different types of techniques: immunoprecipitation analysis, Western analysis, and immunofluorescence analysis. Each has particular advantages. These are presented in the introduction for each procedure described.

Immunoprecipitation

Immunoprecipitation analysis is a very powerful technique. When combined with radiolabeling of cells, immunoprecipitation analysis allows you to monitor the synthesis of a protein and, in pulse-chase experiments, to monitor the kinetics of processing and turnover of the protein as well. Radiolabeling followed by immunoprecipitation is not the method of choice for quantitating the total amount of a given protein in a cell. For that you should turn to Western analysis. An efficient, broadly applicable technique for immunoprecipitation analysis is presented here.

Materials

1. Protein A-Sepharose CL-4B beads (Sigma, cat. # P3391).
2. [^{35}S]-methionine or [^{35}S]-cysteine (New England Nuclear or Amersham).
3. Dialyzed fetal calf serum (Sigma, cat. # F0392).
4. Methionine-minus or cysteine-minus media (MEM Select-Amine Kit, Gibco, cat. # 300-9050AV).
5. Lysis buffer: 20 mM Tris-HCl, pH 8.0, 200 mM LiCl, (Sigma, cat. # L8895), 0.5% NP-40 (Sigma, cat. # N3516), 1 mM EDTA. Filter-sterilize through a 0.2 μm filter, and store at 4°C.
6. Buffer B: 20 mM Tris-HCl, pH 8.0, 100 mM NaCl, 0.5% NP-40.
7. 2X gel-loading buffer: 4% SDS, 15% glycerol, 62.5 mM Tris-HCl, pH 6.8, 0.005% bromophenol blue, 200 mM dithiothreitol (add dithiothreitol just prior to use from a 1 M stock).
8. Nutator (Baxter, cat. # R4187-1).
9. PBS with Ca^{++} and Mg^{++} (Sigma, cat. # D5780).

Method

1. Preparation of Protein A-Sepharose CL-4B beads (hereafter called "Protein A beads")
   a. Fill the bottle containing the beads with enough H_2O to quantitatively transfer the 1.0 gram of the beads to a 15-ml test tube.
   b. Allow the beads to swell for 30 minutes on ice, with occasional agitation to resuspend the beads.
   c. When the beads have settled, the beads and H_2O should be 50:50 (v/v). Adjust the H_2O level if necessary.
   d. Freeze aliquots of the bead slurry at −20°C.
2. Labeling adherent cells
   a. Wash a 60-mm dish of cells (subconfluent) twice with PBS.
   b. Add 1.0 ml of methionine-minus or cysteine-minus medium containing 5% dialyzed FCS. Incubate at 37°C for 30 minutes.
   c. Add 100–200 μCi of the appropriate radioactive amino acid to the medium. Incubate at 37°C for 3 hours, rocking the dish every 30 minutes.
   d. Save the media (if desired) for immunoprecipitation. Store at −70°C if not processing immediately.

e. Add 1.0 ml of lysis buffer to the dish. Incubate at 4°C for 5 minutes. (The plasma membrane is lysed with this buffer. This liberates the cytoplasmic proteins and frees many proteins from the nuclei but does not destroy the ultrastructure of the nuclei, which remain attached to the dish.)
  f. Transfer the lysate to a microfuge tube. Spin at 14 Krpm at 4°C for 5 minutes. Transfer the supernatant to a new microfuge tube. Store at $-70°C$ if not processing immediately.
3. Labeling non-adherent cells
  a. Wash the cells twice with PBS.
  b. Resuspend the cells in methionine-minus or cysteine-minus medium, 5% dialyzed FCS, at a concentration range of $1.0–10.0 \times 10^6$ cells/ml. Pipet cells into a 12-well plate, 1 ml of cell suspension per well. Incubate at 37°C for 30 minutes.
  c. Add 100–200 μCi of the appropriate label, and continue to incubate at 37°C, for 3 hours. Rock the plate every 30 minutes.
  d. Transfer the cell suspension to a microfuge tube. Spin at 14 Krpm for 1 minute.
  e. Save the media supernatant for immunoprecipitation. Store at $-70°C$ if not processing immediately.
  f. Resuspend the cell pellet in 1.0 ml of lysis buffer. Incubate for 5 minutes at 4°C, with occasional rocking.
  g. Spin the tube at 14 Krpm for 5 minutes at 4°C. Transfer the supernatant to a new microfuge tube. Store at $-70°C$ if not processing immediately.
4. Immunoprecipitation
  a. To each sample tube, add the appropriate amount of antibody (determined by antibody titration experiments).
  b. Incubate at 4°C for 60 minutes on the Nutator.
  c. Spin the sample tubes at 14 Krpm for 15 minutes at 4°C.
  d. Transfer the supernatants to new microfuge tubes, and add to each tube 30 μl of a Protein A bead slurry. Use disposable pipetman tips whose tips have been cut off with a clean razor blade to create a larger orifice. The larger orifice ensures quantitative transfer and will keep the beads from being crushed during transfer.
  e. Incubate for 60 minutes at 4°C on the Nutator.
  f. Spin for *15 seconds* in a microfuge and discard the supernatant in a radioactive-waste container.
  g. Rinse the beads three times with 1.0 ml of lysis buffer and twice with 1.0 ml of Buffer B. All spins should be for no longer than 15 seconds.
  h. Remove the supernatant from the final wash and leave the beads undisturbed in a volume of approximately 30 μl. Add 30 μl of 2X sample buffer. Boil for three minutes and load the samples on a polyacrylamide gel. Samples can be stored at $-70°C$.
5. Preparation of SDS-polyacrylamide gels

Upon completion of your immunoprecipitation or prior to Western analysis of a protein mixture or cell lysate, you will want to separate the protein products on a polyacrylamide gel. The protocol presented here is quick because we run relatively small gels, and highly reproducible because we always run the gels under the exact same conditions.

Materials

1. Vertical mini-gel apparatus: Several designs are available. Our laboratory uses a C.B.S. Scientific Company Vertical Mini-Gel Unit (cat. # MGV-100)

with 0.75-mm spacers, 10- or 14-well combs, and notched glass plates, 11.3 cm (width) × 10.0 cm (height).
2. Gel-pouring stand (W.E.P. company, cat. # GEL101).
3. Acrylamide/Bis (N, N'-methylene-bis-acrylamide), 29:1 mixture (Bio-Rad, cat. # 161-0124). Prepare a 50% w/v solution in deionized, warm H_2O. Filter through a 0.45-μ filter and store in dark bottles at 4°C.
4. TEMED (N,N,N',N'-tetramethylethylenediamine, electrophoresis-grade; Bio-Rad, cat. # 161-0801; and other manufacturers).
5. Ammonium persulfate (APS) (Bio-Rad, cat. # 161-0700): Add 1 g to 10 ml of deionized H_2O (10% w/v) and store at 4°C.
6. 10% SDS.
7. 1.5 M Tris-HCl, pH 8.8.
8. 0.5 M Tris-HCl, pH 6.8.
9. 10X Tris-glycine SDS electrophoresis buffer (10X TGS): 250 mM Tris, pH 8.3, 2.5 M glycine (Sigma, cat. # G4392), 1% SDS.
10. Rainbow protein molecular-weight markers (Amersham, available as [^{14}C]-methylated or nonradioactive protein standards). These markers run as discrete colored bands, allowing the experimenter to follow the progress of electrophoresis while the gel is running and to assess the extent of protein transfer from an acrylamide gel to a filter during Western analysis.
11. 2X SDS gel-loading buffer: 4% SDS, 15% glycerol, 62.5 mM Tris-HCl, pH 6.8, 0.005% bromophenol blue, 200 mM dithiothreitol (add dithiothreitol just before use from a 1 M stock).
12. Microcapillary pipette tips (West Coast Scientific, Inc., cat. # MC-50BR).
13. Staining solution: 0.25% (v/v) Coomassie brilliant blue R250 (Sigma, cat. # B0149), 10% (v/v) glacial acetic acid, 45% (v/v) methanol.
14. Destaining solution 10% (v/v) glacial acetic acid, 45% (v/v) methanol.
15. Enlightning (New England Nuclear, cat. # NEF-974).

Method

1. Clean and dry the glass plates.
2. Assemble the glass plates by placing the side spacers between the notched and unnotched plates. Clamp one side with the free clamp and the other side with the clamp mounted to the gel-pouring stand.
3. Add 5 ml of molten 1% agarose in H_2O, at 65°C, to the trough in the gel-pouring stand, and place the bottom of the glass-plate-assembly in the trough to seal the bottom of the assembly.
4. SDS-polyacrylamide mixtures for 12% and 15% gels:

	12%	15%
1.5 M Tris-HCl, pH 8.8	5.0 ml	5.0 ml
50% acrylamide	4.8 ml	6.0 ml
10% SDS	0.2 ml	0.2 ml
H_2O	9.8 ml	8.6 ml
10% APS	0.2 ml	0.2 ml
	20.0 ml	20.0 ml

5. Add 20 μl of TEMED to the SDS-polyacrylamide mixture. Swirl rapidly and pour immediately into the gap between the plates. Leave sufficient space for the stacking gel (length of the teeth, plus 1 cm). With an 18-gauge needle and syringe, carefully overlay with H_2O or isobutanol.
6. After polymerization (30–45 minutes), pour off the overlay and rinse with H_2O to remove unpolymerized acrylamide. Drain as much H_2O as possible. Use a Kimwipe if necessary.

7. 5% Stacker mixture:

2.5 ml	0.5 M Tris-HCl, pH 6.8
1.0 ml	50% acrylamide
0.1 ml	10% SDS
6.3 ml	H$_2$O
0.1 ml	10% APS
10.0 ml	Total

8. Add 25 μl of TEMED to the stacker mixture. Swirl rapidly, and pour into the gap between the plates over the resolving gel. Immediately insert the comb, avoiding trapped bubbles.
9. After polymerization (30–45 minutes), carefully pull out the comb and rinse the wells with H$_2$O from a bent 18-gauge needle fitted to a 5-ml disposable syringe.
10. Mount the gel assembly into the electrophoresis apparatus and add 1X TGS to the upper and lower chambers. Rinse the wells with electrophoresis buffer by gentle squirting with a pasteur pipette.
11. Mix the samples with one volume of 2X SDS gel-loading buffer. Boil the samples for 3 minutes.
12. Load 12–15 μl in each well with a pipetman using the disposable capillary-thin pipette tips. Load an equal volume of 1X SDS gel-loading buffer (diluting with sample buffer) in any unused well.
13. Connect to a power supply. Apply a voltage of 30 volts until the dye enters the resolving gel, then increase the voltage to 75 volts. When the Rainbow markers have migrated the desired distance, turn off the power supply. Remove the gel from the electrophoresis apparatus.
14. a. Proceed to Western analysis, or
    b. Stain the gel with staining solution for 30 minutes; destain with destaining solution; dry, if desired, on a gel dryer; or
    c. Treat with Enlightning for 20 minutes; dry at 60°C on a gel dryer; expose to Kodak XAR-5 or Amersham Hyperfilm-ECL film at −70°C.

Note:

1. Spurious background:
   a. Reduce the amount of antibody used. Include only enough antibody to precipitate all of the protein of interest.
   b. Pre-absorb the samples with Protein A beads. Add the beads to the sample. Incubate for 30 minutes. Pellet the beads and save the supernatant, then, with the primary antibody and fresh beads, begin immunoprecipitation of the supernatant as described above.
   c. Immunoprecipitate the sample with pre-immune serum from the same animal from which the primary antibody was produced (if available), then immunoprecipitate with the primary antibody.
   d. Affinity-purify the antibody on an antigen-affinity column prior to use.
2. Poor incorporation of label: Determine if your medium is deficient in the appropriate amino acid.

Western-Type Techniques

In contrast to immunoprecipitation of radiolabeled proteins, Western analysis is quantitative and serves as a mass determination for a particular protein. This procedure is not useful in evaluating the rate of synthesis, turnover, or processing of a protein. This procedure is considerably quicker than immunoprecipitation analysis, especially in the analysis of a large number of samples, for two reasons. First, you do not need to immunoprecipitate each sample prior to gel loading; rather you just load your samples onto the gel. Second, Western analysis is ideally suited to nonisotopic, enhanced chemiluminescence detection techniques. In our hands these techniques are sufficiently sensitive to detect less than 7 picograms of a 17kD molecule per sample with x-ray film exposure times of just a few seconds. An autoradiographic image of similar intensity of a parallel culture radiolabeled with ^{35}S-cysteine requires an exposure time of 24–48 hours. Western analysis combined with nonisotopic detection is a very powerful technique and is presented here.

Materials

1. 10X stock Tris-glycine buffer, pH 8.5: 0.25M Tris (Sigma, cat. # T1503), 1.92 M glycine (Sigma, cat. # G 7126).
2. 10X stock phosphate buffered saline, pH 6.8 (PBS): 0.2 % KCl (Fisher Scientific, cat. # P217-500), 8.0% NaCl (Fisher Scientific, cat. # S671-500), 0.2% potassium phosphate monobasic (KH_2PO_4; Fisher Scientific, cat. # P285-500), 1.14% sodium phosphate monobasic (NaH_2PO_4; Fisher Scientific, cat. # S369-500).
3. Transfer buffer: 0.025 M Tris, 0.192 M glycine, 20% methanol (Mallinckrodt; Baxter, cat. # 3024-4*NY).
4. Blocking solution (refrigerate if storing): 1.0 M glycine, 5% dry milk (Carnation), 5% fetal calf serum (Sigma, cat. # F4884), 1% ovalbumin (powder; Sigma, cat. # A5253).
5. Wash solution (refrigerate if storing): 1X stock PBS, pH 6.8, 1% fetal calf serum, 0.1% TWEEN 20 (Bio-Rad, cat. # 170-6531), 0.1% dry milk, 0.1% ovalbumin (crystalline; Sigma, cat. # A5378).
6. T-PBS (for ECL wash of Western blots): 1X stock PBS, pH 6.8, 0.05% TWEEN 20.
7. Goat anti-rabbit IgG-horseradish peroxidase (Bio-Rad, cat. # 170-6515).
8. ECL reagents (Amersham, cat. # RPN.2105).
9. Trans-Blot apparatus (Bio-Rad, cat. # 170-3910).

Western Transfer—Protein Transfer from Acrylamide Gels

Having run a polyacrylamide gel with the protein samples to be analyzed on it, you must transfer those protein samples to nitrocellulose for Western analysis. Unlike Southern and Northern transfer techniques, in which DNA and RNA molecules can be transferred by capillary action, in Western transfer you electrophorese the proteins from the polyacrylamide gel onto the nitrocellulose filter. The procedure for this transfer is described here.

Transfer Method

1. Soak one 20 cm × 14 cm sheet of Whatman 3MM paper with transfer buffer and lay on a dry Scotch Brite pad.
2. Place the protein gel on top of the Whatman paper, and remove any air bubbles.
3. Briefly submerge a pre-wet 9.0 cm × 6.0 cm sheet of nitrocellulose in a small volume of transfer buffer and place it on top of the acrylamide gel. Clip the membrane along the top to indicate any lanes to be cut and probed with different antisera later.
4. Remove any air bubbles from the gel/nitrocellulose membrane interface with gloved fingertips, then place a second 20 cm × 14 cm sheet of Whatman paper soaked in transfer buffer on top of the nitrocellulose membrane.

5. Put a dry Scotch Brite pad atop this Whatman paper. Close the unit and submerge it in the Trans-Blot tank containing approx. 2.5 l of transfer buffer.
6. The final arrangement of the gel/nitrocellulose sandwich in the Trans-Blot unit is listed in order:
   a. Cathode (−)   BLACK
   b. Scotch Brite pad
   c. Presoaked Whatman 3MM paper
   d. Protein gel
   e. Nitrocellulose membrane
   f. Presoaked Whatman 3MM paper
   g. Scotch Brite pad
   h. Anode (+)   RED
7. Run the Trans-Blot Cell for 1 hour at a constant voltage of 40 volts (@ 215 mA).

Blocking Method

After Western transfer, prior to exposure with the primary antibody, the nitrocellulose filter bearing the electrophoresed protein samples is treated for an extended period of time with blocking solution. This blocking solution, rich in protein, serves to occupy previously unoccupied protein binding sites on the membrane. This creates a situation in which, when the blocked filter is exposed to the primary antibody, the primary antibody and later the secondary antibody only bind to the antigens on the filter they are directed against and do not bind to the filter nonspecifically. Blocking is the secret of the sensitivity of Western analysis. The extended blocking procedure I describe here works very well.

1. After transfer, block unoccupied sites on the nitrocellulose membrane with ∼ 50 ml of blocking solution for 2 days at 4°C, with agitation, or for 30 minutes at room temperature, with agitation.
2. Wash the blocked filter with wash solution for 5 minutes. Change wash solution and repeat washing twice more. All subsequent antibody incubation and washing steps are done with wash solution unless noted. Perform all incubations and washes with the filter oriented protein-side up.

Primary Antibody Incubation

After the blocking process has been completed, the Western transfer filter is exposed to the antibody directed against the protein of interest. Here is how this is done.

1. Incubate the nitrocellulose membrane at room temperature in a tray with the primary antibody diluted in wash solution for 3 hours, at room temperature, with agitation, or at 4°C overnight, with agitation.
   The dilution of primary antibody should be determined empirically. We routinely use a *1:1000 dilution* of primary antibody: 20 μl primary antibody + 20 ml wash solution.
2. After the primary-antibody incubation, wash the membrane for 5 minutes with wash solution. Repeat this wash step twice more.
3. Wash once for 15 minutes with T-PBS (ECL wash solution).

Secondary-Antibody Incubation

After incubation with the primary antibody, the Western transfer filter is incubated with the secondary antibody. This secondary antibody is a polyclonal antibody directed against the first antibody and is covalently attached to a reporter molecule, in this case horseradish peroxidase. These secondary-antibody molecules bind to the primary-antibody molecules at multiple sites and, as a result, multiple reporter molecules become associated with a single primary-antibody molecule. These reporter molecules serve to catalyze a chemical reaction that oxidizes a special substrate, which, in the case of Enhanced Chemiluminescence, results in the generation of photons that serve to expose x-ray film whenever an antigen/primary antibody/secondary antibody/reporter molecule complex has formed.

1. Incubate the nitrocellulose membrane at room temperature in a tray with the secondary antibody diluted in wash solution for *1 hour* with agitation.
    Use a *1:3000 dilution* of goat-anti-rabbit IgG-HRP complex: $10 \mu l$ goat-anti-rabbit IgG-HRP + 30 ml T-PBS wash solution.
2. Wash the membrane for 15 minutes with T-PBS wash solution. Repeat this wash step twice more.

ECL Method (Prepare the ECL-detection Reagent Fresh)

The final step in the Western analysis is the addition of the horseradish peroxidase substrate to the processed, complexed filter. Here I describe the uses of ECL detection reagent. It is with this reagent that, post incubation, film exposure times of but a few seconds may be all that is necessary to detect a clear signal from the complexed reporter molecules.

1. Mix an equal volume of Reagent 1 and an equal volume of Reagent 2 for a final volume equivalent of 0.125 ml/cm^2 membrane.
2. Place the ECL-detection reagent in a clean tray and agitate.
3. Add the filter and agitate gently for *exactly 60 seconds*.
4. Place the wet filter, protein side up, on the shiny side of Benchcote paper.
5. Wrap the damp filter with Saran Wrap.
6. Expose to hyperfilm-ECL(Amersham, cat. # RPN.2103) in a dark room. Under many circumstances, exposure times of 5–10 seconds will be sufficient.

Immunofluorescence Analysis

Immunofluorescence analysis provides different types of information than Western analysis of transfected or infected cells for the production of a foreign protein. With this technique, one can identify a rare cell in a transfected or infected cell population that produces the protein of interest. Furthermore, this technique may enable the investigator to determine the subcellular localization of the protein of interest, be it the cytoplasm, cell membrane, or nucleus. There are many themes and variations on this technique. A protocol that has broad application is described here. Some fixation options are presented as well. Several fixation methods should be tested to determine which is optimal for the antigen of interest.

Materials

1. PBS.
2. 3.5% formaldehyde (HCHO) in PBS.
3. Ethanol, 70% and 100%.
4. Primary antibody to desired peptide.
5. Fluorescein-conjugated or rhodamine-conjugated second antibody.
6. Mounting media: 80% glycerol, 50 mM N-propyl-gallate (anti-fade reagent, Sigma, cat. # P3130).
7. Microscope equipped for epifluorescence with appropriate filters for excitation and viewing.

8. BSA (Fraction V, Sigma, cat. # A9418).
9. Clear fingernail polish.

Method

1. Aspirate off the media. Wash the plate twice with PBS.
2. Fix the cells with cold 3.5% HCHO on ice for 20 minutes.
3. Wash the plate once with PBS.
4. Add 70% ethanol (at −20°C) to the plate and incubate at −20°C for 7 minutes.
5. Replace with 100% ethanol (at −20°C) and incubate at −20°C for 7 minutes.
6. Replace with 70% ethanol (at −20°C) and incubate at −20°C for 5 minutes.
7. Wash the plate three times with PBS.
8. Circle an area about 1 cm in diameter on the bottom of the plate with a permanent marker.
9. Remove all cells outside the circle with a Kimwipe.
10. Dilute the primary antibody in 3% BSA in PBS. (Appropriate dilution must be determined by titration experiments.)
11. Add 50 µl of diluted antibody to circled area. Incubate for 30 minutes at room temperature.
12. Wash the plate twice with PBS.
13. Add 50 µl of diluted fluorochrome-conjugated second antibody. (Working range must be determined by titration; it is usually 1:10 to 1:300.) Incubate for 30 minutes at room temperature.
14. Wash the plate twice with PBS.
15. Add 5–10 µl of 80% glycerol, 50 mM N-propyl-gallate. Cover with coverslip. Seal edges with clear fingernail polish.
16. Observe by ultraviolet microscopy:

Fluorochrome	Excitation, nm	Emission, nm	Color
Fluorescein	495	525	Green
Rhodamine	552	570	Red

Fixation options include glutaraldehyde and organic solvents such as acetone and methanol.

For glutaraldehyde fixation, prepare a 1% solution (electron microscope grade) in PBS.

Glutaraldehyde Method

1. Aspirate off the media. Wash the plate twice with PBS.
2. Incubate cells in a 1% glutaraldehyde solution, *in a fume hood*, for 1 hour at room temperature.
3. Wash the cells twice with PBS.
4. Permeabilize the fixed cells by incubating in methanol for 2 minutes at room temperature.
5. Rinse gently in PBS with four changes over a period of 5 min.
6. Proceed with step 8 of above formaldehyde procedure.

For fixation with organic solvents, prepare a fresh solution of 50% acetone/50% methanol.

Acetone/Methanol Method

1. Aspirate off the media. Wash the plate twice with PBS.
2. Drain the plate but do not allow the cells to dry.
3. Fill the plate with the 50% acetone/50% methanol solution. Agitate gently and incubate for 2 minutes at room temperature.
4. Remove solvent and rinse with PBS.
5. Proceed with step 8 of above formaldehyde procedure.

ANALYSIS OF POST-TRANSLATIONAL MODIFICATION

The post-translational modification of a protein can dramatically affect its half-life, its antigenicity as well as its biological function. If the gene you have transferred and expressed encodes a protein that is modified post-translationally, it is imperative that you examine the state of the protein regarding the desired modification. Post-translational modification of newly synthesized proteins has been discussed previously (see Part 1, Processing of Proteins Encoded by Transferred Genes). In this section I will present methods for the analysis of these post-translational modifications. These methods include the analysis of N-linked glycosylation, analysis of O-linked glycosylation, analysis of tyrosine sulfation, and analysis of fatty acid acylation (Dorner and Kaufman, 1990). These procedures incorporate other procedures described in Part III of this manual, including radiolabeling of cells in culture, immunoprecipitation analysis, and the running of polyacrylamide gels. To complete your analysis of post-translational modification, you will require the materials listed for these protocols as well as specific materials, listed below, for each type of analysis of post-translational modification.

Analysis of N-Linked Glycosylation

There are two methods available for the analysis of N-linked glycosylation. The first employs glycosidases to remove specific sugar moieties from the protein, thus altering the migration of the glycosidase-treated protein relative to its glycosylated counterpart in a polyacrylamide gel. This procedure provides information on the amount of oligosaccharide associated with the protein. The glycosidase Endo-β-N-acetylglucosaminidase (Endo H) cleaves high mannose and specific types of hybrid carbohydrate moieties at the GlcNacβ1-4G1nNAc linkage, leaving a single GlnNAc residue attached to an asparagine. The glycosidase Peptide-N^4-(N-acetyl-β-glucosaminyl) asparagine amidase (N-gly) cleaves all N-linked carbohydrate regardless of the structure by hydrolyzing the asparaginyl-oligosaccharide bond, thus completely removing the sugar residue from the protein.

Analysis of Glycosylation with Glycosidases

The conditions described here are appropriate for approximately 200 μl of cell extract or conditioned media or 3 μg of a purified glycoprotein.

Materials

1. 0.15 M sodium acetate, pH 6.0 (Sigma, cat. # S8625).
2. 0.1 M PMSF (Sigma, cat. # P7626).
3. Endo H (Sigma, cat. # E6878).
4. 1.0 M sodium phosphate, pH 8.3.
5. NP-40 (Sigma, cat. # N6507).
6. N-gly (Genzyme, N-gly-10).

Method

1. Plate the cells producing the protein of interest and radiolabel as described (see immunoprecipitation).
2. Isolate the radiolabeled protein of interest by the immunoprecipitation procedure.

3. Elute the protein from the protein-A sepharose beads through the addition of 60 µl of a solution containing: 50 mM Tris, pH 6.8, 0.5% SDS, 0.1 M 2-mercaptoethanol. Heat the mixture to 90–100°C for 5 minutes.
4. Pellet the beads and remove the supernatant fraction. Divide the 60 µl supernatant into three 20 µl fractions and react as described here:

Reaction 1 (control)

20 µl	protein sample
20 µl	0.15 M sodium acetate, pH 6.0
0.5 µl	0.1 M PMSF
40.5 µl	Total

Reaction 2 (Endo H digestion)

20 µl	protein sample
20 µl	0.1 U/ml Endo H in 0.15 M sodium acetate, pH 6.0
0.5 µl	0.1 M PMSF
40.5 µl	Total

Reaction 3 (N-gly)

20 µl	protein sample
5 µl	H_2O
8 µl	1.0 M sodium phosphate, pH 8.3
5 µl	14% NP-40
1.5 µl	250 U/ml N-gly
0.5 µl	0.1 M PMSF
40.0 µl	Total

4. Incubate these reactions overnight at 37°C.
5. Terminate the digestion by the addition of an equal volume of 2X gel-loading buffer.
6. Electrophorese the samples on a polyacrylamide gel of the appropriate porosity. Subject dried gel to autoradiography.

Analysis of Glycosylation with Inhibitors of Carbohydrate Addition or Processing

The inhibitors described here either block glycosylation completely or inhibit a particular step in glycosylation (see Figure 3-1). The antibiotic tunicamycin blocks the formation of the dolichol phosphate-linked oligosaccharide donor, completely blocking N-linked glycosylation. The glucose analog deoxynojirimycin (DNJ) inhibits the endoplasmic reticulum glycosidases I and II, thus blocking the removal of the three glucose residues from the oligosaccharide core unit, ultimately preventing further processing of the oligosaccharide. The mannose analogue deoxymannojirimycin (DMJ) inhibits two different mannosidases: golgi complex mannosidase I, responsible for the conversion of the Man_8 oligosaccharide to Man_6, and an endoplasmic reticulum mannosidase of similar specificity. Swainsonine (SWSN) inhibits golgi complex mannosidase II which removes terminal 1,3– and 1,6–linked mannose residues, preventing formation of complex carbohydrates resulting in the accumulation of Man_5-containing oligosaccharides.

Materials

1. Tunicamycin (Boehringer Mannheim, cat. # 1248080).
2. DNJ (Boehringer Mannheim, cat. # 973858).
3. DMJ (Boehringer Mannheim, cat. # 981605).
4. SWSN (Boehringer Mannheim, cat. # 810169).

Methods

1. Plate the cells producing the protein of interest as described (see immunoprecipitation).
2. Wash the cells with medium and then add fresh medium containing the inhibitor at a concentration to be determined empirically (approximately 10 µg/ml for tunicamycin; 5 mM DNJ; 1 mM DMJ; and 0.1 mM SWSN). Incubate the cells 1–2 hours in the presence of the inhibitor prior to the addition of the radioactive amino acids for radiolabeling.
3. Harvest the cells or medium and subject either or both to immunoprecipitation analysis and polyacrylamide gel electrophoresis as described.

Analysis of O-Linked Glycosylation

The enzyme endo-α-N-acetyl-D-galactosiminidase (O-glycanase) cleaves the Galβ1,3galNAc disaccharride unit linked to serine or threonine. The removal of O-linked carbohydrate by this enzyme is inhibited by substituents on the Gal or GalNAc residues. When the substituent is sialic acid, it can be removed by digestion with neuraminidase prior to digestion with O-glycanase. Post digestion the protein is immunoprecipitated and analyzed by polyacrylamide gel electrophoresis. These procedures are described here.

Materials

1. NP-40.
2. 0.4 M calcium acetate (Sigma, cat. # C 1000).
3. 0.4 M Tris-maleate, pH 6.0 (Sigma, cat. # T3128).
4. 0.1 M PMSF (Sigma, cat. # P7626).
5. Neuraminidase (Genzyme, cat. # NSS-1).
6. O-glycanase (Genzyme, cat. # O-ASE-25).

Method

1. Plate the cells producing the protein of interest and radiolabel as described (see immunoprecipitation).
2. Isolate the radiolabeled protein of interest by the immunoprecipitation procedure.
3. Resuspend the protein-A sepharose bead pellet in 40 µl of 0.5% SDS, 0.1 M 2-mercaptoethanol. Denature the protein by heating to 90–100°C for 5 minutes. Pellet the beads, then remove and save the supernatant, and then react as described here:

40 µl	protein sample
10 µl	14% NP-40
2 µl	0.4 M calcium acetate
4 µl	0.4 M Tris-maleate, pH 6.0
22 µl	H_2O
1 µl	0.1 M PMSF
1 µl	Neuraminidase (1 U/ml; Genzyme)
80 µl	Total

Note: D-galactolactone can be added to the reaction to prevent removal of galactose residues from N-linked glycosylation by any contaminating galactosidases. To accomplish this, bring the reaction to a final concentration of 10 mM D-galactolactone (Sigma, cat. # G0375).

4. Incubate this reaction for 1 hour at 37°C.
5. Divide the reaction into two 40 µl aliquots, add O-glycanase to a final concentration of 100–200 U/ml *to one aliquot*, and incubate this reaction overnight at 37°C.
6. Terminate the digestion by the addition of an equal volume of 2X polyacrylamide gel electrophoresis sample buffer.
7. Electrophorese the samples on a polyacrylamide gel of the appropriate porosity. Subject the gel to autoradiography. Compare the migration of the sample treated with neuraminidase and O-glycanase to the sample treated with neuraminidase alone. A decrease in molecular weight or in heterogeneity serves to indicate the presence of O-linked carbohydrate in the sample treated with O-glycanase.

Analysis of Tyrosine Sulfation

The analysis of sulfate addition to tyrosine as an O_4-sulfate begins by labeling cells with [^{35}S]-sulfate. Such labeling leads to the incorporation of sulfate at both tyrosine residues as well as carbohydrate. Differentiate between these two types of modifications either by the addition of an inhibitor of N-linked glycosylation during sulfate labeling or, post-labeling, by treating the protein with N-glycanase to remove N-linked and potentially sulfated carbohydrate.

Materials

1. [^{35}S]-sulfuric acid (Amersham, cat. # ST.44).
2. Sulfate-free media (JRH Biosciences, formerly Hazelton Research Products).
3. Sulfate-free media with reduced levels of cysteine and methionine (2% of normal cysteine and methionine levels; JRH Biosciences, formerly Hazelton Research Products).
4. Dialyzed fetal calf serum (Sigma, cat. # F0392).

Method

1. Plate the cells producing the protein of interest as described (see immunoprecipitation).
2. Labeling cells in sulfate-free medium with reduced levels of cysteine and methionine can increase the rate of incorporation of radiolabeled sulfate but may also interfere with total protein synthesis. To determine if this is the case, radiolabel test cultures with [^{35}S]-methionine in sulfate minus medium and sulfate minus medium with reduced levels of cysteine and methionine and compare the total incorporation of radioactive amino acid into protein, post labeling, by TCA analysis. Select the appropriate medium for your purposes and wash the cells with medium and then add 1.5 ml of sulfate-free medium to a 60 mm plate. Incubate the cells for 30 minutes at 37°C.
3. Remove this medium and add 2 ml of the appropriate sulfate-free medium containing 0.5–1.0 mCi [^{35}S]-sulfuric acid and dialyzed fetal calf serum to the culture. Cells are incubated overnight in this medium.
4. Isolate the radiolabeled protein of interest by the immunoprecipitation procedure.

5. Elute the protein from the protein-A sepharose beads through the addition of 40 µl of a solution of 50 mM Tris, pH 6.8, 0.5% SDS, 0.1 M 2-mercaptoethanol. Heat the mixture to 90–100°C for 5 minutes.
6. Pellet the beads and remove the supernatant fraction. If no inhibitor of N-linked glycosylation was added during the radiolabeling with [^{35}S]-sulfuric acid, digest 20 µl of the 40 µl elute with N-gly as described (see analysis of N-linked glycosylation).
7. Upon completion of these procedures, add an equal volume of 2X polyacrylamide gel electrophoresis buffer and subject the samples to polyacrylamide gel electrophoresis and autoradiography.

Analysis of Fatty Acid Acylation

Materials

1. [^{3}H]-myristic acid (Amersham, cat. # TRK.907).
2. [^{3}H]-palmitic acid (Amersham, cat. # TRK.909).
3. Serum-free medium.
4. Dialyzed serum delipidated by extraction with butanol: diisopropyl alcohol.
5. 1.0 M hydroxylamine (Sigma cat. # H2391).
6. 1.0 M Tris, pH 10.

Method

1. Plate the cells producing the protein of interest as described (see immunoprecipitation).
2. The cells are labeled in either serum-free medium or medium containing dialyzed, delipidated serum containing either [^{3}H]-myristic acid or [^{3}H]-palmitic acid. The labeled fatty acids are first dissolved in dimethyl sulfoxide (DMSO) and then added to the medium for a final concentration of 0.2–0.5 mCi/ml. The cells are incubated for a maximum of 4 hours at 37°C. To prevent the appearance of artifactual fatty acid addition by the conversion of myristate to palmitate and subsequent palmitoylation in a myristoylation analysis, label the cell cultures for no longer than 90 minutes.
3. Harvest the cells or media and subject either or both to immunoprecipitation analysis and polyacrylamide gel electrophoresis as described (see immunoprecipitation analysis).
4. Treatment of the polyacrylamide gels with either hydroxylamine or alkali will reveal the structure of the covalent bond joining the fatty acid to the protein. After fixation soak the gels in either 1 M hydroxylamine, pH 7.0, or 1 M Tris, pH 10, for 16 hours at room temperature. The gels are then dried and exposed to x-ray film. Amide bonds, those by which myristate is coupled to protein, are resistant to hydroxylamine or alkali treatment. Thioester or O-ester linkages, characteristic of palmitoylation, are sensitive to these treatments. Thus, the disappearance of radioactivity from gels so treated is indicative of palmitoylation, not myristoylation.

PART III REFERENCES

CARTER, P., BEDOULLE, H., WAYE, M., and WINTER, G. 1985. Oligonucleotide site-directed mutagenesis in M13. Based on a practical course held at EMBL Heidelberg, 1984. MRC Laboratory of Molecular Biology, Hills Road, Cambridge CB2 2HQ.

DORNER, A. J., and KAUFMAN, R. J. 1990. Analysis of synthesis, processing and secretion of proteins expressed in mammalian cells. *Methods Enzymol.* **185**:557.

HELMUTH, R. 1990. Non-isotopic detection of PCR products. In *PCR Protocols: A Guide to Methods and Applications.* Eds. Innis, M., Gelfand, D., Sninsky, J., and White, T. San Diego: Academic Press, Inc., p. 21.

KAWASAKI, E. 1990. Amplification of RNA. In *PCR Protocols: A Guide to Methods and Applications.* Eds. Innis, M., Gelfand, D., Sninsky, J., and White, T. San Diego: Academic Press, Inc., p. 119.

KELLER, G. H., and MANAK, M. M. 1989. *DNA Probes.* New York: Stockton Press.

LO, Y.-M., MEHAL, W., and FLEMING, K. 1990. Incorporation of biotinylated dUTP. In *PCR Protocols: A Guide to Methods and Applications.* Eds. Innis, M., Gelfand, D., Sninsky, J., and White, T. San Diego: Academic Press, Inc., p. 113.

MANIATIS, T., FRITSCH, E., and SAMBROOK, J. 1982. *Molecular Cloning: A Laboratory Manual.* Cold Spring Harbor, NY: Cold Spring Harbor Laboratory Press.

MCINNES, J., VISE, P., HABILI, N., and SYMONS, R. 1987. Chemical biotinylation of nucleic acids with photobiotin and their use as hybridization probes. *Focus* **9**:1.

MULLIS, K. B., and FALOONA, F. A. 1987. Specific sythesis of DNA *in vitro* via a polymerase-catalyzed chain reaction. *Methods Enzymol.* **155**:335.

MULLIS, K., FALOONA, F., SCHARF, S., SAIKI, R., HORN, G., and ERLICH, H. 1986. Specific enzymatic amplification of DNA in vitro: The polymerase chain reaction. *Cold Spring Harbor Symp. Quant. Biol.* **51**:263.

SAIKI, R. K., GELFAND, D. H., STOFFEL, S., SCHARF, S. J., HIGUCHI, R., HORN, G. T., MULLIS, K. B., and ERLICH, H. A. 1988. Primer-directed enzymatic amplification of DNA with a thermostable DNA polymerase. *Science* **239**:487.

SAMBROOK, J., FRITSCH, E., and MANIATIS, T. 1989. *Molecular Cloning: A Laboratory Manual*, 2nd edition. Cold Spring Harbor, NY: Cold Spring Harbor Laboratory Press.

APPENDIX: SUPPLIERS

Aldrich Chemical Company
P.O. Box 2060
Milwaukee, WI 53201, USA
800-558-9160
Fax 414-273-4979

Amersham Corporation
2636 South Clearbrook Drive
Arlington Heights, IL 60005, USA
708-593-6300
800-323-9750
Fax 800-228-8735

In UK:
Amersham International plc.
Research Products Division
White Lion Road, Amersham
Buckinghamshire HP7 9LL, UK
(44) 494-544000

AMRAD Corporation Ltd.
Level 2, 17-27 Cotham Road
Kew, Victoria 3101, Australia
61+3 853 0022
Fax 61+3 853 0202

Baxter Healthcare Corporation
Scientific Products Division
1210 Waukegan Road
McGraw Park, IL 60085, USA
708-689-8410
800-633-7370
Fax 708-473-2114

Beckman Instruments Inc.
45 Belmont Drive
P.O. Box 6764
Somerset, NJ 08875-6764, USA
908-560-0076
800-742-2345
Fax 908-560-0470

In Europe:
Beckman Instruments International SA
22 Rue Juste-Olivier, CH-1260
Nyon, Switzerland
(41)-22-631181
Fax (41)-22-621810

Bethesda Research Laboratories Inc. (BRL)
(See GIBCO BRL)

BIO 101 Inc.
P.O. Box 2284
La Jolla, CA 92038-2284, USA
619-598-7299
800-424-6101
Fax 619-598-0116

Bio-Rad Laboratories
3300 Regatta Boulevard
Richmond, CA 94804, USA
415-232-7000
800-227-5589
Fax 415-232-4257

In UK:
Bio-Rad Laboratories Ltd.
Bio-Rad House
Maylands Avenue, Hemel Hempstead
Hertfordshire HP2 7TD
(44)-442-232552
Fax (44)-442-59118

Boehringer Mannheim Biochemicals, Orders:
P.O. Box 50414
Indianapolis, IN 46250, USA
317-576-3019
800-262-1640
Fax 317-576-2754

Other information
9115 Hauge Road
Indianapolis, IN 46256, USA

APPENDIX: SUPPLIERS

In Europe:
Sandhoferstrasse 116
Postfach 310120
D6800 Mannheim, FRG
(49)-621-7592890
Fax (49)-621-7598611

In UK:
BCL-Boehringer
Boehringer Mannheim House
Bell Lane, Lewes
East Sussex, BN7 1LG, UK
(44)-273-48044

Brinkman Instruments Inc.
Cantiague Road
Westbury, NY 11590, USA
516-334-7500
Fax 516-334-7506

In Europe:
Eppendorf-Netheler-Hinz GmbH
P.O. Box 65 06 70
D-2000 Hamburg 65, FRG
(49)-40-538-01-0
Fax (49)-40-538-01-556

In UK:
BDH Ltd.
Apparatus Division
P.O. Box 8
Dagenheim, Essex RM 8 1RY, UK
(44)-81-597-8821
Fax (44)-81-597-8300

BTX, Inc.
3742 Jewell Street
San Diego, CA 92109, USA
800-289-2465
619-270-0861
Fax 619-483-3817

In UK:
Biotech Instruments Ltd.
183 Camford Way
Luton, Bedfordshire
LU3 3AN, UK
(44)-582-502388
Fax (44)-582-597091

Calbiochem
10933 North Torrey Pines Road
La Jolla, CA 92037, USA
619-450-9600
800-854-3417
Fax 619-453-3552

In UK:
Novabiochem (UK) Ltd.
3 Heathcoat Building
Highfields Science Park
University Boulevard
Nottingham NG7 2QJ
(44)-602-430840
Fax (44)-602-430951

C.B.S. Scientific
P.O. Box 856
Del Mar, CA 92014, USA
619-755-4959
Fax 619-755-0733

Clontech
4030 Fabian Way
Palo Alto, CA 94303, USA
415-424-8222
800-662-2566
Fax 415-424-1352

Collaborative Research Inc.
2 Oak Park
Bedford, MA 01730, USA
617-275-0004
Fax 617-275-0043

Difco Laboratories
P.O. Box 331058
Detroit, MI 48232-7058, USA
313-462-8500
800-521-0851
Fax 313-462-8517

In UK:
Difco Laboratories
P.O. Box 14B
Central Avenue, East Molesey
Surrey, KT80SE, UK
(44)-81-979-9951
Fax (44)-81-979-2506

Eastman Kodak
Laboratory Research Products Division
343 State Street
Building 701
Rochester, NY 14652-3512
800-225-5352

APPENDIX: SUPPLIERS

In UK:
Phase Separations Ltd.
Unit 19
Deeside Industrial Park, Queensferry
Clwyd, CH5 2NU, UK
(44)-244-816816
Fax (44)-244-821232

Enzo Diagnostics
325 Hudson Street
New York, NY 10013, USA
212-741-3838

Fisher Scientific
52 Fadem Road
Springfield, NJ 07081, USA
201-379-1400
Fax 201-379-7638

ICN-Flow
3300 Hyland Avenue
Costa Mesa, CA 92626
714-545-0100
800-368-3569
Fax 714-557-4872

In UK:
ICN-Flow
Woodcock Hill
Harefield Road, Rickmansworth
Hertfordshire WD3 1PQ, UK
(44)-923-774666
Fax (44)-923-777005

Fluka Chemical Corporation
980 South 2nd Street
Ronkonkoma, NY 11779-7240, USA
516-467-0980
800-358-5287
Fax 516-467-0663

In Europe:
Fluka Chemie AG
Industrie-strasse 25
CH-9470 Buchs, Switzerland
(41)-85-69511
Fax (41)-85-65449

In UK:
Fluka Chemicals Ltd.
Peakdale Road, Glossop
Derbyshire, UK
(44)-457-862518
Fax (44)-457-854307

FMC Marine Colloids
Bioproducts Department
5 Maple Street
Rockland, ME 04841, USA
207-594-3353
800-341-1574
Fax 207-594-3391

Genzyme Corporation
1 Kendall Square
Cambridge, MA 02139
617-876-9404
800-332-1042
Fax 617-252-7700

GIBCO BRL
P.O. Box 68
Grand Island, NY 14072, USA
800-828-6686
Fax 800-331-2286

In UK:
GIBCO-BRL Ltd.
Unit 4, Cowley Mill Trading Estate
Longbridge Way
Uxbridge, UB8 2YG, UK
(44)-895-36355
Fax (44)-895-53159

Haake Buchler Instruments, Inc.
244 Saddle River Road
P.O. Box 549
Saddle Brook, NJ 07662-6001, USA
201-843-2320
800-631-1369

Irvine Scientific
2511 Daimler Street
Santa Ana, CA 92705-5588, USA
714-261-7800
800-437-5706
Fax 714-261-6522

Isolab, Inc.
4350 Drawer
Akron, OH 44303, USA
216-825-4526
800-321-9632
Fax 216-825-8520

APPENDIX: SUPPLIERS

JRH Biosciences (formerly Hazelton Research Products)
P.O. Box 14848
Lenexa, KS 66215-0848, USA
913-469-5500
800-255-6032
Fax 800-441-1561

Micron Separations, Inc.
135 Flanders Road
P.O. Box 1046
Westborough, MA 01581
508-366-8212
800-444-8212
Fax 508-366-5840

Millipore Corporation
80 Ashby Road
Bedford, MA 01730
617-275-9200
800-225-1380
Fax 617-275-8200

In UK:
Millipore (UK) Ltd.
Millipore House, The Boulevard
Blackmoor Lane
Watford, WD1 8YW, UK
(44)-923-816375
Fax (44)-923-818297

New England Biolabs Inc.
32 Tozer Road
Beverly, MA 01915, USA
508-927-5054
800-632-5227
Fax 508-921-1350

New England Nuclear (NEN)
549 Albany Street
Boston, MA 02118, USA
617-482-9595
800-551-2121
Fax 617-482-1380

In Europe:
du Pont de Nemours (Deutschland) GmbH
Biotechnology Systems Division
NEN Research Products
Postfach A01240
6072 Dreieich 4, FRG
(49)-6103-803-115
Fax (49)-6103-85115

Nunc Inc.
2000 North Aurora Road
Naperville, IL 60563, USA
708-983-5700
800-288-6862
Fax 708-416-2556

In UK:
Gibco Ltd.
P.O. Box 35
Trident House, Renfrew Road
Paisley, Scotland PA3 4EF
(44)-41-889-6100
Fax (44)-41-887-1167

In Europe:
Nunc A/S
Postbox 280
Kamstrup, DK 4000
Roskilde, Denmark
(45)-42-359065
Fax (45)-42-350105

Pall BioSupport Company
Pall Conference Center
77 Crescent Beach Road
Glen Cove, N.Y. 11542
607-753-6041
Fax 607-756-7350

In UK:
Pall Process Filtration Ltd.
Europa House
Havant Street
Portsmouth P013PD, UK
(44)-70-5753545
(44)-70-5831324

Perkin Elmer Cetus
761 Main Avenue
Norwalk, CT 06859, USA
800-762-4002
Fax 203-761-9645

Pharmacia LKB Biotechnology Inc.
800 Centennial Avenue
Piscataway, NJ 08854, USA
201-457-8000
800-526-3593
Fax 201-457-8643

APPENDIX: SUPPLIERS

In UK:
Pharmacia LKB Biotechnology Ltd.
Pharmacia House, Midsummer Boulevard
Central Milton Keynes, Buckinghamshire MK9 3HP, UK
(44)-908-661101
Fax (44)-908-690091

In Europe:
Pharmacia LKB Biotechnology AB
P.O. Box 175
Bjorkgatan 30
751 82 Uppsala, Sweden
(46)-18-163000
Fax (46)-18-143820

Promega Biotec
2800 Woods Hollow Road
Madison, WI 53711-5399, USA
608-274-4330
800-356-9526
Fax 608-273-6967

Ratliffe Biochemicals
252 La Cueva
Los Alamos, NM 87544, USA
505-662-7977

Sarstedt Inc.
Research Products Division
Box 468
Newton, NC 28658-0468, USA
704-465-4000
Fax 704-465-4003

Schleicher & Schuell Inc.
10 Optical Avenue
Keene, NH 03431, USA
603-352-3810
800-245-4024
Fax 603-357-3627

Sigma Chemical Company
P.O. Box 14509
St. Louis, MO 63178, USA
314-771-5750
800-325-3010
Fax 314-771-5757

In UK:
Sigma Chemical Company Ltd.
Fancy Road
Poole Dorset BH17 7NH, UK
(44)-202-733114
Fax (44)-202-715460

Stratagene
11099 North Torrey Pines Road
La Jolla, CA 92037, USA
619-535-5400
800-424-5444
Fax 619-535-5430

In Europe:
Stratagene GmbH
Postfach 105466, D-6900
Heidelberg, FRG
(49)-6221-40-06-34
Fax (49)-6221-40-06-39

United States Biochemical Corporation (USB)
P.O. Box 22400
Cleveland, OH 44122, USA
216-765-5000
800-321-9322
Fax 216-464-5075

VWR Scientific
P.O. Box 7900 San Francisco, CA 94120, USA
415-468-7150
Fax 415-330-4185

W.E.P. Company
P.O. Box 55626
Seattle, WA 98155
206-367-4271

West Coast Scientific, Inc.
2542 Barrington Court
Hayward, CA 94545
415-732-1111
800-367-8462
Fax 415-732-1131

Whatman Laboratory Products Inc.
9 Bridewell Place
Clifton, NJ 07014, USA
201-773-5800
800-922-0361
Fax 201-472-6949

In UK:
Whatman Ltd.
Springfield Mill, Maidstone
Kent ME14 2LE, UK
(44)-622-692022
Fax (44)-622-691425

INDEX

A

Adenine phosphoribosyltransferase, 104
Adenosine deaminase, 106–107, 110–111
Adenovirus-based vectors, 31–33
Albumin gene enhancer, 11, 15
α-fetoprotein gene enhancer, 11, 15
Aminoglycoside phosphotransferase, 105
Amplification, 103, 108–111
Aqueous hybridization, 141–146, 149–155
Asparagine synthetase, 107, 111
Aspartate transcarbamylase, 105
Assays for gene expression:
   analysis of post-translational modification, 220–221
      analysis of fatty acid acylation, 226
      analysis of N-linked glycosylation, 222–224
      analysis of O-linked glycosylation, 224–225
      analysis of tyrosine sulfation, 225–226
   analysis of protein, 212
      immunofluorescence analysis, 219–220
      immunoprecipitation analysis, 213–216
      Western-type techniques, 216–219
   analysis of RNA (Northern- and PCR-based techniques), 205–212
Assays for gene transfer, 179–180
   PCR-based techniques, 184–188
   preparation of genomic DNA, 180
   Southern-type techniques, 181–183
   synthesis of nucleic-acid probes for hybridization analysis:
      hybridization conditions for nucleic-acid probes, 201–204
      non-isotopically labeled probes, 192–198
      non-isotopic detection, 199–200
      synthesis of radiolabeled nucleic-acid probes, 189–192

B

β-actin gene enhancer, 11, 14
β-interferon gene enhancer, 11, 13
Bio-Rad Gene Pulser®, 101
Biotin/phenol extraction, 144–146, 149–155

C

CAD gene, 106, 110
Calcium phosphate transfection, 24, 25, 96–98
Carrier DNA, 96
Cells and cell lines:
   assays for colony formation, anchorage-independent growth, and focus formation, 94–95
   basic tissue-culture techniques, 85–86
   propagation of cell lines, 87–90
   propagation of embryonal stem cells, 91–93

Cellular enhancers, 5, 10–11
   additional, 14–16
   immune system, 11–13
c-fos gene enhancer, 11, 15
c-HA-*ras* gene enhancer, 11, 15
Co-amplification, 108
Collagenase gene enhancer, 11, 15
COS-type cell, 26

D

DEAE dextran transfection, 25, 99–100
Dihydrofolate reductase, 106, 109–110
Directional cloning, 27
DNA microinjection, 25
DNA transfer:
   calcium phosphate transfection method, 96–98
   DEAE dextran transfection method, 99–100
   electroporation, 101–102
dsRNA-activated inhibitor (DAI), 21

E

EBO pcDX, 36
Elastase I gene enhancer, 11, 14
Electrocell Manipulator® 600, 101
Electroporation, 24, 101–102
Embryonal stem cells, propagation of, 91–93
Enhancers, 4
   cellular, 5, 10–11
     mechanism of action, 18
     properties that characterize, 5
   viral, 5–10
Epstein-Barr-virus-based vectors, 34–36
ESGRO, 91
Eukaryotic control elements, 3–4
   messenger RNA degradation signals and polyadenylation, 19–20
   promoter and intron interactions, 19
   promoters and enhancers, 4–18
   translational control, 20–21
Expression cloning, 114–116
   cDNA library construction, 121–131
   comparison of construction methodologies, 117–120
   preparation and technical requirements, 117
   solid-state amplification, 131–132
   transfection and SIB selection, 133–135

G

γ-globin/β-globin gene enhancers, 11, 15
Gene-replacement vectors, 56–60
Gibbon ape leukemia virus, 6
Glutamine synthetase, 107, 111

H

Hepatitis B virus enhancer, 6, 9
Herpes simplex thymidine kinase (HSVtk), 4
Herpes-simplex-virus-based vectors, 36–39
Histidinol dehydrogenase, 106
HLA DQ α and β gene enhancers, 11, 13
Homologous recombination and gene-replacement vectors, 56–60
Human cytomegalovirus enhancer, 6, 10
Human immunodeficiency virus enhancers, 6, 9–10
Hydroxylapatite (HAP) chromatography, 141–144, 146–148
Hygromycin-B-phosphotransferase, 105
Hypoxanthine-guanine phosphoribosyltransferase, 105

I

Immunofluorescence analysis, 219–220
Immunoglobulin gene enhancers, 11, 12
Immunoprecipitation analysis, 213–216
Inducible promoters and enhancers, 5
   metallothionein expression elements, 16–17
   mouse mammary tumor virus (MMTV) expression elements, 17–18
Insulin gene enhancer, 11, 16
Interleukin-2 gene enhancers, 11, 13
Interleukin-2 receptor gene enhancers, 11, 13

K

Kinase (see polynucleotide kinase)

L

Linkers/linkering, 14, 27-29, 117-119, 125, 130
Ligase, 149, 154
Library/libraries, 29, 35, 94, 114-123, 125, 128, 131-134, 136-140, 146-147, 149-150, 155-156, 158, 191, 201, 203, 205

M

Messenger RNA degradation signals and polyadenylation, 19–20
Metallothionein, 111
Metallothionein expression elements, 16–17
Metallothionein gene enhancer, 11, 14–15
Moloney murine leukemia virus (MoMuLV), 7
Moloney murine sarcoma virus (MoMuSV), 7
Mouse mammary tumor virus (MMTV) expression elements, 17–18
Multiple drug resistance, 106, 109
Muscle creatine kinase gene enhancer, 11, 14

N

Neural-cell-adhesion molecule (NCAM) gene enhancer, 11, 16
Northern blot analysis, 179
  analysis of RNA using, 209–210

O

Okayama and Berg vector, 120
O-linked glycosylation, 65
Ornithine decarboxylase, 105, 109

P

Packaging cell lines, 51
Papilloma-virus-based vectors, 43–47
Papilloma virus enhancers, 6, 8–9
PA317 packaging line, 52
PCR-based gene assembly, 60–61, 165, 167–171
  gene synthesis, 165–171
PCR-based gene expression, 60–61, 171–172
PCR-based techniques, 179, 184–188
  analysis of RNA using, 210–212
PE501 packaging line, 51
PERT hybridization, 139, 146–148
pMT2 vector, 28
Polyadenylation, role in mRNA stability, 19–20
Polynucleotide kinase, 190-191
Polyoma viral enhancer, 6–7
Polyoma-virus-based vectors, 29–31
Position effect, 21
Prealbumin gene enhancer, 11, 14
Promoters, 4–19
  intron interactions with, 19
Protein(s), post-translational modification, 65
Protein processing, 63
Protein synthesis, 62–63

R

Retroviral enhancers, 6, 7–8
Retroviral vectors, 47–56
Retrovirus-mediated gene transfer:
  generation of high-titer helper-virus-free recombinant-retrovirus stocks, 161–163
  titration and analysis of recombinant-retrovirus stocks, 163–164
RNA polymerase II promoter elements, 4

S

Selection and amplification, 108–111
  negative selection, 112–113
  positive selection, 103–107
Shuttle vectors, 25
SIB selection, 114, 133–135
Solid-state amplification, 116, 131–132
Southern-blot analysis, 179, 181–183
SRα vector, 21, 27, 28, 63, 120
Stable transfection, 24–25
Subtracted libraries, 136, 137
  aqueous hybridization and hydroxylapatite (HAP) chromatography, 149–155
  biotin/phenol extraction, 155–157
Subtracted probes, 136, 137
  aqueous hybridization and

Subtracted probes—*Continued*
   biotin/phenol extraction, 144–146
   aqueous hybridization and hydroxylapatite (HAP) chromatography, 141–144
   phenol-emulsion reassociation-technique (PERT) and HAP chromatography, 146–148
   production of first-strand cDNA, high- or low-specific activity, 139–141
Subtractive hybridization, 136–138
   generation of subtracted probes, 139–148
   generation of subtracted libraries, 149–157
   hybridization analysis of plasmid cDNA libraries, 158–160
Suicide vectors, 55
Suppliers, 229–238
SV40-based vectors, 25–29
SV40 viral enhancer, 5–6, 7

T

TATA box, 4
T-cell receptor gene enhancers, 11, 12–13
Thymidine kinase, 104
Transgenome, 97
Transient transfection, 24
Tryptophan synthetase, 105–106

V

Vaccinia-virus-based vectors, 39–43
Vectors, 23–24
   adenovirus-based, 31–33
   Epstein-Barr-virus-based, 34–36
   herpes-simplex-virus-based, 36–39
   homologous recombination and gene-replacement, 56–60
   papilloma-virus-based, 43–47
   PCR-based expression, 60–61
   polyoma-virus-based, 29–31
   retroviral, 47–56
   stable transfection, 24–25
   SV40-based, 25–29
   transient transfection, 24
   vaccinia-virus-based, 39–43
Viral enhancers, 5, 6
   gibbon ape leukemia, 6
   hepatitis B, 6, 9
   human cytomegalovirus, 6, 10
   human immunodeficiency, 6, 9–10
   papilloma, 6, 8–9
   polyoma, 6–7
   retroviral, 6, 7–8
   SV40, 5–6
Viral infection, 24

X

Xanthine-guanine phosphoribosyltransferase, 105